# HETEROGENEOUS CONTRIBUTIONS TO NUMERICAL COGNITION

# HETEROGENEOUS CONTRIBUTIONS TO NUMERICAL COGNITION

## Learning and Education in Mathematical Cognition

Edited by

**WIM FIAS**

Ghent University, Faculty of Psychology and Educational Sciences, Department of Experimental Psychology, Ghent, Belgium

**AVISHAI HENIK**

Department of Psychology, Ben-Gurion University of the Negev, Beer-Sheva, Israel

ELSEVIER

ACADEMIC PRESS

An imprint of Elsevier

Academic Press is an imprint of Elsevier
125 London Wall, London EC2Y 5AS, United Kingdom
525 B Street, Suite 1650, San Diego, CA 92101, United States
50 Hampshire Street, 5th Floor, Cambridge, MA 02139, United States
The Boulevard, Langford Lane, Kidlington, Oxford OX5 1GB, United Kingdom



**Notices**

Knowledge and best practice in this field are constantly changing. As new research and experience broaden our understanding, changes in research methods, professional practices, or medical treatment may become necessary.

Practitioners and researchers must always rely on their own experience and knowledge in evaluating and using any information, methods, compounds, or experiments described herein. In using such information or methods they should be mindful of their own safety and the safety of others, including parties for whom they have a professional responsibility.

To the fullest extent of the law, neither the Publisher nor the authors, contributors, or editors, assume any liability for any injury and/or damage to persons or property as a matter of products liability, negligence or otherwise, or from any use or operation of any methods, products, instructions, or ideas contained in the material herein.

**Library of Congress Cataloging-in-Publication Data**
A catalog record for this book is available from the Library of Congress

**British Library Cataloguing-in-Publication Data**
A catalogue record for this book is available from the British Library

ISBN: 978-0-12-817414-2



**Cover image:** The Tolle Lege Library in Thagaste Trefpunt Augustijnen in Ghent, Belgium, with permission.

*Publisher:* Nikki Levy
*Acquisitions Editor:* Natalie Farra
*Editorial Project Manager:* Kristi Anderson
*Production Project Manager:* Sujatha Thirugnana Sambandam
*Cover Designer:* Alan Studholme

Typeset by SPi Global, India

# Contents

# Contributors

**Marie Amalric**
Department of Psychology, CAOs Lab, Carnegie Mellon University, Pittsburgh, PA, United States

**Grégoire Borst**
Institut de Psychologie, Université de Paris, LaPsyDÉ, CNRS, Paris; Institut Universitaire de France, France

**Lital Daches Cohen**
Edmond J. Safra Brain Research Center for the Study of Learning Disabilities; Department of Learning Disabilities, University of Haifa, Haifa, Israel

**Bert De Smedt**
Parenting and Special Education Research Unit, Katholieke Universiteit Leuven; Faculty of Psychology and Educational Sciences, KU Leuven, University of Leuven, Leuven, Belgium

**Fien Depaepe**
Centre for Instructional Psychology and Technology, KU Leuven Kulak, Kortrijk; ITEC, KU Leuven, IMEC, Leuven, Belgium

**Annemie Desoete**
Ghent University, Faculty of Psychology and Educational Sciences, Department of Experimental-Clinical and Health Psychology, Ghent, Belgium

**Wim Fias**
Ghent University, Faculty of Psychology and Educational Sciences, Department of Experimental Psychology, Ghent, Belgium

**Roland H. Grabner**
Institute of Psychology, University of Graz, Graz, Austria

**Minna Hannula-Sormunen**
Department of Teacher Education, University of Turku, Turku, Finland

**Avishai Henik**
Department of Psychology, Ben-Gurion University of the Negev, Beer-Sheva, Israel

**Roi Cohen Kadosh**
Department of Experimental Psychology, University of Oxford, Oxford, United Kingdom

**Evelyn H. Kroesbergen**
Behavioural Science Institute, Radboud University Nijmegen, Nijmegen, The Netherlands

**Karin Kucian**
Center for MR-Research, University Children's Hospital Zurich, Zurich, Switzerland

**Karin Landerl**
Institute of Psychology, University of Graz, Graz, Austria

**Erno Lehtinen**
Department of Teacher Education, University of Turku, Turku, Finland; Vytautas Magnus University, Kaunas, Lithuania

**Jake McMullen**
Department of Teacher Education, University of Turku, Turku, Finland

**Kinga Morsanyi**
Centre for Mathematical Cognition, School of Science, Loughborough University, Loughborough, United Kingdom

**Rafael E. Núñez**
Department of Cognitive Science, University of California, San Diego, La Jolla, CA, United States

**Bert Reynvoet**
KU Leuven, Faculty of Psychology and Educational Sciences, Brain & Cognition, Leuven, Belgium

**Miriam Rosenberg-Lee**
Department of Psychology; Behavioral Neural Sciences Graduate Program, Rutgers University, Newark, NJ, United States

**Orly Rubinsten**
Edmond J. Safra Brain Research Center for the Study of Learning Disabilities; Department of Learning Disabilities, University of Haifa, Haifa, Israel

**Delphine Sasanguie**
HOGENT, Research Centre for Learning in Diversity, Ghent, Belgium

**Joke Torbeyns**
Centre for Instructional Psychology and Technology, Katholieke Universiteit Leuven, Leuven, Belgium

**Nienke E.R. van Bueren**
Behavioural Science Institute, Radboud University Nijmegen, Nijmegen, The Netherlands; Department of Experimental Psychology, University of Oxford, Oxford, United Kingdom

**Stijn Van Der Auwera**
Centre for Instructional Psychology and Technology, Katholieke Universiteit Leuven, Leuven, Belgium

**Stefanie Vanbecelaere**
KU Leuven, Faculty of Psychology and Educational Sciences, Centre for Instructional Psychology and Technology; KU Leuven, IMEC Research Group, ITEC, Leuven, Belgium

**Lieven Verschaffel**
Centre for Instructional Psychology and Technology, Katholieke Universiteit Leuven, Leuven, Belgium

**Stephan E. Vogel**
Institute of Psychology, University of Graz, Graz, Austria

**Stella Vosniadou**
College of Education, Psychology and Social Work, Flinders University, Bedford Park, SA, Australia

# Introduction

**Wim Fias[a] and Avishai Henik[b]**
[a]Ghent University, Faculty of Psychology and Educational Sciences, Department of Experimental Psychology, Ghent, Belgium
[b]Department of Psychology, Ben-Gurion University of the Negev, Beer-Sheva, Israel

Although not officially bookmarked as such, this book follows on the book *Heterogeneity of Function in Numerical Cognition,* published in 2018. This book, like the previous book, originated from a small group meeting to which we had invited speakers with different backgrounds, to put their knowledge together in order to come to a more refined and deeper understanding of numerical and mathematical cognition. In the first book, we focused on how cognitive functions, other than the ones that are specifically related to math, contribute to our numerical abilities. Language, performance control and selective attention, spatial processing and imagery, executive functions, and memory were explored. In the present book, we want to bring the field of cognitive (neuro)science of numerical and mathematical cognition in closer contact with the field of education and intervention.

The cognitive (neuro)science of numerical and mathematical cognition studies the neurocognitive basis of how we do mathematics, with the ultimate goal of understanding how numerical and mathematical knowledge are stored and processed in our cognitive systems. Applications and implications for education and intervention are considered, but often in thought exercises that illustrate the hope rather than the fact that basic findings can be translated to practice.

Education and intervention research focuses on developing methods to facilitate the acquisition of mathematical knowledge and skill in children that show a typical developmental trajectory and in children with learning disabilities. Yet, the knowledge and expertise on which the development of training and intervention programs is built remain to a large extent an unexplored territory for cognitive neuroscientists, isolated from the understanding of the cognitive structures and their neural basis that underlie mathematical cognition.

Of course, the above puts it in a black and white fashion, but it is undeniably a fact that mathematical cognition is studied from different backgrounds and approaches in largely separate fields of study. We wanted to bring basic research on mathematical cognition and research on learning and education in mathematics closer together.

Crossing borders in science requires clear and unequivocally defined terminology. Núñez (Chapter 1) makes an important case in this respect, as he describes how the sloppy use of a term as simple as "number" has caused confusion in attempts to understand the nature and origin of mathematics. He explains how formally the terms "number" or "numerical" are not applicable to describe the quantity discrimination abilities observed in animals and preverbal infants. He proposes to use the term "quantical" to refer to these nonsymbolic skills. He argues that explicit learning and education is needed to bridge the gap between quantical and numerical cognition.

In cognitive (neuro)science, there is a strong tendency to focus on basic number processing tasks (like number magnitude tasks or mental arithmetic) because they allow high levels of experimental control. Yet, this comes with a cost, as it narrows the scope to these basic levels of knowledge and skill, while, of course, math knowledge and expertise is far more complex and extends to geometry, algebra, measurement theory, problem solving, and so on. The distance between the basic levels of mathematical skill that receive most attention of researchers and the advanced levels of mathematics that are the ultimate goal of formal education is enormous. Bridging that gap is a major challenge for the field. In the first section of the book, higher level mathematics is addressed.

Borst (Chapter 2) starts from the observation that complex cognitive tasks can be solved by logical algorithms, but that at the same time, people employ heuristics. As opposed to the slow and effortful nature of logical algorithms, heuristic strategies are fast and effortless. Because such intuitive heuristics are sometimes misleading, it is crucial for learning higher level mathematics to be able to obtain inhibitory control. This progressive inhibitory ability is crucial, not only for mathematical understanding, but also for logical reasoning. A close connection between mathematics and reasoning skill can be expected.

Morsanyi (Chapter 3) explores the connection between logical reasoning and mathematical ability in more detail. She highlights that mathematical ability does not only depend on number magnitude and other types of numerical knowledge but also involves reasoning abilities. In particular, she shows that transitive and conditional inferences are specifically related to mathematical skill. Transitive inference is related to number line tasks and plays a role in measurement, class inclusion, and comparative reasoning and may therefore, through further development, help in scaffolding several types of mathematical knowledge (geometry, etc.). Conditional reasoning, on the other hand, is related to number order tasks and mathematical fluency.

Amalric (Chapter 4) reports a summary of a number of neuroimaging experiments that looked at the neural basis of high-level mathematical expertise in the domains of geometry, algebra, analysis, and so on. A math-responsive network is established in the bilateral intraparietal sulcus and the inferior temporal cortex. This network is distinct from language processing networks but overlaps with visual and particularly spatial processing regions. Interestingly, this network overlaps with the neural correlates of deductive reasoning, again confirming an intrinsic link between mathematics and reasoning.

Complex forms of mathematics and reasoning having been addressed in the first section of the book, the second section concentrates more on education.

Verschaffel (Chapter 6) makes the transition from higher levels of mathematical understanding (i.e., knowledge that goes further than understanding number magnitude and knowing the arithmetic tables) to education. He focuses on subtraction by addition as an interesting case of higher level knowledge and how it can create strategic opportunities to solve math problems by different strategies. Once higher levels of mathematical knowledge are addressed, which supersede the mere understanding of numerical magnitudes and the rote verbal learning of arithmetic tables, the importance of strategy repertoire and strategy use becomes more important. To understand the cognitive basis of how strategies are selected (cf. the work of Patrick Lemaire) and how efficient strategy use should be instructed are important topics for further investigation. Advancement can be made by conjoining the cognitive and educational approaches.

Understanding and working with rational numbers is a crucial stage in mastery of mathematical knowledge. This is realized within both cognitive neuroscience and education. Cognitive neuroscience brings in that the whole number bias needs to be inhibited (Rosenberg-Lee, Chapter 7; Borst, Chapter 2). Yet, cognitive neuroscience so far hasn't concentrated much on what comes in place, that is, how the rational numbers are cognitively and neurally represented. The latter, which constitutes rational number knowledge, is something that is brought by education. Vosniadou (Chapter 5) explains the framework of conceptual change, which emphasizes how initial knowledge about numbers (whole number bias) has to be changed and restructured to allow the understanding of rational numbers. Rosenberg-Lee (Chapter 7) accepts the challenge to think about how knowledge about whole numbers ought to be expanded to allow understanding of rational numbers. She argues that the new knowledge structures that are constructed when rational numbers are learnt could be built on magnitude representations.

Learning studies in which learning is experimentally manipulated are rare in the domain of math. Yet, if we want to make the connection between cognitive neuroscience and education, it is imperative to conduct such studies, as argued by De Smedt (Chapter 8). Natural experiments in which educational contexts are compared may constitute a further tool to bring mathematical cognition and education closer together.

The following four chapters (i.e., Chapters 9–12) focus on intervention. Intervention studies vary in their aims, target populations, age of participants, and other aspects. The first three of these chapters present specific research and methods of interventions and their results, and the last chapter (Chapter 12) in this section presents a review of the intervention literature.

Numerical abilities start developing very early in life and preschool children exemplify individual differences in such abilities. Preschool children learn the counting principles and learn to work with magnitude and quantities. This learning helps them understand math operations (i.e., addition, subtraction, multiplication, and division) and learn to master arithmetic facts (e.g., $3 \times 4 = 12$). Not surprisingly, Kucian (Chapter 10) found that deficiencies in basic magnitude-numerical competencies, found in kindergarten, predicted calculation abilities later on in school. Importantly, Kucian reports that prevention training in kindergarten could help reduce children's difficulties and help them reach average levels of numerical abilities. Similarly, Hannula-Sormunen, McMullen, and Lehtinen (Chapter 9) discuss spontaneous focusing on numerosity (SFON). They suggest that the tendency to focus on numerosity varies among preschool children and that variation in SFON predicts the development of numerical abilities in school. Henik (Chapter 16) suggests that SFON, studied by Hannula-Sormunen and colleagues, requires an ability similar to what that needs to develop in order for Piaget's number conservation to appear. This is the ability to pay attention to numerosity and at the same time ignore distracting features (e.g., density, object identity) that are associated with numerosity and may hinder focusing on numerosity.

van Bueren, Kroesbergen, and Cohen Kadosh (Chapter 11) present research on brain stimulation that aims to improve numerical skills. They focus on noninvasive brain stimulation (NIBS), and in particular, on transcranial electrical stimulation (tES). This type of intervention uses one or more electrodes, placed on the scalp, to deliver an electrical current (in most cases 1–2 mA) to a specific area of the brain. Effective brain stimulation is dependent on studies (e.g., fMRI) that reveal the brain structures involved in performing various mathematical tasks. It has been suggested that the parietal lobes, and especially the intraparietal sulcus (IPS), are involved in basic mathematical tasks. Accordingly, tES has been found to be efficient in improving math learning. Moreover, parietal tES may facilitate transfer of improved mathematical performance.

Reynvoet, Vanbecelaere, Depaepe, and Sasanguie (Chapter 12) surveyed reviews and metaanalyses that targeted interventions aimed to improve numerical proficiency and basic mathematical abilities. The surveyed studies were directed at children and adolescents characterized as typical or atypical (or at risk) achievers. Interestingly, the number of intervention studies increased dramatically in recent years. Figure 2 in Reynvoet and colleagues' chapter shows that the number of intervention studies was small in the last decade of the 20th century and the first decade of the current century. The number of studies grew dramatically in the last decade. This impressive expansion of studies led to efforts to synthesize multiple studies and these efforts have been conveyed by systematic reviews and metaanalytic studies, which have increased in recent years also (see Figure 3 in Reynvoet and colleagues' chapter). Forty-one reviews/metaanalyses were included in Reynvoet and colleagues' work. Their metareview provides information on the effectiveness of interventions, especially regarding the preferred age of interventions, preferred skills to be trained, type of instruction, and the characteristics of the participants (who should be targeted by the intervention). For example, it seems that interventions could be effective at any age. Yet, Reynvoet and colleagues recommend to start at a young age and suggested that early interventions could induce more favorable attitudes toward numbers and math. In line with Reynvoet and colleagues' recommendations, Hannula-Sormunen and colleagues (Chapter 9) found that training could improve focusing on numerosity and, as a result, lead to improved numerical abilities.

Chapters 13–16 focus on math deficiencies and difficulties. As is evident already from the previous section (on intervention studies), math difficulties could be related to domain-general and to domain-specific factors. Obviously, difficulties vary according to age. The fact that children of different ages tackle different specific math problems is not very surprising. However, it seems that the domain-general difficulties may also vary by age. For example, it is not clear whether the executive functions required in preschool are the same as those required for the more advanced, higher levels of mathematics, presented in Chapters 2–4. In this sense, the current discussions raise a question regarding how unitary our conceptual framework is. Is the inhibition discussed in research of preschool children the same as the inhibition discussed in research of the adult population? With this question in mind, we can turn to the four last chapters of this book.

Daches Cohen and Rubinsten (Chapter 13) discuss the influence of math anxiety on math achievements and on numerical cognition in general. The influence of math anxiety is also discussed by Desoete (Chapter 14) and Landerl, Vogel, and Grabner (Chapter 15). Note that a recent summary of the Program for International Student Assessment (PISA) suggested that 30%–60% of the participants reported signs of math anxiety. Math anxiety relates to a tension, distress, or fear triggered by facing or thinking of numerical information. Such a feeling is detrimental to performance in academic and daily situations. Daches Cohen and Rubinsten suggest that math anxiety is the result of the interplay between environmental factors and an individual's intrinsic characteristics. It seems that there may exist a vicious circle of influences. Low achievements may lead to negative feelings regarding numbers and math, which may interact with internal characteristics of individuals to develop math anxiety. Importantly, an early tendency toward math anxiety may interfere with math performance, which may exacerbate the negative feeling toward math. Interestingly, Hannula-Sormunen and colleagues (Chapter 9), Kucian (Chapter 10), Desoete (Chapter 14), and Henik (Chapter 16) discuss preschool difficulties in acquiring the very basic concepts of numerical cognition. Accordingly, the framework put forward by Daches Cohen and Rubinsten suggests that such early difficulties might have broader effects. Even if such difficulties do not lead to full-blown math anxiety, they may still produce some negative feelings and attitudes that may interfere and hinder progress of numerical processing.

Desoete (Chapter 14) presents longitudinal studies that examined the prediction of typically achieving and low achieving children. Desoete examines basic numerical processing like subitizing, counting, and magnitude comparisons. It seems that it is easier to predict typical achievement than to identify which children would develop arithmetic difficulties. However, we should still remember Kucian's (Chapter 10) suggestion that children with a risk for dyscalculia could be identified already in kindergarten. Importantly, Desoete reports that quite a few (about 46%) children who present arithmetic difficulties at the age of 6–7 years old show average math performance later at the age of 10 years old. In line with this report, Kucian suggested that prevention training in kindergarten could improve performance of children at risk for dyscalculia to an average level. These changes from atypical to typical performance suggest that children might overcome arithmetic difficulties. This is an important conclusion to bear in mind when one thinks of math education.

Landerl, Vogel, and Grabner (Chapter 15) present an overview of the factors involved in early development of dyscalculia. Dyscalculia is a neurodevelopmental disorder reflected in children's difficulties in arithmetic and their falling behind their classmates. Predominately by this and similar descriptions and definitions, dyscalculia is diagnosed in school. However, Landerl and colleagues point out that early signs for the risk to develop dyscalculia could be found in preschool years (see also Kucian, Chapter 10, and Desoete, Chapter 14). Landerl and colleagues suggest that in addition to domain-specific factors directly related to math (e.g., symbolic and non-symbolic representations of numerosities), there exist domain-general factors related to various other aspects of learning (e.g., working memory, executive functions) that may be involved in the development of dyscalculia. Interestingly, among the domain-general factors, Landerl and her colleagues discuss the role of verbal abilities. Note that Desoete also presents results that language abilities in kindergarten predicted arithmetical abilities in the second grade.

Alongside the development of specific math abilities, children develop domain-general abilities like attention and executive functions. Most, if not all, of the authors of this book stress the intimate connection between the development of numerical abilities and the development of more general abilities. These general abilities like language, attention, working memory, and other executive functions are important for math performance and math development. Moreover, it seems that these heterogenous abilities are crucial for proper early development of the very basic numerical competencies. Hannula-Sormunen and colleagues (Chapter 9) and Henik (Chapter 16) stress that attention to numerosity varies among preschool children and might depend not only on the ability to focus on numerosity but also on the ability to inhibit features of the scene that are irrelevant to numerosity. This ability to inhibit irrelevant information might be specific to the numerical and magnitude domain or might be related to the development of more general executive abilities. Accordingly, it is important that teachers and parents (see Desoete on the influence of mothers' focus on numerosity and Henik on parents' contributions to the development of self-regulation) pay attention to the development of both specific and general abilities very early in life for the proper development of children in general and math ability in particular.

CHAPTER 1

# From quantical to numerical cognition: A crucial passage for understanding the nature of mathematics and its origins

**Rafael E. Núñez**
Department of Cognitive Science, University of California, San Diego, La Jolla, CA, United States

What is the nature of mathematics? What are its foundations and origins? Many readers would perhaps agree that a reasonable place to begin such quest is with the concept of "number." But where do numbers come from? The field of "numerical cognition," which gathers efforts from multiple disciplines, ranging from experimental psychology to cognitive linguistics, and from animal cognition to cognitive neuroscience, has studied these questions for several decades. The field has made progress in many directions (see, for example, Campbell, 2005; Cohen Kadosh & Dowker, 2015; Henik, 2016), but it has also proposed various accounts that are deeply incompatible and inconsistent. For instance, while it is very well established that there are hundreds of human cultures that do not exhibit arithmetic and lack words for quantities beyond three or four (Bowern & Zentz, 2012; Epps, Bowern, Hansen, Hill, & Zentz, 2012; Everett, 2017), studies in nonhuman animal cognition, in stark contrast, unproblematically report that there are "mathematical abilities" in monkeys (Beran, Perdue, & Evans, 2015), "numerical and arithmetic abilities in non-primate species" (Agrillo, 2015), "spontaneous number representation in mosquitofish" (Dadda, Piffer, Agrillo, & Bisazza, 2009), and that "numerical cognition in honeybees enables addition and subtraction" (Howard, Avarguès-Weber, Garcia, Greentree, & Dyer, 2019). And while historians of mathematics document with painstaking detail the struggles and difficulties that were involved in the (relatively recent) development of the concept of "zero" (Ifrah, 1985; Menninger, 1969), a recent article in *Science* magazine states that bees, despite their relatively simple nervous system and small cortex-less brains, are not prevented "from knowing how to understand numbers, including zero" (Nieder, 2018). How can

*Heterogeneous Contributions to Numerical Cognition*
https://doi.org/10.1016/B978-0-12-817414-2.00004-X

insects with tiny cortex-less brains and nonsymbolic language be involved in "counting" and understanding "numbers" and "zero" while healthy humans from ancestral, environmentally adapted cultures, with brains that have about 90 billion more neurons, sophisticated symbolic languages and cultural practices do not? Clearly, there is something fundamental that does not square.

Such profound incompatibilities hurt the field of "numerical cognition" and lead to confusing and deeply inconsistent theoretical developments, especially concerning its underlying biological dimensions, the role of language, and the complex interplay between biological endowment and cultural developments. Here, I will argue that this current impasse is largely due to a notoriously sloppy and confusing use of crucial technical concepts in the field—primarily involving the terms "number," "counting," "arithmetic," and even "mathematics"—which allows for accommodated interpretations that defend preferred views. Good science requires clear concepts, sharp distinctions, and rigorous terminology in order to avoid blurry interpretations of data and the proliferation of misleading accounts and theories.

To look for clarity, a crucial question we—practitioners in the field—should ask is the following: Are the quantity-related capacities that have been observed in human infants and many nonhuman animals really about *numbers*, and about *arithmetic*, let alone about *mathematics*? (Núñez, 2017a). And if not, how can we incorporate those valuable data into a naturalistic theory of the nature of number and arithmetic (and eventually, mathematics)? As mentioned, a huge obstacle for developing clear theories in this domain relies on the fact that most of us working directly or indirectly in the field of "numerical cognition" have been immersed in the (bad) habit of being extremely sloppy in the use of crucial terms, beginning with the term "number." And this practice is made worse due to the fact that the term "number" is, in both, everyday and technical domains, highly polysemous (Núñez & Marghetis, 2015)—i.e., it has many meanings. As an extension, its adjectival form—"numerical"—can apply to many instances, which generate lots of confusion when the various meanings are blended or used interchangeably. To practice good science, we must clearly specify these meanings in a sound and unambiguous manner to avoid confusions that would lead to unwarranted conclusions, and eventually to bad science. For example, consider the current COVID-19 pandemic. Clearly distinguishing the various types of coronaviruses is crucial for understanding the nature and etiology of the diseases they produce in humans, and the development of potential vaccines and treatments. Blurring them in one conceptual coronavirus blob—e.g., referring to human, porcine, or avian coronaviruses interchangeably or equating a potentially deadly human coronavirus (SARS-CoV-2 that has led to the current pandemic) with one that produces mild symptoms of the common cold (e.g., HCoV-OC43)—leads to terrible science with catastrophic consequences for humanity. Keeping the proportions, similar conceptual distinctions must be made in the field of "numerical cognition" with respect to quantity-related phenomena, including the fundamental conceptual "number."

## The problems of using blurry concepts and sloppy terminology in numerical cognition

To address the question of the nature of "number," "arithmetic," and eventually, of "mathematics," one must be clear about what is meant by "number." As we will see below, this specification will determine what mental, behavioral, or neurological processes can be properly qualified as "numerical," so one can unambiguously analyze whether infants and animals operate with "*numerical* competence," "*numerical* cognition," "*numerical* judgment," "*numerical* abilities," "*numerical* capacities," "*numerical* reasoning," "*numerical* representation," and so on. Clarifying what is meant by "number" is crucial because the term, being highly polysemous, has many different (but somewhat related) meanings. And this, not only in everyday language, but also in the academic fields of mathematical cognition, developmental psychology, comparative (animal) studies, the neuroscience of numerical cognition, and even in mathematics itself.

Consider, for instance, the following English expressions: "five is a prime *number*," "this is my passport *number*," "an infinitesimal *number* does not have Archimedean properties," "there is a *number* of things I want to talk to you about," "fourth is an ordinal *number*," "$\aleph_0$ is a transfinite *number*, not a real number." The meanings of "number" in these expressions are related, and are well accepted, but cannot be captured by a unique definition. One might think that this is due to "vagueness" in natural language. But this is not so. In academic literature, whether it is neuroscience or animal psychology, we can also find cases of polysemy which are potentially confusing. And this is the case even in pure mathematics, where "numbers" are abstract entities in their own right, governed by precise properties (e.g., the Peano axioms for natural numbers, or completeness, for the real numbers). In mathematics, "numbers" are referred to *symbolically*, represented by specific written (or spoken) signs—called *numerals*—and can be used to perform operations and calculations. Depending on their properties, specific collections of these numbers are designated as natural numbers, ordinal numbers, negative numbers, whole numbers, rational numbers, irrational numbers, real numbers, imaginary numbers, complex numbers, infinitesimal numbers, hyperreal numbers, surreal numbers, transfinite cardinal numbers, and so on, all of which designate types of "numbers" but in each case with somewhat different meanings. Thus evaluating a statement like "each number has a successor" or "each number is either equal to, or greater than, or smaller than, zero" cannot be done properly, unless it is specified what is meant by "number." If by "number" what is meant is "whole number" (i.e., integer), then both these statements are true. But if what is meant is "complex number," then both statements are false. With focus and clarity, it is possible to constrain and specify the meanings such that we end up with the appropriate and desired denotations that are necessary for a transparent evaluation of claims. A similar minimal level of rigor should apply to any scientific endeavor, which includes, for instance, claims about the origins of numbers and the hypothesized biological endowment of "numerical abilities" in infants, nonhuman primates, birds, fish, and insects.

## "Terminological chaos" and "misapplications of terms" in numerical cognition

Calling for the (now urgent) need for clarifying basic concepts in the field is by no means new. Indeed, already three decades ago researchers in the troubled domain of "numerical competence" in nonhuman animals and children spoke of the "terminological chaos" (Davis & Pérusse, 1988, p. 562), the lack of "clarification of terms" (Von Glasersfeld, 1988, p. 601), and the unnecessary suffering "from the misapplication of terms" (Boysen, 1988, p. 580) that existed in the field when dealing with terms as (seemingly) simple as "counting," "subitizing," and "estimation." Scholars worked hard to propose taxonomies and definitional clarifications for such terms but their discrepancies were multiple and their success limited (see, for example, Davis & Pérusse, 1988, and related commentaries). Unfortunately, the situation is not better today. In fact, to some extent it is even worse. At least in the 1980s, beyond the disagreements, scholars were aware of the "terminological chaos." Nowadays, however, scholars (including myself) seem to be doing business as usual, rarely aware of an existing "contemporary terminological chaos" in the field of "numerical cognition."

Today, number-related terms that were once relatively precise, such as "numeral," "numerosity," or "numerousness"—some of which we inherited from the careful and rigorous psychophysics of the mid-20th century (e.g., Stevens, 1939/2006, 1951)—are blurred and often used as synonyms or as interchangeable terms. For example, in scholarly articles in numerical cognition, the term "number" is sometimes used to mean "numeral" (the *sign* for a number: e.g., the Hindu-Arabic digit "5" or the Roman "V"). Or sometimes "number" is used instead of "numerousness" (e.g., Rugani, Regolin, & Vallortigara, 2011), despite the warnings that "numerousness discrimination ... represents a simple perceptual ability that bears no obvious relation to number" (Davis, Albert, & Barron, 1985, p. 1222). Importantly, the term "number" is often loosely used in place of "numerosity"—a technical term utilized by psychophysicists in the 1950s to specifically designate a scale of measurement for evaluating the "numerousness" of a stimuli (e.g., a collection of objects) and by means of which an experimenter establishes "the cardinal attribute of physical collections of objects" (Stevens, 1951, p. 5; see also Stevens, 1939/2006). While "numerousness"—"a property or attribute we are able to discriminate when we regard a collection of objects" (Stevens, 1939/2006)—is a property or attribute of a stimulus, which can be measured by the investigator in "numerosity" units, "number" is a (usually exact) quantifier referred to symbolically that remains in the mind of the formally schooled experimenter. Thus we often find scholarly articles that while properly referring to "numerosity" in their titles, describe the studies and their conclusions in terms of "numbers." For instance, in an influential study in developmental psychology by Xu (2003), the title reads "Numerosity discrimination in infants." But the abstract goes on to say that "these studies provide the first direct comparison between discrimination of small and large numbers in infants ... providing evidence for the existence of two systems of number representation in infancy" (Xu, 2003, p. B15). Or, for example, while referring to the appropriately titled report in comparative cognition "Ordering of the numerosities 1–9 by monkeys" (Brannon & Terrace, 1998), the main author in a subsequent publication analyzes the findings under "abstract number" stating that "number was the only valid cue" monkeys relied on (Brannon, 2005, p. 86). Similarly, in a recent neuroscience review by Nieder (2016) entitled "The neuronal code for number," the author analyzes how "numerosity-selective neurons were present both in the ventral intraparietal (VIP) area of the posterior parietal cortex (PPC) and the lateral prefrontal cortex (lPFC) of the primate brain without numerical training" and concludes that "collectively, these findings suggest that numbers reflect a natural perceptual category" (Nieder, 2016, p. 370).

In sum, the "terminological chaos" and "misapplications of terms" described more than 30 years ago are still very much alive today. The relatively precise definitions of various quantity-related terms used by the rigorous psychophysicists of the mid-twentieth century are routinely blurred. Occasionally, one might find some timid reminders in the background of a few texts pointing to this sloppy and misleading use of concepts and terminology. For example, when contrasting human and nonhuman animal "numerical abilities," Dehaene writes in his book *The Number Sense*: "The difference with our [human] verbal counting is so enormous that we should perhaps not talk about 'number' in animals at all, because by number we often imply a discrete symbol. This is why scientists, when they describe perception of numerical quantities, speak of 'numerosity' or 'numerousness' rather than number" (Dehaene, 1997, p. 35). A similar remark was made by comparative psychologists M. Beran and A. Parrish, who pointed out that "quantity judgments ... are subsumed under the research domain known as numerical cognition. However, they should more accurately be considered as cases of quantitative cognition because they do not have to rely on the use of abstract numerical concepts or a 'number sense'" (Beran & Parrish, 2016, p. 176). Unfortunately, however, such rare constructive recommendations are in practice largely ignored and by-passed as authors proceed with a business-as-usual attitude where it is taken at face value that newborn babies, salamanders, fish, and bees perceive (and even operate with) "numbers" and monkeys de facto have "mathematical abilities." As a result, the loose and overinclusive use of the term "number" (and "arithmetic") is now pervasive in the literature in infant development, comparative psychology, and cognitive neuroscience, leading, as mentioned earlier, to titles that unproblematically read: "Rhesus monkeys spontaneously compute addition operations over large numbers" (Flombaum, Junge, & Hauser, 2005), "Spontaneous number discrimination of multi-format auditory stimuli in cotton-top tamarins" (Hauser, Dehaene, & Dehaene-Lambertz, 2002), "Spontaneous number representation in semi-free-ranging rhesus monkeys" (Hauser, Carey, & Hauser, 2000), "Large number discrimination in 6-month-old infants" (Xu & Spelke, 2000), "Chicks with a number sense" (Brugger, 2015), "Evidence for counting in insects" (Dacke & Srinivasan, 2008), and so on.

But the lack of rigor and loose terminology hasn't just created random noise and confusion. The "terminological chaos" and "misapplication of terms" have worked primarily in one direction: to support the nativist view that biological evolution alone endowed many species with capacities *specific* for "number" and "arithmetic." While the good news is that the amount of studies investigating the biologically endowed abilities for discriminating small and large quantities in many species has increased exponentially, the bad news is that the conceptual clarity and rigor for interpreting those results have not followed. With the sloppy and overinclusive terminology, the interpretations of these results have led to fallacious conclusions. There is no question that many nonhuman animals exhibit basic quantity-related behaviors that in many respects are similar to that of humans, so putting the investigation of such behaviors in an evolutionary context is the right thing to do. However, this has to be done with conceptual rigor and clarity. Unfortunately, unlike the cautious and rigorous psychophysicists of the mid-twentieth century, the field of "numerical cognition" has brought the term "number" itself (i.e., the very *explanandum*—what is to be explained) into the working theoretical constructs themselves (i.e., in the *explanans*—the explanation). Such is the case, for instance of the theoretical construct "Approximative Number System," or "ANS," which following Weber-Fechner's law is claimed to be an "evolutionarily ancient innate system for approximate number" (Halberda, Mazzocco, & Feigenson, 2008; Piazza, 2010). This has facilitated the development of nativists proposals, since the hypothesized biological "system" (i.e., the ANS) has "number" already built in. Thus the "terminological chaos" and consistent "misapplications of terms" have ended up reifying teleological claims about evolved capacities *specific* for number and arithmetic.

## The issue is not merely about words

Why are these terminological distinctions important and why should anyone care? The issue is not just a matter of lexical pickiness. Questions of conceptual clarity and terminology are not "just about words" since they often convey profound theoretical implications. Keeping the terms straight makes all the difference in the world when addressing the question of the nature of number, arithmetic, and mathematics. Not doing so can seriously mislead the very theories one is trying to develop. After all, researchers in evolution of language, for instance, have been for a long time promptly corrected when loosely referring to the "grammar" of gorillas or the "language" of bees (Hockett, 1960). Why should the use of the terms "number" and "numerical" be any different? The use of these terms for the scientific understating of their nature and origins should demand nothing less. Thus when a researcher observes that a pigeon, a monkey, a chicken, a bee, or a human infant is able, with a certain degree of success, to discriminate two stimuli that differ in quantity or "numerousness"—measured by the experimenter as having, say, "numerosity 2" and "numerosity 5," respectively—she doesn't have to assume that the organism is operating with some kind of "number" concept, or "number sense." With these important distinctions in her toolkit, the researcher has an adequate language for properly expressing what has been observed: She can safely claim that the individual (a pigeon, a bee, or an infant) has been able to discriminate the two stimuli on the basis of their "numerosity" (which she has measured), without having to assume—if there is no further evidence—that it was done on the basis of their "number." "Number" is a symbolically sustained concept that remains in the mind of the experimenter who has a life of schooling and training behind her, all of which crucially mediated by rich linguistic, notational, and cultural practices.

The point is that this unprecise, overinclusive, and sloppy use of terminology is not neutral and it is harmful. It blurs the question of how genuine "number" (as an exact symbolic quantifier, with properties to be listed below) is brought to being, and by extension, how arithmetic and mathematics are created. This confusion must be avoided at all costs. We must rescue some of the healthy and rigorous practices of the psychophysicists of the mid–20th century, and with a clearer and more theory-neutral concepts and terminology describe and analyze the existing data with finer distinctions. This is why so far in this text I have been purposefully referring to the remarkable capacities observed in infants and nonhuman animals as relating to *quantity*, not (necessarily) to *number*. By "quantity" I have meant a property or attribute an organism is able to discriminate when perceiving a collection (or sequence) of objects (or sounds) (and this, borrowing a slightly modified definition of "numerousness" due to the renowned experimental psychologist Stevens (1939/2006)). I thus have explicitly avoided characterizing the infant and animal data as being "numerical" and about "number," precisely because there is no evidence that these quantity-related capacities actually possess the properties attributed to the prototypical notion of "number" that we are about to see. "Quantity," on the other hand, being a far more generic notion does not presuppose the presence of those more elaborated properties and allows us to refer to the baby, nonhuman primate, bird, and insect findings with suitable terms that explicitly leave out those elaborated properties. "Quantity" makes us avoid the risk of overinterpreting the data in a way that pushes us to teleologically pack arithmetic and mathematics straight and unproblematically into biological evolution.

## Characterizing "number," and what is to be qualified as "numerical"

How are we to understand the well-established human and nonhuman animal quantity-related data? To begin, given that many human cultures do not operate with languages that have numerals beyond the subitizing range (Bowern & Zentz, 2012; Epps et al., 2012; Everett, 2017) and that children present difficulties in learning numbers and counting procedures (Cheung, Rubenson, & Barner, 2017; Piaget, 1952; Sarnecka & Carey, 2008), we must not assume that "number" is a "natural domain of competence" (Antell & Keating, 1983, p. 695) or "a natural perceptual category" (Nieder, 2016, p. 370). We want to explain the *emergence* of number; we definitely do not want to just take it for granted, as preexisting or pregiven. And, importantly, we must be precise about the use of the term "number" and of the associated adjective "numerical." What then is "number" for the purposes of investigating its origins and for evaluating the claim that there is a biologically evolved capacity *specific* for "number"? And, more generally, what is it, if we want to understand the nature and foundations of mathematics?

## Minimal criteria for number

Beyond the natural polysemy of the term "number," a reasonable step for characterizing "number" for research purposes is to begin with an empirical, functional, and pragmatic approach, i.e., to begin with the most prototypical and fundamental properties ascribed to "number" when evoking the familiar counting sequence "one, two, three, ..."—the crux of "numerical cognition." Accordingly, (prototypical counting) number:

- **(i)** quantifies in an exact and discrete manner
- **(ii)** is abstract in the sense that it transcends the quantification of specific commodities or specific types of stimuli
- **(iii)** has a cardinal sense (produced by counting)
- **(iv)** has an ordinal sense (required for enumerating and counting)
- **(v)** is relational
- **(vi)** is combinative, operable
- **(vii)** is referred to symbolically

This is a minimal—certainly nonexhaustive—collection of properties. It doesn't cover all instances of "number," but it covers an important core of the most prototypical and fundamental cases, as in the familiar counting numbers "one," "two," "three," etc.—the *prototypical counting numbers* (Núñez & Marghetis, 2015). For example, the number "nine" quantifies in an exact manner collections with numerosity 9. "Nine" is exact, as it precisely designates a specific discrete quantity that is not just "most of the time" different from "eight" or "ten" (as seen in behavioral responses of most animal studies when corresponding numerosities are presented), but it is *absolutely* and *categorically* distinct from them (property i). "Nine" is also abstract in the sense that it transcends what psychologists call "sensory modalities" (Davis & Pérusse, 1988); it can be used to quantify (exactly) a visually perceived collection of nine balls, or a collection of acoustically perceived nine beeps (property ii). The precise quantification of those collections can be assured by counting the items in them, which gives its cardinal sense of "nine" (property iii). The prototypical counting process involves order, which gives "nine" an "ordinal sense," as when we designate the "ninth" element in a row or in a counting sequence (property iv). Moreover, the meaning of the concept "nine" is inherently intertwined with that of other numbers, and it is specified relationally, as, for instance, by being the successor of "eight" (property v) and by being "one" unit less than "ten." Incidentally, it is worth mentioning that these cardinal (iii) and relational (v) properties are usually taken, unproblematically, to be sufficient for children to learn their numbers. However, although apparently simple, they should not be taken for granted. In fact, the learning of the "cardinality principle" by children is more than just learning "words" and it is far from trivial (Sarnecka & Carey, 2008). Indeed, it has been shown that the generalization of the successor function in children is achieved only years after learning to count (Cheung et al., 2017).

Another important prototypical property of "nine" is that it can be combined to generate other numbers via specific operations (property vi), such as adding two to it to yield exactly "eleven," not just a quantity "around eleven-ish," but exactly, categorically, and distinctively "eleven." Finally, the exact quantification of a collection with numerosity 9 is referred to symbolically (property vii) via specific signs (numerals) such as the spoken/written word "nine," the digit "9," the Roman "IX," or the string of characters "$1001_2$" or "$13_6$" (when "nine" is written in base 2 or in base 6, respectively). The concept of "nine" of course doesn't get reduced to the properties just listed. Many other properties can be mentioned. For example, "nine" is the positive square root of 81, "nine" is not a prime number; it is the base 10 logarithm of 1,000,000,000, and so on. The list is endless. But what matters for the argument is that the most essential and *prototypical properties* of "nineness" (or of most prototypical "numbers") are covered by the properties listed above.

These very properties—being exact, abstract, cardinal, ordinal, relational, combinative-operable, and referred to symbolically—make Arithmetic possible. Critically, property vii—being referred to symbolically—is by nature a conventional *cultural* feature, a signature of *Homo sapiens* (Deacon, 1998). And this has some important entailments. Developing socio-historically, symbolic reference puts "number" in a qualitatively separate realm from the quantity-related phenomena observed in animals (and in humans from many nonindustrialized cultures not immersed in counting, measuring, or writing traditions). It puts "number" outside of the continuum of biological evolution via natural selection because symbolic reference is not instantiated via genetic mutations that get selected under evolutionary pressures. As biological anthropologist Deacon puts it: "symbolic reference must be acquired by learning, and lacks both the natural associations and trans-generational reproductive consequences that would make such references biologically evolvable" (Deacon, 2011, p. 393). Exact quantity-related symbolic reference is not just about "written symbols;" it's not only observed in the familiar "0–9" Hindu-Arabic numerals (digits) or in the number words we use today. Exact quantity-related symbolic reference was already present among trained individuals in societies that developed writing practices and basic accounting techniques in Mesopotamia (Robson, 2008; Schmandt-Besserat, 1992). And, in the absence of writing technology, exact symbolic reference manifests, for instance, in the use of body parts as numerals common in native groups in Papua New Guinea (Saxe, 1981; Wassmann & Dasen, 1994), in the use of knot-based artifacts such as the khipu of the Incas in the Andes (Urton, 2003), in the development of sophisticated linguistic constructions to support complex arithmetic, such as the binary calculating system in the Polynesian Mangarevan (Beller & Bender, 2003; Bender & Beller, 2014), and possibly even in those meticulously human-trained nonhuman animals such as stars chimpanzees Ai (Matsuzawa, 1985) and Sheba (Boysen & Berntson, 1996), and gray parrot Alex (Pepperberg, 1987). Written or not, exact symbolic reference places "number" outside the reach of biological evolution via natural selection. The moral is that if the quantity-related phenomena observed in nonhuman animals and infants do not exhibit these prototypical properties of "number," labeling them as "numerical" is not just inappropriate and misleading but also paves the way for fallacious teleological arguments that can claim that capacities *specific* for number are to be found in biological evolution alone. Such misqualification of "numerical" gives the impression that the underlying biologically endowed capacities are more than what they actually are: inexact perceptual quantity-related phenomena.

Now, having specified the prototypical properties of "number," we can see that if a baby, a bee, or a salamander discriminates two physical (or depicted) collections of objects based on the underlying quantities (or "numerousness"), this *does not* imply that the organism is operating with cardinal or ordinal properties of number (properties iii and iv, respectively) let alone presuppose the involvement of any categorically distinct symbolic-referential activity (property vii, i.e., not just proportions of behavioral responses driven by associative learning). In fact, this is precisely why already in the 1940s experimental psychologist E. Taves carefully and deliberately referred to "the perception of visual *numerousness*" (Taves, 1941), not to the perception of "number" as we, unfortunately, see it expressed in a large proportion of scholarly papers today. The reigning behavioral psychology of the mid-twentieth century may have been simplistic and overreductionistic in many respects when it came to the study of the complexities of the mind, but at least it gave us rigor and terminological precision at the service of the scientific method. Unfortunately, with the rejection of behavioral psychology, we seem to have thrown the baby with the bathwater, and no longer have the same standard of conceptual rigor. If the goal is to advance fruitful understanding, we must raise the standards again and regain the conceptual and terminological rigor that the study of quantity-related phenomena requires.

## Quantical vs numerical cognition: A crucial distinction

The issue to address next is the following: If the quantity-related capacities we observe in infants and nonhuman animals are *not* about *numbers* but about *quantity*, then we should not qualify them as *numerical*. In "numerical cognition," the adjective "numerical" is decidedly overinclusive: any cognition or behavior relating to quantity in babies, monkeys, rats, fish, or bees—whether exact or inexact, treated symbolically or nonsymbolically—is labeled as being "numerical." And, as mentioned earlier, this sloppy overinclusiveness opens the door for stating that thousands of species, from bees to humans, by virtue of being able to discriminate quantities de facto have "number representations" as a result of biological evolution.

To avoid sloppiness, confusion, and fallacious reasoning, the adjective "numerical" should be reserved for those phenomena that exhibit the prototypical properties listed above: exactness, cardinality, ordinality, operativity, and, crucially, symbolic reference. Therefore in the name of scientific precision, when we are, for instance, simply referring to capacities such as subitizing and large approximate quantity discrimination, we should not unproblematically speak of the "numerical" capacities of infants and nonhuman animals, or of their "numerical" cognition, "numerical" representation, "numerical" abilities, "numerical" behavior, and so on. But how should we refer to these biologically endowed phenomena? The English language (like many other languages) unfortunately does not have a common adjective for qualifying the capacities, behaviors, reasoning, and cognition that being *quantity-related* lack the seven properties described above. Moreover, we in academia have already a long-lasting (misleading) habit of using the term "numerical" to qualify a whole range of phenomena that do not distinguish between whether there is, or not, the presence of the relevant prototypical properties of "number."

If we want to address the question of whether there is a biologically evolved capacity for "number" and arithmetic, we must avoid using over-inclusive misleading concepts and confusing terms and refine our scientific terminology. How are we to adjectivally qualify these biologically endowed capacities for *quantities* that are *not numerical* (much less arithmetical)? An option, following a comment by M. Beran and A. Parrish mentioned above, would be to call them *quantitative* capacities (Beran & Parrish, 2016). But, unfortunately, "quantitative" is a term that already has a common meaning—usually contrasted with "qualitative"—which relates to measurements and their numerical and mathematical treatment (as in the analysis of financial markets). So, the term "quantitative" does not seem appropriate. But while keeping its Latin origin "quantitas" (quantity), I propose to refer to subitizing, large quantity discrimination (LQD), and other such biologically endowed capacities as *quantical*—in contrast with "numerical." Note that the adjective quantical is deliberately exclusive. To avoid the misleading overinclusiveness of "numerical," "quantical" explicitly excludes the prototypical properties of number listed above, while preserving basic quantity-related biologically endowed phenomena (see Fig. 1).

Thus quantical cognition, quantical behavior, quantical abilities, quantical discrimination, and so on, although dealing with *quantity*, are not required to meet the more elaborated and complex criteria of being exact, abstract, cardinal, ordinal, combinatorial, operable, or referred to symbolically. Only if such properties are observed in some nonhuman species or human infants, we can qualify them as "numerical." Thus subitizing may be exact within a range (numerosity 1–3), but it is not combinatorial/operative

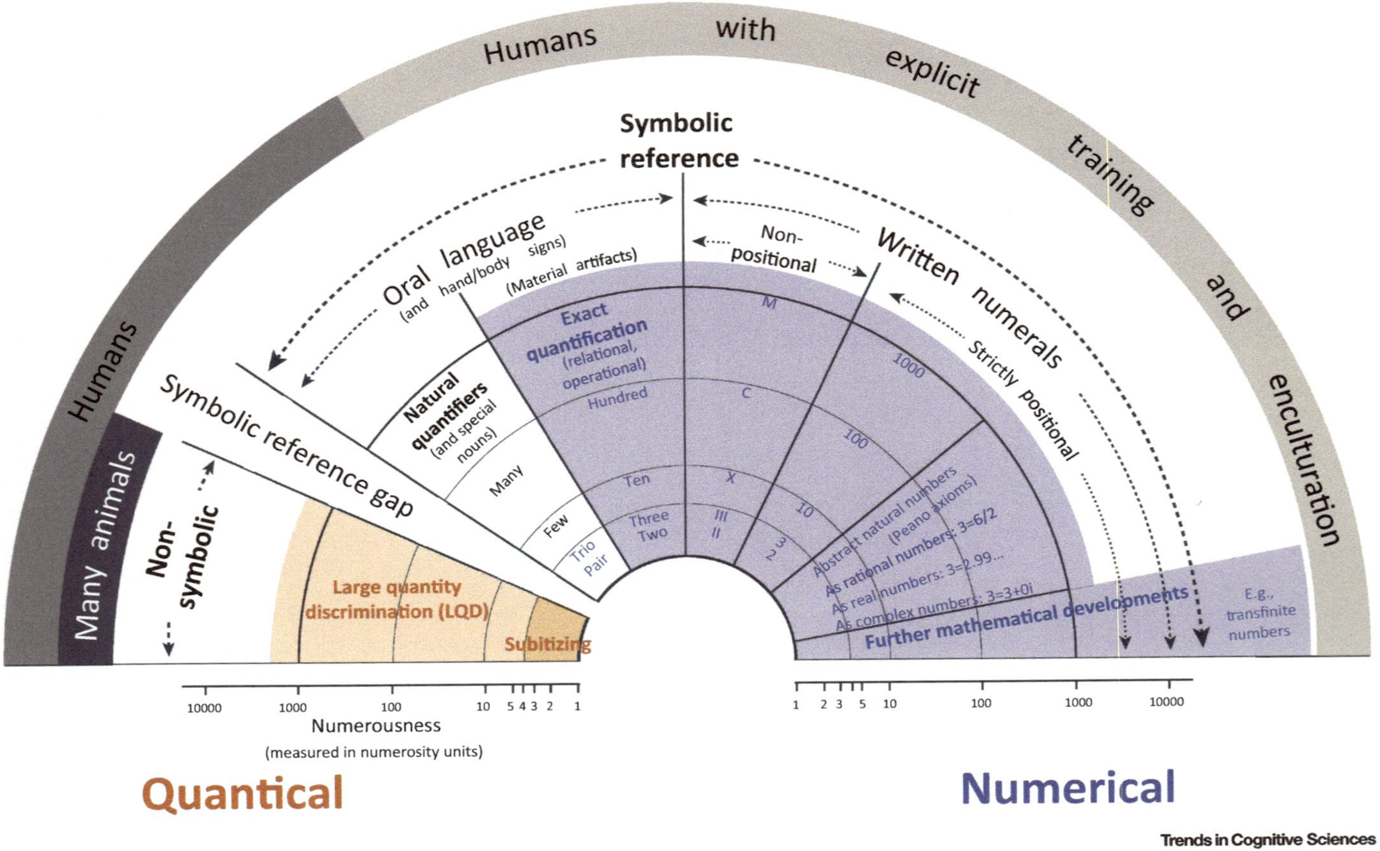


**Fig. 1** Quantical vs numerical cognition: relevant distinctions involved in the treatment of discrete quantity. *Reprinted from Núñez, R. (2017). Is there really an evolved capacity for number? Trends in Cognitive Sciences, 21, 409–424; used with permission from Trends in Cognitive Sciences, Elsevier.*

or referred to symbolically, therefore is not a numerical capacity, but rather a quantical capacity. Similarly, the capacity for discriminating large quantities in an inexact manner (given a particular ratio between their numerosities) is not a numerical capacity either, but a quantical capacity.

The quantical-numerical distinction not only is conceptually more precise, but it also reveals how misleading certain theoretical constructs in the field of "numerical cognition" can be. For example, the construct *Approximate Number System* (ANS) mentioned earlier is believed (not without criticisms; Gevers, Cohen Kadosh, & Gebius, 2016; Gebius, Cohen Kadosh, & Gevers, 2016; Leibovich, Katzin, Harel, & Henik, 2016; Leibovich, Kallai, & Itamar, 2016) to neurally handle quantities above the subitizing range. As evidence, it is often mentioned that there are similarities of ratio effects between nonsymbolic and symbolic stimuli (Dehaene, 1997; Piazza, 2010). However, these similarities can be explained by different underlying mechanisms when it comes to nonsymbolic (quantical) and symbolic (numerical) processing (Krajcsi, 2016; Lyons, Nuerk, & Ansari, 2015), which supports the idea that quantical and numerical cognition are not only fundamentally distinct, but they relate to each other in nonobvious ways. Therefore such findings cast serious doubts on the claim put forward by S. Dehaene that "when we learn number symbols, we simply learn to attach their arbitrary shapes to the relevant non-symbolic quantity representations" (Dehaene, 2007, p. 552). Furthermore, the term "ANS" implicitly takes "numbers" as preexisting primitives to which noisy mental representations are taken to be "approximate" to. But if a prototypical property of "number" is to be exact, then the term "ANS" really is a misleading teleologically driven oxymoron. It makes the hypothesized neural system to be about "numbers" and misleadingly puts "number"—with all its complexities—directly in the realm of what is biologically endowed (i.e., in the explanans—in the very "explanation" of number).

To capture the discontinuity brought by exact symbolic quantification, the quantical-numerical distinction is crucial. The adjective "quantical" does not take preexisting numbers as primitives (placed in the explanans) with representations being approximate to them. Rather, it labels large quantity discrimination, for instance—upon which quantifications such as "few" and "many" build—without being "approximate" to anything. Indeed, *quantical cognition* may serve as a basis for explaining three important features of human languages worldwide:

- The (nearly) universal presence of exact quantification within the subitizing range (numerosities 1–3).

- The nonuniversality of exact quantification beyond the subitizing range.
- The universal presence of natural quantifiers that capture fuzzy and imprecise distinctions made in the range of large quantity discrimination.

Natural quantifiers are universal linguistic manifestations of seemingly meaningful experience with medium-to-large quantities built on quantical cognition, which would explain why native speakers don't need scholastic (or explicitly concocted) training to achieve proficiency in using them. Critically, although very functional in dealing with *quantities* in everyday life, quantifiers in natural language, like quantical competences in animals, cannot scale up to build an exact number system and an exact arithmetic, that is, an arithmetic that yields specific results for specific operations (e.g., "few" plus "several" doesn't yield an exact and unambiguous quantifier). For this, the concept of number—exact, abstract, cardinal, ordinal, combinatorial, operable, and symbolic—is needed. And this seems to require extra ingredients and cultural preoccupations and practices that are not shaped via biological evolution properly. Not surprisingly, lacking precise cardinality, combinatoriality, and operativity, the field of "numerical cognition" has not devoted much attention to natural quantifiers and the role they might play in the development of numerical cognition.

If we want to scientifically investigate the nature of mathematics and its biological constraints, we must disentangle the various components, properties, and processes in a clear manner. And for this, the quantical-numerical distinction is needed. We do not want to blur the neurological and behavioral processes involved in treating imprecise quantity with exact symbolic number, considering them under the same umbrella term of "numerical cognition." Quantical cognition is biologically endowed, but numerical cognition is not. Quantical cognition may be the manifestation of biologically evolved preconditions (BEPs) for numerical cognition and arithmetic but in itself is not *about* number or arithmetic (let alone mathematics). Indeed, quantical processing seems to be about many sensorial phenomena and dimensions other than number per se (Cantrell & Smith, 2013; Gebius et al., 2016; Leibovich & Ansari, 2016; Leibovich, Katzin, et al., 2016). Quantical capacities are not precursors that inevitably lead to numerical capacities, but rather biologically evolved preconditions (BEPs) for them that only lead to numerical capacities when the relevant cultural traits and practices intervene. Crucially, quantical cognition does not, by itself, scale up to produce number, and eventually arithmetic, but numerical cognition does.

## Number: Not just language and symbols, but evolving cultural preoccupations and practices that generate them

To account for the observed discontinuity between human and nonhuman numerical and arithmetical capacities, human language and symbols are usually invoked as *sufficient* conditions for explaining the gap (e.g., Leibovich, Katzin, et al., 2016; Wiese, 2003a, 2003b). However, the underemphasized study of small-scale nonindustrialized human cultures reveals that language and symbolization may be *necessary* conditions for the formation of number but alone do not produce numbers and arithmetic: they are necessary, but not sufficient conditions for explaining the discontinuity. As mentioned, all human cultures and societies have developed and employ spoken language, but not all have developed a system of exact numbers beyond the subitizing range. Rather, they have all developed natural quantifiers, which appear to symbolize universal quantity-related experiences (e.g., the perception of "numerousness") brought by biologically endowed quantical cognition. Critically, although very functional in dealing with *quantities* in everyday life, quantifiers in natural language—like quantical competences in animals—can not scale up to build an exact number system and arithmetic (for complementary arguments along these lines, see Overmann & Coolidge, 2013). To yield exact results for specific operations, the concept of number is needed—symbolic exact quantifiers. This seems to require the crucial contribution of particular cultural preoccupations and practices (Beller et al., 2018; Coolidge & Overmann, 2012; Robson, 2008; Schmandt-Besserat, 1992), which for the case of numbers have been well documented for groups in Papua New Guinea (Saxe, 1981, 2012) and Polynesia (Beller & Bender, 2003), for instance. Sustained by language (Bender & Beller, 2014), symbolic reference (Bender & Beller, 2014; Núñez, Cooperrider, & Wassmann, 2012; Saxe, 1981) (e.g., body count systems; Saxe, 1981, 2012; Wassmann & Dasen, 1994), imagination, and metaphor (Lakoff & Núñez, 2000; Núñez, 2009; Núñez & Marghetis, 2015), these preoccupations and practices are not shaped via biological evolution proper, but via cultural evolution and the enculturation of the human brain (Ansari, 2008; Núñez, 2017b).

In humans, the learning of the simplest aspects of numbers and arithmetic requires extensive training and cultural scaffolding. Importantly, this enculturation process affects even the most fundamental aspects of neural processing of number. For example, an fMRI study compared the brain activation of schooled native speakers of Chinese and English while they performed simple addition and made relative magnitude judgments (Tang et al., 2006). Although both groups were presented with the same symbols—Hindu-Arabic numerals—different brain activation and functional connectivity between brain regions were found. Native Chinese speakers showed more activation in premotor regions, whereas native English speakers showed more activation in the left supplemental motor and perisylvian regions, including Broca's and Wernicke's areas, which are usually associated with language production and comprehension. These findings support the claim that the neural circuits and brain regions that are recruited to sustain even the most fundamental aspects of exact symbolic number processing are critically mediated by cultural factors, such as writing systems, educational organization, and enculturation.

## Concluding remarks with a challenging scientific question: How do we get from quantical to numerical cognition?

In order to begin investigating the nature and origin of mathematics, this chapter addressed the following fundamental guiding questions: Are the quantity-related capacities that have been observed in human infants and many nonhuman animals really about *numbers*, and about *arithmetic*? And, relatedly, is there a biologically evolved capacity *specific* for number and arithmetic? We have seen that humans and many nonhuman species do have biologically endowed abilities for perceiving and discriminating *quantities*, at least in an imprecise manner and in certain formats (e.g., food items). But, although robust, these observations do not support claims that there is an evolved capacity specific for number and arithmetic. Such claims are based on an implicit teleological rationale guided by sloppy and misleading terminology. They build on an inaccurate conception of biological evolution that takes "number" and "arithmetic" as preexisting entities. Through this lens, influential views in developmental psychology, animal cognition, and cognitive neuroscience have failed to consider crucial human data from nonindustrialized millennia-old cultures and have overinterpreted results with trained animals in captivity, resulting in an overstated role for biological evolution in the origins of numbers, arithmetic, and mathematics.

"Number" is a complex and polysemous concept, which must be specified in detail, especially when evaluating claims about evolution. Accordingly, the behaviors, capacities, and cognition that are to be labeled as "numerical" must be carefully delineated. To understand the existing human and nonhuman data in light of evolution, we must disentangle biologically endowed quantical capacities—which are imprecise and nonsymbolic—from strictly speaking numerical capacities that exhibit exact symbolic quantification. Importantly, the distinction makes clear that quantical capacities don't scale up to numerical capacities via natural selection alone.

Equipped with more precise and appropriate concepts and terminology, we can state that the impressive biologically endowed capacities observed in nonhuman animals are not *numerical*, but *quantical*. And this applies also to human infants, and to healthy human adults who belong to cultures with limited number lexicon and no writing traditions. Exact, systematic, and symbolic quantification beyond the subitizing range has materialized via cultural preoccupations, necessities, and practices that involve language and symbolic reference. The entailment is that number, arithmetic and, by extension, mathematics, did not arise from a *specific* "mathematical core" that evolved biologically over millions of years, manifesting in monkeys, birds, fish, and insects. Instead, building on domain-general biologically evolved preconditions (BEPs) that sustain quantical cognition, developed from much more recent cultural developments that took place outside of natural selection.

Currently, the field of "numerical cognition" asks questions such as how an "approximate number system" becomes increasingly more precise, and how with symbolic language, training, learning, and expertise, precise concepts of "number" come to being. The question, however, implicitly takes as point of departure that there are preexisting "numbers"—specific reference points such as "six" or "fourteen"—to which noisy and fuzzy mental representations are "approximate" to. With the quantical-to-numerical approach, the question is of a different nature. What is taken as point of departure is not that there are preexisting "numbers," like "six" or "fourteen," with approximate representations. But rather that, supported by quantical cognition, there are systematic perceptual experiences that lead to imprecise symbolic quantifications such as "few" and "many" which have no specific reference points at all. They are not "approximate" to anything. They just quantify numerousness in an imprecise manner. Via natural quantifiers in human languages, rough quantical cognition can become abstract and symbolic. But the human language capacity by itself is not enough to lead to numerical cognition. All cultures around the world have languages, and as far as we know, all of them are able to quantify in an imprecise manner. But precise quantification beyond the subitizing range seems to require extra cultural preoccupations, conventions, training, and learning, all of which abound in our industrialized society with writing technologies, systematic and organized educational systems, and complex societal commercial exchanges.

In sum, in order to investigate the question of the foundations and origin of mathematics, an important group of questions must be addressed: What does it take to move from nonsymbolic biologically evolved *quantical* cognition to symbolic and precise *numerical* cognition? How do the two forms relate to each other? What selective pressures exactly may have given rise to quantical cognition in so many different species? What are the necessary conditions for human cultures to develop exact quantification and number systems? What exactly is the biological enculturation process that supports them? What are the neurological underpinnings of the learning of quantifiers in natural language? and so on. Although ubiquitous today in the industrialized world, "numerical" cognition should not be taken at face value. To understand the origins of arithmetic and mathematics, and the processes through which these disciplines are learned, we must understand the complex passage from quantical to numerical cognition. A passage that made possible via specific forms of learning and education, only some humans in certain socio-historic contexts have realized.

## References


Agrillo, C. (2015). Numerical and arithmetic abilities in non-primate species. In R. Cohen Kadosh, & A. Dowker (Eds.), *The Oxford handbook of numerical cognition* (pp. 214–236). Oxford, UK: Oxford University Press.

Ansari, D. (2008). Effects of development and enculturation on number representation in the brain. *Nature Reviews Neuroscience*, *9*, 278–291.

Antell, S. E., & Keating, D. P. (1983). Perception of numerical invariance in neonates. *Child Development*, *54*, 695–701.

Beller, S., & Bender, A. (2003). The limits of counting: Numerical cognition between evolution and culture. *Science*, *319*, 213–215.

Beller, S., Bender, A., Chrisomalis, S., Jordan, F., Overmann, K., Saxe, G., et al. (2018). The cultural challenge in mathematical cognition. *Journal of Numerical Cognition*, *4*(2), 448–463. https://doi.org/10.5964/jnc.v4i2.137.

Bender, A., & Beller, S. (2014). Mangarevan invention of binary steps for easier calculation. *Proceedings of the National Academy of Sciences of the United States of America*, *111*(4), 1322–1327.

Beran, M., & Parrish, A. (2016). Going for more: Discrete and continuous quantity judgments by nonhuman animals. In A. Henik (Ed.), *Continuous issues in numerical cognition: How many or how much* (pp. 175–192). Amsterdam: Elsevier.

Beran, M., Perdue, B., & Evans, T. (2015). Monkey mathematical abilities. In R. C. Kadosh, & A. Dowker (Eds.), *The Oxford handbook of numerical cognition* (pp. 237–257). Oxford University Press.

Bowern, C., & Zentz, J. (2012). Diversity in the numeral systems of Australian languages. *Anthropological Linguistics*, *54*(2), 133–160.

Boysen, S. (1988). Kanting processes in the chimpanzee: What (and who) really counts? *Behavioral and Brain Sciences, 11*, 580.
Boysen, S., & Berntson, G. (1996). Quantity-based inference and symbolic representation in chimpanzees (pan troglodytes). *Journal of Experimental Psychology: Animal Behavior Processes, 22*, 76–86.
Brannon, E. (2005). What animals know about numbers. In J. Campbell (Ed.), *Handbook of mathematical cognition* (pp. 85–107). New York: Psychology Press.
Brannon, E., & Terrace, H. (1998). Ordering of the numerosities 1 to 9 by monkeys. *Science, 282*(5389), 746–749.
Brugger, P. (2015). Chicks with a number sense. *Science, 347*, 477–478.
Campbell, J. (Ed.). (2005). *Handbook of mathematical cognition*. New York: Psychology Press.
Cantrell, L., & Smith, L. B. (2013). Open questions and a proposal: A critical review of the evidence on infant numerical abilities. *Cognition, 128*(3), 331–352.
Cheung, P., Rubenson, M., & Barner, D. (2017). To infinity and beyond: Children generalize the successor function to all possible numbers years after learning to count. *Cognitive Psychology, 92*, 22–36.
Cohen Kadosh, R., & Dowker, A. (Eds.). (2015). *The Oxford handbook of numerical cognition*. Oxford, UK: Oxford University Press.
Coolidge, F. L., & Overmann, K. A. (2012). Numerosity, abstraction, and the emergence of symbolic thinking. *Current Anthropology, 53*(2), 204–225.
Dacke, M., & Srinivasan, M. V. (2008). Evidence for counting in insects. *Animal Cognition, 11*(4), 683–689.
Dadda, M., Piffer, L., Agrillo, C., & Bisazza, A. (2009). Spontaneous number representation in mosquitofish. *Cognition, 112*, 343–348.
Davis, H., Albert, M., & Barron, R. (1985). Detection of number or numerousness by human infants. *Science, 228*, 1222.
Davis, H., & Pérusse, R. (1988). Numerical competence in animals: Definitional issues, current evidence, and a new research agenda. *Behavioral and Brain Sciences, 11*, 561–615.
Deacon, T. W. (1998). *The symbolic species: The co-evolution of language and the brain*. New York: WW Norton & Company.
Deacon, T. W. (2011). The symbol concept. In K. R. Gibson, & K.R., & Tallerman, M. (Eds.), *The Oxford handbook of language evolution* (pp. 393–405). Oxford University Press. https://doi.org/10.1093/oxfordhb/9780199541119.013.0043.
Dehaene, S. (1997). *The number sense: How the mind creates mathematics*. New York: Oxford University Press.
Dehaene, S. (2007). Symbols and quantities in parietal cortex: Elements of a mathematical theory of number representation and manipulation. In P. Haggard, & Y. Rossetti (Eds.), *Attention & performance XXII. Sensori-motor foundations of higher cognition* (pp. 527–574).
Epps, P., Bowern, C., Hansen, C. A., Hill, J., & Zentz, J. (2012). On numeral complexity in hunter-gatherer languages. *Linguistic Typology, 16*(1), 41–109.
Everett, C. (2017). *Numbers and the making of us: Counting and the course of human cultures*. Cambridge MA: Harvard University Press.
Flombaum, J. I., Junge, J. A., & Hauser, M. D. (2005). Rhesus monkeys (Maccaca mulatta) spontaneously compute addition operations over large numbers. *Cognition, 97*, 315–325.
Gebius, T., Cohen Kadosh, R., & Gevers, W. (2016). Sensory-integration system rather than approximate number system underlies numerosity processing: A critical review. *Acta Psychologica, 171*, 17–35.
Gevers, W., Cohen Kadosh, R., & Gebius, T. (2016). Sensory integration theory: An alternative to the approximate number system. In A. Henik (Ed.), *Continues issues in numerical cognition: How many or how much* (pp. 405–418). Amsterdam: Elsevier.
Halberda, J., Mazzocco, M., & Feigenson, L. (2008). Individual differences in non-verbal number acuity correlate with maths achievement. *Nature, 455*, 665–668.

Hauser, M. D., Carey, S., & Hauser, L. B. (2000). Spontaneous number representation in semi–free–ranging rhesus monkeys. *Proceedings of the Royal Society of London B: Biological Sciences*, *267*(1445), 829–833.

Hauser, M. D., Dehaene, S., & Dehaene-Lambertz, G. (2002). Spontaneous number discrimination of multi-format auditory stimuli in cotton-top tamarins (Saguinus Oedipus). *Cognition*, *86*(2), B23–B32.

Henik, A. (Ed.). (2016). *Continuous issues in numerical cognition: How many or how much*. Amsterdam: Elsevier.

Hockett, C. (1960). The origin of speech. *Scientific American*, *203*, 88–111.

Howard, S. R., Avarguès-Weber, A., Garcia, J. E., Greentree, A. D., & Dyer, A. G. (2019). Numerical cognition in honeybees enables addition and subtraction. *Science Advances*, *5*(2), eaav0961.

Ifrah, G. (1985). *From one to zero: A universal history of numbers*. New York: Viking Press.

Krajcsi, A. (2016). Numerical distance and size effects dissociate in indo-Arabic number comparison. *Psychonomic Bulletin Review*. https://doi.org/10.3758/s13423-016-1175-6. (October 17 epub ahead of print.

Lakoff, G., & Núñez, R. (2000). *Where mathematics comes from: How the embodied mind brings mathematics into being*. New York: Basic Books.

Leibovich, T., & Ansari, D. (2016). The symbol-grounding problem in numerical cognition: A review of theory, evidence, and outstanding questions. *Canadian Journal of Experimental Psychology*, *70*(1), 12–23.

Leibovich, T., Kallai, A. Y., & Itamar, S. (2016). What do we measure when we measure magnitudes? In A. Henik, & A. (Eds.), *Continuous issues in numerical cognition: How many or how much* (pp. 355–374). Amsterdam: Elsevier.

Leibovich, T., Katzin, N., Harel, M., & Henik, A. (2016). From 'sense of number' to 'sense of magnitude'—The role of continuous magnitudes in numerical cognition. *Behavioral and Brain Sciences*. https://doi.org/10.1017/S0140525X16000960 (published online 17 august 2016).

Lyons, I. M., Nuerk, H.-C., & Ansari, D. (2015). Rethinking the implications of numerical ratio effects for understanding the development of representational precision and numerical processing across formats. *Journal of Experimental Psychology: General*, *144*(5), 1021–1035.

Matsuzawa, T. (1985). Use of numbers by a chimpanzee. *Nature*, *315*, 57–59.

Menninger, K. (1969). *Number words and number symbols: A cultural history of numbers*. Cambridge, MA: MIT Press.

Nieder, A. (2016). The neuronal code for number. *Nature Reviews Neuroscience*, *17*, 366–382.

Nieder, A. (2018). Honey bees zero in on the empty set. *Science*, *360*, 1069–1070.

Núñez, R. (2009). Numbers and arithmetic: Neither hardwired nor out there. *Biological Theory*, *4*, 68–83.

Núñez, R. (2017a). Is there really an evolved capacity for number? *Trends in Cognitive Sciences*, *21*, 409–424.

Núñez, R. (2017b). Number—Biological enculturation beyond natural selection. *Trends in Cognitive Sciences*, *21*, 404–405. https://doi.org/10.1016/j.tics.2017.03.013.

Núñez, R., Cooperrider, K., & Wassmann, J. (2012). Number concepts without number lines in an indigenous Group of Papua new Guinea. *PLoS One*. https://doi.org/10.1371/journal.pone.0035662.

Núñez, R., & Marghetis, T. (2015). Cognitive linguistics and the concept(s) of number. In R. Cohen Kadosh, & A. Dowker (Eds.), *The Oxford handbook of numerical cognition* (pp. 377–401). Oxford, UK: Oxford University Press.

Overmann, K., & Coolidge, F. L. (2013). On the nature of numerosity and the role of language in developing number concepts. *Current Anthropology*, *54*(1), 83–84.

Pepperberg, I. M. (1987). Evidence for conceptual quantitative abilities in the African Grey parrot: Labelling of cardinal sets. *Ethology, 75*, 37–61.
Piaget, J. (1952). *The Child's conception of number*. New York: Routledge.
Piazza, M. (2010). Neurocognitive start-up tools for symbolic number representations. *Trends in Cognitive Sciences, 14*(12), 542–551.
Robson, E. (2008). *Mathematics in ancient Irak*. Princeton, NJ: Princeton University Press.
Rugani, R., Regolin, L., & Vallortigara, G. (2011). Summation of large numerousness by new-born chicks. *Frontiers in Psychology (Comparative Psychology)*. https://doi.org/10.3389/fpsyg.2011.00179.
Sarnecka, B., & Carey, S. (2008). How counting represents number: What children must learn and when they learn it. *Cognition, 108*, 662–674.
Saxe, G. (1981). Body parts as numerals: A developmental analysis of numeration among the Oksapmin in Papua New Guinea. *Child Development, 52*, 306–316.
Saxe, G. (2012). *Cultural development of mathematical ideas: Papua New Guinea studies*. New York: Cambridge University Press.
Schmandt-Besserat, D. (1992). *Before writing: From counting to cuneiform*. Austin: University of Texas Press.
Stevens, S. S. (1939/2006). On the problem of scales for the measurement of psychological magnitudes. *Proceedings of Fechner Day, 22*(1), 23–27.
Stevens, S. S. (1951). Mathematics, measurement, and psychophysics. In S. S. Stevens (Ed.), *Handbook of experimental psychology* (pp. 1–49). New York: John Wiley.
Tang, Y., Zhang, W., Chen, K., Feng, S., Ye Ji, Y., Shen, J., et al. (2006). Arithmetic processing in the brain shaped by cultures. *Proceedings of the National Academy of Sciences, 103*(28), 10775–10780.
Taves, E. H. (1941). Two mechanisms for the perception of visual numerousness. *Archives of Psychology (Columbia University), 265*, 47.
Urton, G. (2003). *Signs of the Inka Khipu: Binary coding in the Andean knotted string records*. Austin, TX: Texas University Press.
Von Glasersfeld, E. (1988). Difficulties of demonstrating the possession of concepts. *Behavioral and Brain Sciences, 11*, 601.
Wassmann, J., & Dasen, P. (1994). Yupno number system and counting. *Journal of Cross-Cultural Psychology, 25*, 78–94.
Wiese, H. (2003a). Iconic and non-iconic stages in number development: The role of language. *Trends in Cognitive Sciences*, 7(9), 385–390.
Wiese, H. (2003b). *Numbers, language, and the human mind*. Cambridge, UK: Cambridge University Press.
Xu, F. (2003). Numerosity discrimination in infants: Evidence for two systems of representations. *Cognition, 89*, B15–B25.
Xu, F., & Spelke, E. (2000). Large number discrimination in 6-month-old infants. *Cognition, 74*, B1–B11.

SECTION I

# Advanced mathematics, reasoning and problem solving

# CHAPTER 2

# Toward an inhibitory control theory of the reasoning brain in development

**Grégoire Borst**[a,b]
[a]Institut de Psychologie, Université de Paris, LaPsyDÉ, CNRS, Paris, France
[b]Institut Universitaire de France, Paris, France

## The paradoxes of cognitive development

In one of the most famous developmental psychology tasks—the number-conservation task, designed by Jean Piaget at the beginning of the 20th century (Piaget, 1983)—two alignments of tokens of equal length are presented, and children are asked whether the two alignments contain the same number of tokens. At age 5, children acknowledge the so-called initial equivalence, stating that the two alignments indeed contain the same number of tokens. Then, the experimenter transforms the length of one of the two alignments and asks once again whether the two alignments contain the same number of tokens. Piaget observed that children under 7 years of age stated that the two alignments did not contain the same number of tokens after the transformation of the length of one of the two alignments, disregarding the fact that the transformation affected one physical quantity (i.e., length) without affecting the other (i.e., numerosity). Surprisingly, in the 1980s, studies demonstrated that infants already understood at age 4 months that length did not covary with number (Antell & Keating, 1983). This type of paradox, in which older children commit systematic errors in contexts that are supposed to require the same knowledge already mastered by younger children (or even infants), is not limited to one cognitive domain or one age range. Adults, for instance, are systematically biased in very simple logical problems that require logical reasoning similar to the one displayed by 7-year-old children (Kahneman, 2011). In one such problem, the so-called Linda conjunction fallacy problem, Linda is described as a woman who was a feminist activist in the 1970s, and participants are asked whether it is more likely that Linda would today be a banker or a banker and a feminist. Most adults think that Linda is more likely to be a banker

*Heterogeneous Contributions to Numerical Cognition*
https://doi.org/10.1016/B978-0-12-817414-2.00006-3

and a feminist, disregarding the principles of probabilities (i.e., the probability of a conjunction of events is always lower than the probability of a single event). Strikingly, children have already mastered the class-inclusion logic needed to succeed in the Linda conjunction fallacy problem at age 7, as demonstrated by their ability to succeed in the class-inclusion task designed by Piaget (1983). In this task, children are presented with eight daisies and two roses and asked whether there are more daisies than flowers. To succeed at this task, children must understand that there are more elements in the superordinate class (i.e., flower) than in the subordinate classes (i.e., daisies), just as in the Linda problem, one must understand that there are always more bankers (i.e., superordinate class) than bankers and feminists (i.e., subordinate class).

In the present article, we first argue that these paradoxes of cognitive development can be best accounted for by assuming that cognitive development is akin to a nonlinear dynamic system (Smith & Thelen, 2003), in contrast with incremental models of development, such as Piaget's constructivist (Piaget, 1983), or innate models of development, such as Spelke's core knowledge (Spelke, 2000) models. We then present converging evidence that systematic errors can appear at each age and in specific contexts both in the lab and at school when one fails to inhibit a misleading strategy to activate the appropriate one (Houdé & Borst, 2014). Finally, we discuss potential future research avenues to further understand the role of inhibitory control in cognitive development.

## From linear to nonlinear dynamic models of cognitive development

Cognitive development is generally conceived as linear and incremental, with more complex knowledge progressively acquired with age. For instance, in the seminal constructivist theory, Piaget (1983) argues that children progressively acquire a logical understanding of the principles that govern the physical world by directly acting in it. According to this theory, the logical structures of children's mind become increasingly more complex throughout four fundamental shifts (or stages) during which the logical structures at a given stage are combined to create more complex structures at the next stage. These shifts occur at about the same age for all children. They start with a shift from the sensorimotor stage (from birth to 2 years of age), during which perception and motor processes are prevalent, to the preoperational stage (from 2 to 7 years of age), during which children are mainly illogical, according to Piaget. Shifting to the concrete operational stage (from 7 to 12 years of age), children start to reason logically but only on concrete objects (i.e., the success in the number-conservation and class-inclusion tasks presented above are hallmarks of the shift to this stage), while the shift to the formal operational stage (after 12 years of age) allows children to reason on abstract propositions.

Piaget's constructivist theory has been widely criticized in part because it underestimates the rich conceptual knowledge of infants on the physical principles of their environment (Spelke, 2000). Indeed, 4-month-old infants seem already to possess an understanding of the physical properties of objects, of agents and their actions, of the numerical properties of sets of objects, and of the spatial relations of objects and their geometrical organization in the environment (Spelke, 2000). Evidence that knowledge was similar in infants and in adults led some developmental psychologists to argue that cognitive development essentially consists of the refinement of these four core knowledge systems (Spelke, 2000). A clear limitation of these two models is their inability to account for nonlinear (i.e., U-shaped) developmental patterns in which older children commit systematic errors in a situation that supposedly engages the same knowledge mastered by infants, and adults fail to solve problems requiring knowledge mastered by children and adolescents (Houdé & Borst, 2014).

Some have argued that the nature and degree of the explicitness of the knowledge tested in tasks administered to infants and children can explain the apparent paradox of early competencies followed by late reasoning errors. According to this theoretical account, tasks administered to infants mainly test their implicit knowledge, whereas tasks administered to children, particularly Piagetian tasks, test explicit knowledge of the same principles (Karmiloff-Smith, 1992). It is easier to succeed on the former than the latter, which would explain early competencies in infants followed by later reasoning errors in older children. This model assumes that a domain-general process of representational redescription allows children to shift from implicit to explicit knowledge regarding the principles that govern the physical world. Although this model explains the U-shaped developmental pattern observed in various domains between infanthood and childhood, it cannot account for the late reasoning errors observed in adults in tasks that engage the explicit knowledge already mastered by children.

In contrast to previous models, we argue that cognitive development should be conceived as a nonlinear dynamic system (Siegler, 1999; Smith & Thelen, 2003) in which errors can appear at any age and in any context when children, adolescents, or adults rely on prepotent responses, illogical intuitions, or misleading strategies (such as heuristics) to solve a task rather than on deliberate, logical strategies adapted to the context. Importantly, we assume that inhibitory control is one of the core mechanisms of intelligence that allows one to overcome systematic errors in a given task by inhibiting misleading prepotent responses and strategies (Houdé & Borst, 2014). This assumption is consistent with studies showing that the ability of children to succeed in theory of mind (Benson, Sabbagh, Carlson, & Zelazo, 2013), counterfactual reasoning (Beck, Riggs, & Gorniak, 2009), and strategic reasoning (Apperly & Carroll, 2009) tasks might be in part related to the growing inhibitory control ability between the ages of 3 and 5 years. Inhibitory control is conceived in our model and these studies as a process that allows children and adults to resist habits or automatisms, temptations, distractions, or interference and to adapt to conflicting situations (Diamond, 2013).

As in Siegler's model (Siegler, 1999), we postulate that several strategies with different degrees of complexity and sophistication can be used to solve a task, but as Kahneman (2011) does, we assume that individuals rely spontaneously on heuristics (i.e., fast, effortless, global or holistic strategies) rather than logical algorithms (i.e., slow, effortful, analytical strategies) to perform such tasks. In most contexts, heuristics are the most widely adapted strategies and provide a faster way to reach a solution than algorithms, but in other contexts, they can be misleading when they interfere with the activation of the proper adapted algorithms. Thus systematic errors do not necessarily reveal a lack of knowledge; rather, they reveal a lack of ability to express a given knowledge in a context in which a misleading heuristic constructed during childhood interferes with it. According to our theoretical framework, the progressive ability to inhibit misleading heuristics sustains cognitive development by allowing children, adolescents, and adults to understand the domain of the validity of algorithms by succeeding in activating those algorithms in various contexts in which they compete with different heuristics (Houdé & Borst, 2014). In addition, the ability to inhibit misleading heuristics in a given context is directly related to the ability to detect the conflict between the misleading heuristics and the logical algorithms (De Neys, 2012). A growing number of studies provided evidence that children, adolescents, and adults when committing errors typically display conflict sensitivity as reflected in longer response times and lower confidence in their response in the conflict condition (in which the heuristic is misleading and interferes with the algorithm) than in the no-conflict condition (in which the heuristic is adapted to the context and congruent with the algorithm) (De Neys, 2012). Importantly, conflict detection occurs under time pressure and cognitive load suggesting that it could be an automatic process (De Neys, 2017).

We acknowledge that the development of inhibitory control ability is necessary but might not be sufficient to produce conceptual development during childhood and adolescence. Indeed, the development of other executive functions such as working memory (Halford, Cowan, & Andrews, 2007) might also be critical to support conceptual development and reasoning (Morsanyi & Handley, 2008). Finally, we note that inhibiting misleading heuristics might remain challenging throughout childhood, adolescence, and adulthood because of the slow maturation (until 25 years of age) of the prefrontal cortex (Casey, Tottenham, Liston, & Durston, 2005), an area of the brain devoted to inhibitory control (Aron, Robbins, & Poldrack, 2014) and more generally to executive functions (Diamond, 2013).

In what follows, we present evidence that success in three seminal Piagetian tasks (i.e., the A-B, number-conservation, and class-inclusion tasks), in reasoning tasks, and in perspective-taking tasks reflects not a new stage of development but the progressive ability to inhibit misleading heuristics.

## Inhibitory control and misleading heuristics in the lab

### The A-B problem

In the A-B problem designed by Piaget (Piaget, 1983) to test infants' (8–12 months old) understanding of object permanence (i.e., understanding that an object continues to exist even when out of sight), an infant is placed in front of two equally reachable screens (A and B), and an object is hidden behind A. Infants as young as 8 months can easily find the object under A. After a few repetitions of hiding and finding the object under A, the object is transferred from A to B. Infants up to 12 months old commit the famous A-not-B error, seeking the object under A while the object is hidden under B. This error is surprising in light of the fact that 4-month-old infants have already mastered several principles regarding objects and their displacement in the environment (Spelke, 2000). Several studies have demonstrated that the A-not-B error reflects not a lack of object permanence but a lack of maturation of the prefrontal cortex, allowing infants before 12 months of age to inhibit the prepotent gesture (i.e., a motor heuristic) of reaching for the object under A, reinforced by the repetition of hiding the object under A (Diamond, 1998). Convergent evidence for the role of inhibitory control in succeeding in the A-B problem was provided by an electroencephalographic study showing a relation between infants' improvement in this task and an increase in the electric activity of the frontal areas of the brain (Bell & Fox, 1992).

## The number-conservation task

As in Piaget's A-B problem, several studies have demonstrated that success in the number-conservation task for both children and adults relies on inhibiting a misleading heuristic—here, the "length-equals-number" heuristic (Borst, Simon, Vidal, & Houdé, 2013; Houdé et al., 2011; Houdé & Guichart, 2001). Indeed, children under 7 years of age erroneously think that the longer alignment contains more tokens after tokens have been spread apart in the number-conservation task. The "length-equals-number" heuristic is most likely created during childhood because our brain detects in our visual environment that longer alignments tend to have more objects (Houdé & Borst, 2014). Interestingly, this heuristic is sometimes reinforced at school when teaching preschool children to count: illustrations in the classroom or in textbooks depict the 1-to-10 Arabic numbering series with rows of animals of increasing length, reinforcing the "length-equals-number" heuristic. Using a negative priming paradigm, Houdé and Guichart (2001) provided the first evidence that solving the number-conservation task is rooted in the ability to inhibit the "length-equals-number" heuristic. The negative priming paradigm rests on the logic that if a heuristic is inhibited on a given item (i.e., the prime), then the subsequent activation of that heuristic in the next item (the probe) will be disrupted, as revealed by slower or less accurate responses (Tipper, 2001). In the negative priming adaptation of the number-conservation task, fourth graders (10 years of age) typically require more time to determine whether two rows of tokens contain the same number of tokens in items in which length and number covary (e.g., the longer row contains more tokens than the shorter one, a context in which the "length-equals-number" heuristic is automatically triggered) after performing number-conservation-like tasks in which length and number interfere (e.g., the longer row contains the same number of tokens as the shorter one) than after performing items in which length and number do not covary or interfere (i.e., tokens are not aligned along the horizontal axis, a context that does not automatically trigger the "length-equals-number" heuristic). This negative priming effect suggests that the ability to perform the number-conservation task is directly related to the ability to inhibit the misleading "length-equals-number" heuristic in children. A functional magnetic resonance imagery (fMRI) study on

60 children aged 5–9 provided convergent evidence for the role of inhibitory control in the number-conservation task by showing that the progressive ability to perform the number-conservation task elicits increasing activation not only in the intraparietal sulcus (an area dedicated to number and length processing, Dormal & Pesenti, 2009) but also in the right inferior frontal gyrus (an epicenter of inhibitory control, Aron et al., 2014; Houdé et al., 2011). Finally, a high-density event-related potential (ERP) study demonstrated that adults still need to inhibit the misleading "length-equals-number" heuristic to perform the number-conservation task (Borst, Simon, et al., 2013).

Interestingly, one study recently demonstrated that 6-year-old children are more prone to accept the numerical equivalence in an adaptation of the number-conservation task when the difference in length of the two rows increased (Viarouge, Houdé, & Borst, 2019). In this adaptation of the number-conservation task, children were asked to perform a computerized same-different numerical estimation task with horizontally arranged dots (as in Piaget's number-conservation task after the length of one row is transformed). The spacing of the dots in one row (and thus the length ratio between the two rows) varied systematically. Children's ability to make a "same" judgment when the two rows contained an equal number of dots decreased progressively as the difference in the lengths of the two rows increased. Thus children's performance on this task decreased as the interference between the nonnumerical and numerical magnitudes increased. In addition, children who orient more spontaneously toward numerosity in a card-sorting task were more likely to succeed the number-conservation task. Spontaneous orientation toward different dimensions of magnitude (i.e., numerosity spacing, or item size in a set of dots) was measured by using a card-sorting task (i.e., the SOMAG, see Viarouge et al., 2018). In each trial, children were presented with three images depicting arrays of dots: Two cards (targets) were horizontally aligned at the top of the screen, while the third card (reference) was centered at the bottom of the screen. Children were asked to spontaneously pick one target card to go with the reference card. The collections of dots shown on the cards could be sorted according to the number of dots, aspects of the dot size, or of the spacing of the dots (the size and spacing dimensions were defined following DeWind, Adams, Platt, & Brannon, 2015). Taken together, these findings suggest that it becomes easier for children to succeed the number-conservation task (a) as the misleading "length-equals-number" heuristic becomes less salient (by increasing the difference in length between the two rows) and/or (b) as they spontaneously rely on the numerical dimension of the task. We speculate that success in the number-conservation task increases in both cases because the misleading heuristic interferes to a lesser extent with the proper algorithm to activate which in turns allows to inhibit the misleading heuristic more easily (see De Neys & Pennycook, 2019 for a similar rational for conflict detection processes in logical reasoning task).

Finally, De Neys, Lubin, and Houdé (2014) studied whether children before 7 years of age fail the number-conservation task because they fail to inhibit the misleading "length-equal-number" heuristic or because they fail to detect the conflict between the misleading heuristic and the algorithm to activate in such context. To do so, De Neys et al. (2014) asked preschoolers to perform the number-conservation task in the conflict and no-conflict conditions. In the conflict condition, 5.7-year-old preschoolers performed the seminal version of the number-conservation task. In the no-conflict condition, the number-conservation task was reversed starting with two rows containing the same number of tokens but with different lengths. The length of one of the two rows was then transformed to display two rows of similar length. As expected, preschoolers were less efficient in the conflict than in the no-conflict condition. Preschoolers who committed errors in the conflict condition of the number-conservation task (i.e., non-conservers) were less confident in their response in the conflict than in the no-conflict condition. In line with previous studies (De Neys, 2012; De Neys & Pennycook, 2019), the difference in the confidence of the response of preschoolers in the conflict and no-conflict condition suggests that they detected the conflict between the length-equals-number misleading strategy and the appropriate logical strategy to use in that context (i.e., reversibility). Thus errors of 5.7-year-old preschoolers in the number-conservation task do not reflect a failure to detect the conflict between the heuristic and the logical strategy but probably a specific difficulty to inhibit the misleading heuristic (Lubin, Simon, Houdé, & De Neys, 2015).

## The class-inclusion task

Using negative priming adaptations of the class-inclusion task, two studies revealed that the inhibition of a misleading heuristic (here, the direct perceptual comparison of the number of daisies and roses) was also critical for success in this Piagetian task for both school-aged children and adults (Borst, Poirel, Pineau, Cassotti, & Houdé, 2013; Perret, Paour, & Blaye, 2003). Indeed, children under 7 years of age erroneously think that there are more daisies than flowers in the class-inclusion task because children spontaneously compare the visuospatial extensions (the number of elements) of the two subordinate classes (e.g., daisies and roses) rather than comparing the superordinate class (flowers) to its subordinate class (daisies).

## Deductive reasoning tasks

The progressive ability to inhibit misleading heuristics is critical to succeed not only in Piagetian problems but also in deductive reasoning problems (Houdé et al., 2000; Prado & Noveck, 2007). For instance, the results from a positron emission tomography (PET) micro-longitudinal (pretests—inhibitory control training—posttest) study (Houdé et al., 2000) and an fMRI study (Prado & Noveck, 2007) converged in showing that key prefrontal regions of the inhibitory control network are activated when falsifying (or verifying) conditional rules, such as "If there is not a red square on the left, then there is a yellow circle on the right" (Evans, 1998). In this type of task, inhibitory control is needed to overcome the perceptual matching bias consisting of selecting the two elements quoted in the rule to falsify it. To provide evidence for the role of inhibitory control in overcoming deductive reasoning errors, Houdé et al. (2000) asked participants to perform the perceptual matching bias task (Evans, 1998) before and after one of two types of interventions. Participants were either taught to inhibit the heuristic or to apply the logical algorithm to use in such task. In both groups, the intervention was based on the Wason task (Wason, 1968). Participants who learned to inhibit the heuristic were the only ones who were able to overcome their deductive reasoning errors. Comparison of the brain activation before and after intervention revealed that brain activation shifted from the posterior perceptual regions preintervention to prefrontal executive regions postintervention including the left-middle-frontal gyrus, the left-inferior-frontal gyrus, and the right ventromedial prefrontal cortex.

Prado and Noveck (2007) provided convergent evidence that the resolution of such deductive reasoning tasks relies in part on the ability to inhibit misleading. Participants determined on each trial whether a conditional rule such as "if there is not a H then there is a square" was falsified (or verified) by a given item (e.g., B and circle). They reported increased activation in the right mid-dorsolateral prefrontal cortex (mid-DLPFC), the medial frontal areas (including the anterior cingulate area), the presupplementary motor area, and the parietal cortices as the perceptual mismatch increased between the conditional rule and the item (i.e., stronger bias). Critically, a psychophysiological interaction analysis revealed that the integration between the visual areas and mid-DLPFC decreased when the perceptual mismatch increased. Thus overcoming deductive reasoning errors in this context might be rooted in part on the inhibitory control exerted by prefrontal regions (i.e., mid-DLPFC and the medial frontal cortex) on lower level visual regions.

## Perspective-taking and theory-of-mind tasks

The development of cognitive abilities in several other domains seems also to be related to the progressive ability to inhibit misleading heuristics (Aïte et al., 2016; Apperly & Carroll, 2009; Beck et al., 2009; Benson et al., 2013; Dumontheil, Apperly, & Blakemore, S-J., 2010; Qureshi, Apperly, & Samson, 2010). For instance, several studies have provided convergent evidence that the inhibition of a first-person (i.e., egocentric) perspective was necessary not only to adopt a third-person (i.e., heterocentric) perspective (Aïte et al., 2016; Dumontheil et al., 2010) but also to succeed on theory of mind tasks (Qureshi et al., 2010). A recent study provided further support for the role of inhibitory control in the development of perspective-taking abilities by showing that the progressive efficiency to adopt a heterocentric perspective from childhood to adulthood was directly related to the progressive ability to inhibit an egocentric perspective with age (Aïte et al., 2016).

Taken together, these findings suggest that transitioning from committing errors to succeeding in various cognitive tasks is rooted in the progressive ability to inhibit misleading heuristics constructed during childhood that are never fully overcome, even in adulthood. In what follows, we present studies showing that, like in laboratory cognitive tasks, a number of systematic difficulties faced by typically developing children and adolescents in fundamental academic learning could be attributed to difficulties in inhibiting misleading scholarly heuristics.

## Inhibitory control and misleading heuristics in the classroom

A growing number of studies suggest that in numeracy (Lubin et al., 2016; Lubin, Vidal, Lanoë, Houdé, & Borst, 2013; Roell, Viarouge, Hilscher, Houdé, & Borst, 2019; Roell, Viarouge, Houdé, & Borst, 2017, 2019; Rossi, Vidal, Létang, Houdé, & Borst, 2019), literacy (Ahr, Houdé, & Borst, 2016, 2017; Borst, Ahr, Roell, & Houdé, 2015; Brault Foisy, Ahr, Masson, Houdé, & Borst, 2017; Lanoë, Lubin, Vidal, Houdé, & Borst, 2016), and science (Masson, Potvin, Riopel, & Foisy, 2014; Potvin, Masson, Lafortune, & Cyr, 2015), the acquisition of new concepts and abilities at school and at the university could be related to the growing ability to inhibit misleading heuristics reinforced in previous years of schooling.

## Numeracy

For instance, comparing fractions with common numerators (e.g., 3/7 vs 3/5) is at all ages more challenging than comparing fractions with common denominators (e.g., 3/4 vs 6/4). When comparing fractions with common numerators, children and adolescents erroneously think that 3/7 is larger than 3/5 because they tend to consider a fraction as two separate whole numbers rather than as one rational number (Stafylidou & Vosniadou, 2004). Thus 3/7 is larger than 3/5 because 7 is larger than 5. The whole number bias (Vamvakoussi, Van Dooren, & Verschaffel, 2012), i.e., a tendency to rely on whole number knowledge when processing rational numbers (Ni & Zhou, 2005), could be at the root of systematic errors when comparing fractions with common numerators. Using a negative priming paradigm, a study provided evidence that 14-year-old adolescents and young adults must inhibit their tendency to rely on the "greater the whole number, the greater the fraction" strategy to compare fractions with common numerators (Rossi et al., 2019).

Similarly, children and even educated adults have difficulty to compare the magnitude of decimal numbers when the smallest decimal number has the greatest number of digits (7.457 vs 7.6) (e.g., Lortie-Forgues, Tian, & Siegler, 2015). As for the comparison of the magnitude of fractions with common numerators, the difficulty to compare the magnitude of decimal numbers when the smallest decimal number has the greatest number of digits could be related to a whole number bias, i.e., an overgeneralization of whole number properties to rational numbers (Ni & Zhou, 2005; Vamvakoussi et al., 2012). In the context of decimal number comparison, the whole number bias consists of using a heuristic adapted to the context of whole number magnitude comparison, such as "the greater the number of digits, the greater its magnitude" to compare decimal numbers. Using such heuristic leads one to state erroneously that 7.457 is larger than 7.6 because 457 is larger than 6. Two negative priming studies on 12-year-old adolescents (Roell et al., 2017) and adults (Roell, Viarouge, Hilscher, et al., 2019) provided evidence that the comparison of the magnitude of two decimal numbers in which the smallest decimal number has the greatest number of digits (7.457 vs 7.6) at all ages relies on the ability to inhibit the "greater the number of digits, the greater its magnitude" heuristic.

Fourth graders have also systematic difficulties in solving arithmetic word problems, such as "John has 20 marbles. He has 10 more marbles than Bill. How many marbles does Bill have?" in which the relational term ("more than") is incongruent with the arithmetic operation to perform ("subtraction"). Most errors in this type of problem are reversal errors: children add the numbers instead of subtracting them (Stern, 1993). Using a negative priming paradigm, two studies showed (a) that the progressive ability to solve this type of problem from childhood to adulthood is related to the progressive ability to inhibit the "add if more, subtract if less" heuristic with age (Lubin et al., 2013) and (b) that experts in mathematics (master's students in mathematics) are more efficient in solving this type of problems than nonexperts (master's students in humanities) because they more efficiently inhibit the "add if more, subtract if less" heuristic (Lubin et al., 2016). Finally, one study provided evidence that third to fifth graders are less confident in their answers in arithmetic word problems in which the "add if more, subtract if less" heuristic interferes with the logical strategy (conflict condition) than in arithmetic word problems in which this heuristic does not interfere with the logical strategy (no-conflict condition). This finding suggests (a) that third to fifth graders detect the conflict between the misleading heuristic and the logical strategy even when they commit errors in solving arithmetic word problems and (b) that these errors are likely due to a difficulty to inhibit the "add if more, subtract if less" heuristic (Lubin, Houdé, & De Neys, 2015).

## Literacy

In literacy, one study on fourth graders, ninth graders, and young adults demonstrated that overcoming systematic verb inflection errors in sentences such as "the dog of neighbors eats," in which children and even adolescents tend to inflect the verb with the nearest noun (i.e., "the dog of the neighbors eat"), is rooted on the growing ability to inhibit the "inflect the verb with the nearest noun" heuristic (Lanoë et al., 2016). Similarly, studies on expert readers (Ahr et al., 2017; Borst et al., 2015; Brault Foisy et al., 2017) and novice readers (first to fifth graders, Ahr et al., 2016) have demonstrated that overcoming mirror errors in reading (i.e., confusing b for d and p for q) is related in part to the ability to inhibit mirror generalization, an intrinsic property of the neurons of the visual system that allows one to identify a visual stimulus regardless of its presentation side (Pegado, Nakamura, Cohen, & Dehaene, 2011). While well adapted for the visual recognition of objects, this heuristic is detrimental for the discrimination of letters with lateral mirror-image counterparts (such as b/d and p/q).

## Scientific reasoning

Conceptual changes in science seem also to be supported by the growing ability to inhibit misconception regarding the physical principles governing our environment (Masson et al., 2014; Potvin et al., 2015). For instance, the misconception that "heavy objects sink more than lighter objects" affects high school students' scientific reasoning regarding flotation (Potvin et al., 2015). Moreover, scientific experts (i.e., physics students) need to inhibit misconceptions related to electricity to reason about electric circuits correctly (Masson et al., 2014).

## Conclusion

A number of studies now show that inhibitory control might be one of the core mechanisms producing cognitive change in various domains, including numerical cognition, categorization, reasoning, perspective taking, and theory of mind. Interestingly, inhibitory control seems also to be involved in fundamental academic learning. In all of these domains and in fundamental academic learning, systematic errors occur when individuals tend to rely on misleading heuristics in a context in which these heuristics interfere with the proper strategies to activate. These heuristics are most likely constructed during childhood, when our "statistical brain" detects systematic regularities in our environment, such as the regularity between the length of an alignment and the number of objects in that alignment. According to our theoretical framework, at each age and in each context, transitioning from error to success in a given context (from Piagetian to theory of mind problems) is the hallmark not of a new logical or structural stage of cognitive development, as Piaget hypothesized (Piaget, 1983), but of the progressive ability to inhibit a misleading heuristic in that context. This theoretical framework can thus account for nonlinear patterns of development defined by the rich conceptual knowledge of infants, followed by systematic errors of children in a context requiring similar conceptual knowledge and by children mastering logical rules that adults fail to use in some contexts. Indeed, errors should be viewed no longer as age specific but as context specific. The probability of an error in a given context is essentially defined by three parameters: the frequency of use of the misleading heuristic, the availability of the logical algorithm, and inhibitory control efficiency.

As research on the role of inhibitory control in conceptual change proceeds, studies will need to determine the extent to which inhibitory control is a domain-general or domain-specific process and whether this generality (or specificity) changes with age. Studies have already shown that inhibitory control can generalize (a) in children during the Color Word Stroop task and the number-conservation task (Linzarini, Houdé, & Borst, 2015) and (b) in adults during ambiguous sentence reading and the Color Word Stroop tasks and during the Necker cubes and Color Word Stroop tasks (Kan et al., 2013), but a more systematic investigation of this issue is needed.

Another critical avenue of research is to determine the type of pedagogical interventions that could potentially improve inhibitory control (see for a review, Diamond & Lee, 2011) and the degree to which improving inhibitory control in one domain could transfer to another one. This is especially critical for education given that cognitive control efficiency, including inhibitory control efficiency in childhood, is a strong predictor of school readiness and academic success (Moffitt et al., 2011).

## References

Ahr, E., Houdé, O., & Borst, G. (2016). Inhibition of the mirror-generalization process in reading in school-aged children. *Journal of Experimental Child Psychology*, *145*, 157–165. https://doi.org/10.1016/j.jecp.2015.12.009.

Ahr, E., Houdé, O., & Borst, G. (2017). Predominance of lateral over vertical mirror errors in reading: A case for neuronal recycling and inhibition. *Brain and Cognition*, *117*, 1–8.

Aïte, A., Berthoz, A., Vidal, J., Roell, M., Zaoui, M., Houdé, O., et al. (2016). Taking a third-person perspective requires inhibitory control: Evidence from a developmental negative priming study. *Child Development*. https://doi.org/10.1111/cdev.12558.

Antell, S., & Keating, D. (1983). Perception of numerical invariance in neonates. *Child Development*, *54*, 695–701. https://doi.org/10.2307/1130057.

Apperly, I. A., & Carroll, D. J. (2009). How do symbols affect 3- to 4-year-olds' executive function? Evidence from a reverse-contingency task: How symbols affect 3- to 4-year-olds' executive function. *Developmental Science*, *12*, 1070–1082. https://doi.org/10.1111/j.1467-7687.2009.00856.x (Houdé & Borst, 2014).

Aron, A. R., Robbins, T. W., & Poldrack, R. A. (2014). Inhibition and the right inferior frontal cortex: One decade on. *Trends in Cognitive Sciences*, *18*, 177–185. https://doi.org/10.1016/j.tics.2013.12.003 (Smith & Thelen, 2003).

Beck, S. R., Riggs, K. J., & Gorniak, S. L. (2009). Relating developments in children's counterfactual thinking and executive functions. *Thinking & Reasoning*, *15*, 337–354. https://doi.org/10.1080/13546780903135904 (Smith & Thelen, 2003).

Bell, M., & Fox, N. (1992). The relations between frontal brain electrical activity and cognitive development during infancy. *Child Development*, *63*, 1142–1163. https://doi.org/10.2307/1131523.

Benson, J. E., Sabbagh, M. A., Carlson, S. M., & Zelazo, P. D. (2013). Individual differences in executive functioning predict preschoolers' improvement from theory-of-mind training. *Developmental Psychology*, *49*, 1615–1627. https://doi.org/10.1037/a0031056.

Borst, G., Ahr, E., Roell, M., & Houdé, O. (2015). The cost of blocking the mirror-generalization process in reading: Evidence for the role of inhibitory control in discriminating letters with lateral mirror-image counterparts. *Psychonomic Bulletin & Review*, *22*, 228–234. https://doi.org/10.3758/s13423-014-0663-9.

Borst, G., Poirel, N., Pineau, A., Cassotti, M., & Houdé, O. (2013). Inhibitory control efficiency in a Piaget-like class-inclusion task in school-age children and adults: A developmental negative priming study. *Developmental Psychology*, *49*, 1366–1374. https://doi.org/10.1037/a0029622 (Karmiloff-Smith, 1992).

Borst, G., Simon, G., Vidal, J., & Houdé, O. (2013). Inhibitory control and visuospatial reversibility in Piaget's seminal number-conservation task: A high-density ERP study. *Frontiers in Human Neuroscience*. https://doi.org/10.3389/fnhum.2013.00920.

Brault Foisy, L.-M., Ahr, E., Masson, S., Houdé, O., & Borst, G. (2017). Is inhibitory control involved in discriminating pseudowords that contain the reversible letters b and d? *Journal of Experimental Child Psychology*, *162*, 259–267.

Casey, B., Tottenham, N., Liston, C., & Durston, S. (2005). Imaging the developing brain: What have we learned about cognitive development? *Trends in Cognitive Sciences*, *9*, 104–110. https://doi.org/10.1016/j.tics.2005.01.011.

De Neys, W. (2012). Bias and conflict: A case for logical intuitions. *Perspectives on Psychological Science*, 7, 28–38.

De Neys, W. (2017). Bias, conflict, and fast logic: Towards a hybrid dual process future? In W. De Neys (Ed.), *Dual process theory 2.0* (pp. 47–65). Oxon, England: Routledge.

De Neys, W., Lubin, A., & Houdé, O. (2014). The smart non-conserver: Preschoolers detect their number conservation errors. *Child Development Research.*. https://doi.org/10.1155/2014/768186.

De Neys, W., & Pennycook, G. (2019). Logic, fast and slow: Advances in dual-process theorizing. *Current Directions in Psychological Science*, *28*, 503–509.

DeWind, N. K., Adams, G. K., Platt, M. L., & Brannon, E. M. (2015). Modeling the approximate number system to quantify the contribution of visual stimulus features. *Cognition*, *142*, 247–265.

Diamond, A. (1998). Understanding the A-not-B error: Working memory vs. reinforced response, or *active* trace vs. latent trace. *Developmental Science*, *1*, 185–189. https://doi.org/10.1111/1467-7687.00022.

Diamond, A. (2013). Executive functions. *Annual Review of Psychology*, *64*, 135–168. https://doi.org/10.1146/annurev-psych-113011-143750.

Diamond, A., & Lee, K. (2011). Interventions shown to aid executive function development in children 4 to 12 years old. *Science*, *333*, 959–964. https://doi.org/10.1126/science.1204529.

Dormal, V., & Pesenti, M. (2009). Common and specific contributions of the intraparietal sulci to Numerosity and length processing. *Human Brain Mapping*, *30*, 2466–2476. https://doi.org/10.1002/hbm.20677.

Dumontheil, I., Apperly, I. A., & Blakemore, S- J. (2010). Online usage of theory of mind continues to develop in late adolescence. *Developmental Science*, *13*(2), 331–338. https://doi.org/10.1111/j.1467-7687.2009.00888.x.

Evans, J. (1998). Matching bias in conditional reasoning. *Thinking & Reasoning*, *4*, 45–82.

Halford, G. S., Cowan, N., & Andrews, G. (2007). Separating cognitive capacity from knowledge: A new hypothesis. *Trends in Cognitive Sciences*, *11*(6), 236–242.

Houdé, O., & Borst, G. (2014). Measuring inhibitory control in children and adults: Brain imaging and mental chronometry. *Frontiers in Psychology*, *5*. https://doi.org/10.3389/fpsyg.2014.00616.

Houdé, O., & Guichart, E. (2001). Negative priming effect after inhibition of number/length interference in a Piaget-like task. *Developmental Science*, *4*, 119–123.

Houdé, O., Pineau, A., Leroux, G., Poirel, N., Perchey, G., Lanoë, C., et al. (2011). Functional magnetic resonance imaging study of Piaget's conservation-of-number task in preschool and school-age children: A neo-Piagetian approach. *Journal of Experimental Child Psychology*, *110*, 332–346. https://doi.org/10.1016/j.jecp.2011.04.008 (Kahneman, 2011).

Houdé, O., Zago, L., Mellet, E., Moutier, S., Pineau, A., Mazoyer, B., et al. (2000). Shifting from the perceptual brain to the logical brain: The neural impact of cognitive inhibition training. *Journal of Cognitive Neuroscience, 12*, 721–728. https://doi.org/10.1162/089892900562525.

Kahneman, D. (2011). *Thinking fast and slow*. London: Allen Lane.

Kan, I. P., Teubner-Rhodes, S., Drummey, A. B., Nutile, L., Krupa, L., & Novick, J. M. (2013). To adapt or not to adapt: The question of domain-general cognitive control. *Cognition, 129*, 637–651. https://doi.org/10.1016/j.cognition.2013.09.001.

Karmiloff-Smith, A. (1992). *Beyhond modularity: A developmental perspective on cognitive science*. Cambridge, MA: MIT Press.

Lanoë, C., Lubin, A., Vidal, J., Houdé, O., & Borst, G. (2016). Inhibitory control is needed to overcome written verb inflection errors: Evidence from a developmental negative priming study. *Cognitive Development, 37*, 18–27. https://doi.org/10.1016/j.cogdev.2015.10.005.

Linzarini, A., Houdé, O., & Borst, G. (2015). When Stroop helps Piaget: An inter-task positive priming paradigm in 9-year-old children. *Journal of Experimental Child Psychology, 139*, 71–82. https://doi.org/10.1016/j.jecp.2015.05.010.

Lortie-Forgues, H., Tian, J., & Siegler, R. S. (2015). Why is learning fraction and decimal arithmetic so difficult? *Developmental Review, 38*, 201–221. https://doi.org/10.1016/j.dr.2015.07.008.

Lubin, A., Houdé, O., & De Neys, W. (2015). Evidence for children's error sensitivity during arithmetic word problem solving. *Learning and Instruction, 40*, 1–8.

Lubin, A., Rossi, S., Lanoë, C., Vidal, J., Houdé, O., & Borst, G. (2016). Expertise, inhibitory control and arithmetic word problems: A negative priming study in mathematics experts. *Learning & Instruction, 45*, 40–48. https://doi.org/10.1016/j.learninstruc.2016.06.004.

Lubin, A., Simon, G., Houdé, O., & De Neys, W. (2015). Inhibition, conflict detection, and number conservation. *ZDM Mathematics Education, 47*, 793–800.

Lubin, A., Vidal, J., Lanoë, C., Houdé, O., & Borst, G. (2013). Inhibitory control is needed for the resolution of arithmetic word problems: A developmental negative priming study. *Journal of Educational Psychology, 105*, 701–708. https://doi.org/10.1037/a0032625.

Masson, S., Potvin, P., Riopel, M., & Foisy, L.-M. B. (2014). Differences in brain activation between novices and experts in science during a task involving a common misconception in electricity. *Mind, Brain, and Education, 8*, 44–55. https://doi.org/10.1111/mbe.12043.

Moffitt, T. E., Arseneault, L., Belsky, D., Dickson, N., Hancox, R. J., Harrington, H., et al. (2011). A gradient of childhood self-control predicts health, wealth, and public safety. *Proceeding of the National Academy of Sciences of the United States of America, 108*, 2693–2698. https://doi.org/10.1073/pnas.1010076108.

Morsanyi, K., & Handley, S. J. (2008). How smart do you need to be to get it wrong? The role of cognitive capacity in the development of heuristic-based judgment. *Journal of Experimental Child Psychology, 99*(1), 18–36.

Ni, Y., & Zhou, Y.-D. (2005). Teaching and learning fraction and rational numbers: The origins and implications of whole number bias. *Educational Psychologist, 40*, 27–52. https://doi.org/10.1207/s15326985ep4001_3.

Pegado, F., Nakamura, K., Cohen, L., & Dehaene, S. (2011). Breaking the symmetry: Mirror discrimination for single letters but not for pictures in the visual word form area. *NeuroImage, 55*, 742–749.

Perret, P., Paour, J.-L., & Blaye, A. (2003). Respective contribution of inhibition and knowledge levels in class inclusion development: A negative priming study. *Developmental Science, 6*, 283–288. https://doi.org/10.1111/1467-7687.00284.

Piaget, J. (1983). Piaget's theory. In P. H. Mussen (Ed.), *Handbook of child psychology* (pp. 103–128). New York: Wiley.

Potvin, P., Masson, S., Lafortune, S., & Cyr, G. (2015). Persistence of the intuitive conception that heavier objects sink more: A reaction time study with different levels of interference. *International Journal of Science and Mathematics Education, 13*, 21–43. https://doi.org/10.1007/s10763-014-9520-6.

Prado, J., & Noveck, I. (2007). Overcoming perceptual features in logical reasoning: A parametric fMRI study. *Journal of Cognitive Neuroscience, 19*, 642–657. https://doi.org/10.1162/jocn.2007.19.4.642.

Qureshi, A. W., Apperly, I. A., & Samson, D. (2010). Executive function is necessary for perspective selection, not Level-1 visual perspective calculation: Evidence from a dual-task study of adults. *Cognition, 117*, 230–236. https://doi.org/10.1016/j.cognition.2010.08.003 (Antell & Keating, 1983).

Roell, M., Viarouge, A., Hilscher, E., Houdé, O., & Borst, G. (2019). Evidence for a visuo-spatial bias in decimal number comparison in adolescents and in adults. *Scientific Reports, 9*, 14770.

Roell, M., Viarouge, A., Houdé, O., & Borst, G. (2017). Inhibitory control and decimal number comparison in school-aged children. *PLoS One, 12*(11). https://doi.org/10.1371/journal.pone.0188276, e0188276.

Roell, M., Viarouge, A., Houdé, O., & Borst, G. (2019). Inhibition of the whole number bias in decimal number comparison: A developmental negative priming study. *Journal of Experimental Child Psychology, 177*, 240–247.

Rossi, S., Vidal, J., Létang, M., Houdé, O., & Borst, G. (2019). Adolescents and adults need inhibitory control to compare fractions. *Journal of Numerical Cognition, 5*, 314–336.

Siegler, R. (1999). Strategic development. *Trends in Cognitive Sciences, 3*, 430–435. https://doi.org/10.1016/S1364-6613(99)01372-8.

Smith, L. B., & Thelen, E. (2003). Development as a dynamic system. *Trends in Cognitive Sciences, 7*, 343–348. https://doi.org/10.1016/S1364-6613(03)00156-6.

Spelke, E. (2000). Core knowledge. *American Psychologist, 55*, 1233–1243. https://doi.org/10.1037/0003-066X.55.11.1233.

Stafylidou, S., & Vosniadou, S. (2004). The development of students' understanding of the numerical value of fractions. *Learning and Instruction, 14*, 503–518. https://doi.org/10.1016/j.learninstruc.2004.06.015.

Stern, E. (1993). What makes certain arithmetic word problems involving the comparison of sets so difficult for children? *Journal of Educational Psychology, 85*, 7–23. https://doi.org/10.1037/0022-0663.85.1.7.

Tipper, S. P. (2001). Does negative priming reflect inhibitory mechanisms? A review and integration of conflicting views. *Quarterly Journal of Experimental Psychology: Human Experimental Psychology, 54*(A), 321–343. https://doi.org/10.1080/02724980042000183.

Vamvakoussi, X., Van Dooren, W., & Verschaffel, L. (2012). Naturally biased? In search for reaction time evidence for a natural number bias in adults. *The Journal of Mathematical Behavior, 31*, 344–355. https://doi.org/10.1016/j.jmathb.2012.02.001.

Viarouge, A., Courtier, P., Hoppe, M., Melnik, J., Houdé, O., & Borst, G. (2018). Spontaneous orientation towards irrelevant dimensions of magnitude and numerical acuity. *Learning and Instruction, 54*, 156–163.

Viarouge, A., Houdé, O., & Borst, G. (2019). The progressive 6-year-old conserver: Numerical saliency and sensitivity as core mechanisms of numerical abstraction in a Piaget-like estimation task. *Cognition, 190*, 137–142.

Wason, P. C. (1968). Reasoning about a rule. *Quarterly Journal of Experimental Psychology, 20*, 273–281.

CHAPTER 3

# Maths and logic: Relationships across development

**Kinga Morsanyi**
Centre for Mathematical Cognition, School of Science, Loughborough University, Loughborough, United Kingdom

It has long been suggested that mathematics skills and logical reasoning are intrinsically linked. Nevertheless, the reasons for this link are not well researched or well understood. The aim of this chapter is to answer some basic questions about these relations. In particular, I will focus on the following issues: What types of logical reasoning are related to mathematics (and what types of mathematics skills are related to logic)? Are these links stable across development? Is the link between maths and logic mediated by general cognitive factors, such as intelligence, working memory, or some other basic processes that are known to be involved in mathematics, such as the ability to create a linear mental representation of item order? Another intriguing question is whether training in mathematics can improve logical reasoning skills, and vice versa. Answering these questions is important for both theoretical and practical purposes.

The idea that mathematical skills might be linked to skills in other disciplines originated in Ancient Greece. The philosopher, Plato, proposed that people with a natural talent for calculation are generally also good at other disciplines, and that training in arithmetic can improve people's skills in other areas, even if a person does not have a natural talent for mathematics. This idea (i.e., that training in a specific skill might generalize to other areas of knowledge) is known as the Theory of Formal Discipline (TFD), and it has long influenced educational policy and societal practices (see Attridge & Inglis, 2013; Inglis & Attridge, 2017 for further discussion of the TFD regarding mathematics education). For example, in the UK, workers who have studied post-compulsory mathematics earn more than those with similar ability levels and qualifications (Dolton & Vignoles, 2002). Moreover, mathematics graduates have the fifth highest employment rate of any degree subject in the UK, and the highest of all the "pure" sciences. These statistics indicate that mathematical knowledge has a high value in the job market. Indeed, mathematics is not only considered the universal language of science, but

*Heterogeneous Contributions to Numerical Cognition*
https://doi.org/10.1016/B978-0-12-817414-2.00010-5

also a powerful tool for analysis and problem solving. Nevertheless, it is not very clear whether mathematical training improves people's thinking skills, or if people with good analytical skills are more likely to study mathematics—or maybe reasoning and mathematics skills are based on some common cognitive factors. The idea that mathematical skills might be linked to skills in other disciplines originated in Ancient Greece. The philosopher, Plato, proposed that people with a natural talent for calculation are generally also good at other disciplines, and that training in arithmetic can improve people's skills in other areas, even if a person does not have a natural talent for mathematics. This idea (i.e., that training in a specific skill might generalize to other areas of knowledge) is known as the Theory of Formal Discipline (TFD), and it has long influenced educational policy and societal practices (see Attridge & Inglis, 2013; Inglis & Attridge, 2017 for further discussion of the TFD regarding mathematics education). For example, in the UK, workers who have studied post-compulsory mathematics earn more than those with similar ability levels and qualifications (Dolton & Vignoles, 2002). Moreover, mathematics graduates have the fifth highest employment rate of any degree subject in the UK, and the highest of all the "pure" sciences. These statistics indicate that mathematical knowledge has a high value in the job market. Indeed, mathematics is not only considered the universal language of science, but also a powerful tool for analysis and problem solving. Nevertheless, it is not very clear whether mathematical training improves people's thinking skills, or if people with good analytical skills are more likely to study mathematics—or maybe reasoning and mathematics skills are based on some common cognitive factors.

Logical reasoning involves reaching a valid conclusion on the basis of multiple premises that are considered to be true (e.g., Johnson-Laird, 1999). This definition implies that although the premises might be consistent with people's knowledge, beliefs, or perceptions, they might also be inconsistent with beliefs, knowledge, or perceptions, or they might have an "abstract" content. Apart from these differences, logical arguments also differ in the types of inferences that they elicit. In the current chapter, I will focus on two particular forms of logical inference: transitive and conditional inference, both of which have been shown to relate to mathematical skills.

## Transitive inferences and linear representations of order

Transitive reasoning is one of the most basic but crucial inferential abilities. It builds on the ability to form relations between concepts, which might also rely on perceptual processes. For example, if Peter and Paul are standing next to each other, we might immediately see that Peter is taller than Paul. Or if we consider two items and four items, if these are presented together, we can easily see that four items are more than two items. Transitive inference can be used to go beyond perceptually available information. For example, if I know that Peter is taller than Paul, and Paul is taller than John, I can infer that John is shorter than Peter, even if I have never seen John and Peter together.

Studies with animals that live in social groups, which are governed by linear dominance hierarchies, have demonstrated that, in the context of social dominance, animals are able to draw transitive inferences. For example, Hogue, Beaugrand, and Laguë (1996) reported that chickens behaved subordinately toward an unfamiliar bird that they observed to defeat another chicken dominant to themselves, although they did not behave subordinately to unfamiliar birds that they observed defeating another unfamiliar bird. This indicates that they did not simply react to dominant behavior, but they were able to infer dominance relationships. A study with infants between the ages of 10 and 13 months (Gazes, Hampton, & Lourenco, 2017) indicated that babies are also able to draw transitive inferences in the context of dominance interactions.

In their study, Gazes et al. (2017) presented babies with a video recording of two dominance interactions between three puppets (bear > elephant; hippo > bear), where the dominant puppet took a toy from the other puppet after a struggle. Subsequently, the babies were presented with interactions between the two puppets that had not interacted during familiarization that were either congruent (hippo > elephant) or incongruent (elephant > hippo) with the inferred hierarchy. The infants looked longer at incongruent than congruent displays, which suggested that they were able to use transitivity to infer dominance relationships. To exclude the possibility that babies simply reacted to a change in dominance behavior in the incongruent condition, Gazes et al. (2017) also included a control condition in which either the hippo or the elephant interacted with a new puppet (a giraffe) in a way that was either congruent with their earlier dominance position (hippo > giraffe or giraffe > elephant) or it was incongruent with their earlier dominance position (giraffe > hippo or elephant > giraffe). The infants looked longer to the incongruent than the congruent scenario in the test condition, but this was not the case in the control condition.

Gazes et al. (2017) noted that in their study, the three puppets in the test condition appeared simultaneously onscreen during the familiarization phase, and they were arranged along a physical line, which was consistent with their dominance position. Moreover, the interactions between the puppets unfolded sequentially along this linear path. This spatial arrangement might have provided the infants with crucial support in forming a single linear representation of the dominance relations among the three puppets, which could have aided their inferences about their relations. Additionally, the fact that, in some contexts, both human infants and nonhuman animals are able to draw transitive inferences suggests that these inferences do not require verbal representations.

Piaget and Inhelder (1967) proposed that children gradually acquire the ability to rely on logically underpinned transitive inferences (i.e., transitive inferences that are based on abstract strategies and do not rely on direct perceptual support). In particular, Piaget proposed that young children struggle with understanding that the middle item in a series can be both smaller than and larger than another item, and that the ability to draw transitive inferences only fully emerges around 7 years of age. This view was challenged by Bryant and Trabasso (1971), who demonstrated that 4-year-olds were able to infer length relations at an above-chance level when training ensured memory for each of the premise pairs. Specifically, children were taught the relative sizes of five colored rods until they reliably selected the longer rod in each of the four adjacent pairs. Subsequently, transitive reasoning ability was tested by presenting the children with nonadjacent pairs that they have never seen together before. In these test trials, the colors of the rods were visible but their lengths were obscured. Children were able to select the longer rod at an above-chance level, demonstrating the ability to rely on transitivity. Given that the children required extensive training to perform this task, these results have been explained by proposing that they had created an internal representation of the premise information in the form of a seriated (potentially spatial) array.

It could be argued that once children can rely on an internal representation of an ordinal set that they can retrieve from memory, they do not need to engage inferential processes, because they can simply read off the relative position of items from their representation of the whole series. At the same time, young children might be genuinely able to draw transitive inferences, but this skill might be obscured by the memory requirements of the reasoning tasks. In order to overcome this problem, Pears and Bryant (1990) created transitive reasoning tasks about the spatial relations *higher* and *lower*, where the children were presented with direct representations of the premises in the form of two-block towers (e.g., blue block on top of a red block, red block on top of a green block, etc.—Fig. 1A). These representations

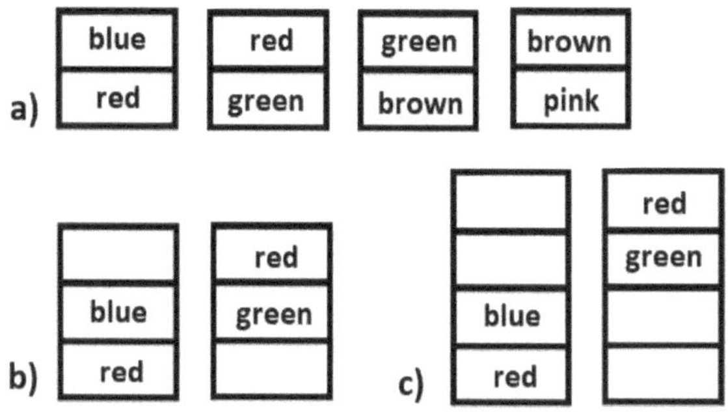


**Fig. 1** Examples of towers used by Pears and Bryant (1990; display A) and Markovits et al. (1995, display B—ambiguous perceptual information and display C—contradictory perceptual information).

could be continuously consulted during the test phase, which eased memory load, but they did not provide a simple answer to questions regarding the relative position of items that were not part of the same premise block. Children were told that based on the information that they were given, they have to build a single tower. But before doing this, they were asked some questions about the relative position of certain items (for example, *is red higher than brown?*). Pears and Bryant (1990) specifically focused on the position of the middle items in the displays, because these appeared both in the higher and lower position in the blocks, which made it necessary for children to rely on inferential (rather than perceptual) processes when they answered these questions. Pears and Bryant (1990) found that 4-year-olds were able to answer such questions at a level well above chance.

Good performance on Pears and Bryant's task made it necessary for children to understand that when they were making a decision about the relative position of the red and brown blocks, they also had to consider the position of the green block, which connected the other two blocks. Once children focused on the blocks that included only the three relevant items, it was easy for them to answer the higher/lower question, because in these two towers red appeared on top and brown appeared on the bottom. In order to test further the inferential abilities of children, Markovits, Dumas, and Malfait (1995) developed further Pears and Bryant's method by presenting children with towers where direct perceptual information could not be used to support transitive inferences. They did this by introducing some neutral (white) blocks, which were used to manipulate the relative positions of the target items (i.e., blue and green) within the premise blocks. In one condition (Fig. 1B), the perceptual information regarding the position of the blue and green items was ambiguous, and in another condition (Fig. 1C), it was contradictory relative to their positions within the towers that had to be built. Markovits et al. (1995) found that 6-year-olds and 8-year-olds performed very well on Pears and Bryant's (1990) version of the task, but 6-year-olds struggled with the more difficult versions of the task, whereas 8-year-olds performed significantly better than 6-year-olds on the tasks where they had to overcome perceptual distraction.

It has been proposed that transitive inferences rely on spatial mental models of the premise information, even when the materials are verbal, and the relations that are described are not spatial. In fact, the premises might also be inconsistent with people's beliefs or they might be abstract. Consider the following problem:

Dogs are stronger than elephants.

Cats are stronger than dogs.

Are elephants stronger than cats?

According to Goodwin and Johnson-Laird (2005), when people are presented with problems like this, based on the meaning of the relations described in the premises, they construct mental models (in the form of schematic spatial arrangements) that incorporate all of the premise information in an integrated way. The logical consequences of relations emerge from these models, which means that people do not need to rely on explicit logical reasoning to solve these problems. Children around the age of 10 are able to reason about problems with belief-inconsistent content when they are provided with instructions to *imagine* that the premises are true, even if they are not true in real life (e.g., Handley, Capon, Beveridge, Dennis, & Evans, 2004; Morsanyi, Devine, Nobes, & Szűcs, 2013). Nevertheless, even 10-year-olds struggle with transitive inference problems with abstract or hard-to-visualize content (Morsanyi et al., 2013).

In summary, transitive inference is a basic form of logical inference, which appears to rely on a (spatial) mental representation of ordered relationships. In some contexts, even nonhuman animals are able to spontaneously draw transitive inferences (e.g., Hogue et al., 1996). Babies also appear to be capable of transitive reasoning (Gazes et al., 2017), at least when the premises are presented to them in the form of a linear sequence, which provides perceptual support for drawing the correct inferences. Young children from around the age of 4 are able to draw transitive inferences relating to the spatial arrangement of items (Bryant & Trabasso, 1971; Pears & Bryant, 1990), and by the age of 7 or 8, they are also able to reason about more complex transitive reasoning problems, even when their inferences conflict with their perceptions (Markovits et al., 1995; Piaget & Inhelder, 1967). By the age of 10, children are also able to solve transitive reasoning problems with belief-inconsistent, verbal materials (Handley et al., 2004; Morsanyi et al., 2013), presumably by building a spatial mental model of the premises. In these cases, we can expect that children need to inhibit the effect of readily available perceptual information or the effect of their beliefs. Nevertheless, the ability to reason about transitive reasoning problems with abstract/hard-to-visualize content (e.g., "*Tom is taller than Bob*") is still challenging at the age of 10 (Morsanyi et al., 2013).

## Conditional inferences and thinking about possibilities

As described in the previous section, transitive reasoning skills are basic and develop early. In fact, these inferences are so simple that people might be inclined to infer transitive relations, even when none exist. For example, Goodwin and Johnson-Laird (2008) found that people often erroneously assume transitivity when they think about the following statements: "*Fred is a blood relative of Bob's, Bob is a blood relative of Ann's*" and "*Dr. Avarado knows Dr. McNamara, Dr. McNamara knows Dr. Villalobos.*"

In addition to using transitive reasoning problems, another common way of investigating logical reasoning skills is to administer conditional inference tasks, which involve reasoning on the basis of statements in the form of "if *p* then *q*" (for example, "*if it is raining, you will get wet,*" or "*if the letter is A, the number is 6*"). Research has established that evaluating conditional statements of the same logical form is more difficult if the content is unbelievable or abstract (e.g., Markovits & Lortie-Forgues, 2011). Indeed, conditional reasoning skills develop gradually during childhood and adolescence, and even educated adults struggle with conditional reasoning (De Neys & Everaerts, 2008; Gauffroy & Barrouillet, 2009; Klaczynski, Schuneman, & Daniel, 2004; Markovits & Barrouillet, 2002).

Suppose that you have a pack of cards, and each card has a letter and a number on it. You are then given a rule in the form of "if *p* then *q*," together with a premise. You must assume that both the rule and the premise are true. Your task is to decide if a conclusion necessarily follows from these statements (i.e., that it always follows, with no exception). Examples of the most commonly investigated conditional inferences are presented in Table 1. Based on formal logic, modus ponens (MP) and modus tollens (MT) are considered valid inferences (that is, the conclusion necessarily follows), and denial of the antecedent (DA) and affirmation of the consequent

**Table 1** Examples of conditional inference tasks (correct responses based on formal logic are in bold).

| | **Modus ponens (MP)** | **Modus tollens (MT)** |
|---|---|---|
| **Valid inferences** | Rule: *If the letter is H then the number is 1*<br>Premise: *The letter is H*<br>Conclusion: *The number is 1*<br><br>**YES (it follows)**<br>No (it does not necessarily follow) | Rule: *If the letter is D then the number is 8*<br>Premise: *The number is not 8*<br>Conclusion: *The letter is not D*<br><br>**YES (it follows)**<br>No (it does not necessarily follow) |
| | **Denial of the antecedent (DA)** | **Affirmation of the consequent (AC)** |
| **Invalid inferences** | Rule: *If the letter is C then the number is 6*<br>Premise: *The letter is not C*<br>Conclusion: *The number is not 6*<br><br>YES (it follows)<br>**No (it does not necessarily follow)** | Rule: *If the letter is X then the number is 4*<br>Premise: *The number is 4*<br>Conclusion: *The letter is X*<br><br>YES (it follows)<br>**No (it does not necessarily follow)** |

(AC) are considered invalid inferences. Nevertheless, the typical finding is that most people do not give a logic-based response when they are evaluating conditional arguments. More specifically, almost all adults (about 97%) endorse MP (i.e., consider that it is a logically valid argument), and about 74% endorse MT, but about 64% also endorse AC, and 56% endorse DA (Schroyens, Schaeken, & d'Ydewalle, 2001). That is, although, to some extent, people are able to discriminate between logically valid and invalid arguments, they generally struggle with recognizing when an argument is logically invalid. In practice, this means that people tend to believe that all of these conclusions follow from the premises.

As mentioned above, people are generally better at reasoning about conditionals when the problems include real-life (thematic) content. For example, consider the following (AC) inference: "*If the radio is turned on, then you will hear music. You hear music. Does it follow that the radio is turned on?*" When people reason about problems with thematic content, they are better able to reject incorrect arguments, due to their ability to rely on counterexamples (e.g., Cummins, 1995; De Neys, Schaeken, & d'Ydewalle, 2005;

Quinn & Markovits, 1998; Thompson, 1994). For example, in the case of the above conditional argument, we could rely on our knowledge that we might hear music because there is a music program on TV or because our neighbor is listening to loud music, and, on this basis, we can argue that when we hear music, it is not necessarily true that the radio is turned on. Although we could rely on a similar process of considering counterexamples when we reason about conditionals with an abstract content, this is more difficult, because our reasoning is not supported by counterexamples that can be readily retrieved from memory. For example, if we consider the AC inference presented in Table 1, we could argue that the conclusion does not follow, because the statement "if the letter is X then the number is 4" is compatible with the alternative statement that "if the letter is not X then the number is 4."

It should be noted that the same process of generating counterexamples can lead to responses that are not in line with formal logic in the case of MP and MT inferences. Consider the following example of an MT inference: "*If it is raining, John gets wet. John doesn't get wet. Does it follow that it is not raining*?" Based on our everyday experiences, it is easy to think of situations where this conclusion does not follow. For example, John could use an umbrella, or he could stay indoors, and then he would not get wet in the rain. Nevertheless, this sort of reasoning does not take into account the instruction that we should suppose that if it is raining, John *necessarily always* gets wet. Similarly, in the case of an MP inference, one might think of disabling conditions. Consider the following example: "*If you throw a chair at a window, the window will break. You throw a chair at a window. Does it follow that the window will break?*" In this case, one could imagine that the chair was very light, or that the window was made of a special type of glass that is very hard to break. Nevertheless, people almost never block (i.e., reject) the MP inference.

Handley, Newstead, and Trippas (2011) demonstrated that drawing an MP inference is even easier than telling whether the conclusion of a conditional argument is believable or unbelievable (although giving responses on the basis of beliefs is considered an effortless, heuristic process—e.g., Evans, Barston, & Pollard, 1983). Handley et al. (2011) presented their participants with problems like: "*When a child is crying, then he is happy. Suppose a child is crying, is the child happy*?" The participants were either instructed to give a response on the basis of their beliefs or on the basis of logic (they were given a detailed explanation of what a logical response means in this case). When logic and belief conflicted, as in the example above, people were faster to give logic-based responses than belief-based responses and also had fewer errors. This suggests that, in the case of the MP inference, correct reasoning is supported by some kind of "intuitive logic," which does not necessitate effortful processing.

Regarding the MT inference, given that people's dominant response to conditional inferences in general is that they think that the conclusion necessarily follows, it is not very easy to interpret response patterns to MT inferences. Indeed, a person might say that the conclusion follows, because they do not consider the problem very carefully, but it is also possible that a person engages in the laborious process of considering counterexamples, but then rejecting them on the basis that they conflict with the rule that was given.

Overall, people are not generally very good at conditional reasoning (at least in the sense of following the rules of normative logic). Moreover, although conditional reasoning is considered a form of complex, logical inference, people tend to rely on memory retrieval, rather than on the application of formal rules, when they reason about invalid inferences. Several studies investigated the involvement of working memory (WM) resources in conditional reasoning, which are thought to play a key role in activating relevant knowledge, as well as inhibiting the retrieval of knowledge when it is not warranted. De Neys, Schaeken, and d'Ydewalle (2003) reported that verbal WM capacity was negatively associated with endorsing the invalid AC and DA inferences, whereas WM capacity showed a quadratic, U-shaped trend, suggesting that acceptance rates of the logically valid MP and MT inferences were the lowest in the case of people at the higher and lower end of the WM capacity continuum. Newstead, Handley, Harley, Wright, and Farrelly (2004) found similar trends when they investigated the relations between conditional reasoning performance and intelligence in the case of the invalid DA and AC inferences, but a somewhat different pattern in the case of the valid MP and MT inferences. Newstead et al. (2004) reported that whereas correct reasoning about the MP, DA, and AC inferences was positively related to intelligence in an adult sample, there was a nonsignificant trend for a negative relationship between correct MT inferences and intelligence. Correct MT inferences were also negatively related to correct reasoning about the DA and AC inferences. A potential explanation of this pattern is that people who relied on the retrieval of counterexamples in the case of DA and AC were also more likely to (erroneously) reason about MT by considering possible counterexamples.

Handley, Capon, Copp, and Harper (2002) investigated the role of verbal and spatial WM resources (including both simple and complex span tasks) in conditional reasoning. They have also included the Tower of Hanoi (ToH) task as an additional reasoning task for a comparison. Handley et al. (2002) found that performance on the conditional reasoning task correlated most strongly with a complex verbal span task, whereas solution time on the ToH correlated most strongly with simple and complex spatial WM span, suggesting that verbal WM resources are more strongly involved in conditional reasoning than spatial WM resources.

Regarding the development of conditional reasoning skills, it should be noted that the ability to consider counterexamples develops gradually, and that the quality of reasoning is strongly influenced by the content of the problems (e.g., De Neys & Everaerts, 2008; Gauffroy & Barrouillet, 2009; Markovits & Barrouillet, 2002). The ability to reason about false or abstract premises is particularly difficult. Markovits and Lortie-Forgues (2011) reported that 6–7-year-olds were unable to reason logically about conditionals with false premises (e.g., *If a towel is dropped in the bath then it becomes dry*), even when these premises were presented in a fantasy context, which generally makes it easier for children to reason about belief-inconsistent content (e.g., Dias & Harris, 1988). By contrast, 9–11-year-olds were able to reason logically about belief-inconsistent content under certain circumstances, and 12–15-year-olds were also able to reason logically about abstract premises when they were primed to rely on the retrieval of counterexamples.

Taken together, the literature on conditional inferences suggests that the MP inference is based on effortless "intuitive logic" (cf., Handley et al., 2011), and correct reasoning about the DA and AC inferences is based on the retrieval of counterexamples from memory. In the case of MT, correct reasoning is more difficult to interpret, as it might reflect default responding (i.e., accepting the conclusion as valid) without considering counterexamples, or the laborious process of first retrieving and then inhibiting counterexamples. In terms of its developmental trajectory, the ability to rely on counterexamples to support conditional inferences continues to develop into early adulthood, although even children can rely on this process in some cases.

It is notable that although both transitive and conditional inference are considered to reflect logical reasoning skills, their developmental trajectories, as well as the strategies that are used to solve these problems, are notably different. Based on these considerations, we can expect that how and why they might link to mathematical skills will also differ. Specifically, we can expect that mathematical operations that utilize spatial representations (and, in particular, the activation of a mental number line) should be most strongly related to transitive reasoning. By contrast, conditional reasoning appears to be related to tasks that require the activation and inhibition of relevant knowledge, and it relies most strongly on verbal WM resources.

## Transitive reasoning and mathematics skills

Earlier, I provided a brief review of the literature on the development of transitive reasoning skills. Wright (2001, p. 376) defined transitivity as "one of the emergent properties of an ordered set when the items in that set can be placed along a single relative continuum." Based on this definition, it is easy to recognize that transitivity is an important property of the number system. That is, reasoning about numbers often involves taking into account transitive relations. Indeed, transitive inference is considered to play an important role in the learning and understanding of various concepts that are important for mathematics, such as measurement (e.g., length, height), class inclusion (e.g., *if all As are Bs, and all Bs are Cs then all As are Cs*), and comparative reasoning (see examples above; Bryant & Kopytynska, 1976; Kallio, 1988; Newstead, Keeble, & Manktelow, 1985; Piaget, Inhelder, & Szeminska, 1960; Rabinowitz & Howe, 1994).

Transitive reasoning skills emerge early in development, and it has been argued that they build crucially on the ability to create a (probably spatial) mental representation of the ordinal relations between items. A similar representation has been proposed to exist in the domain of numbers, which has been termed the mental number line (Moyer & Landauer, 1967; Restle, 1970). Indeed, some phenomena, which have been considered as supporting evidence for the existence of the mental number line (i.e., the distance effect and the SNARC effect[a]), have also been demonstrated in the case of transitive inferences (Potts, 1974; Prado, Noveck, & Van Der Henst, 2010; Prado, Van Der Henst, & Noveck, 2008).

The neural underpinnings of the distance effect in the case of number comparison and transitive inferences have been investigated by Prado, Noveck, and Van Der Henst (2010). The participants in their study were presented with the following information:

[a]The SNARC effect (i.e., spatial-numerical association of response codes; Dehaene, Bossini, & Giraux, 1993) is considered to provide evidence that people represent relative numerical magnitudes along a mental number line. When presented with smaller numbers, people tend to respond faster if those stimuli are associated with a response on the left-hand side (for example, they have to press a key with their left hand), and when presented with larger numbers, people respond faster if those stimuli are instead associated with a right-hand side response.

*The rectangle is to the left of the oval, the star is to the left of the triangle, the oval is to the left of the hexagon, the circle is to the left of the square, the arrow is to the left of the rectangle, the square is to the left of the star, and the triangle is to the left of the arrow.*

Subsequently, the participants were asked to decide whether some statements relating to the above description were true or false (e.g., "*The square is to the left of the triangle*"). The participants also performed a number comparison task, involving one-digit numbers. In terms of behavioral results, Prado, Noveck, and Van Der Henst (2010) found similar distance effects in both tasks with participants responding more quickly as the distance between items increased. Additionally, Prado et al. found similar activity in the horizontal segment of the intraparietal sulcus (hIPS) during the number comparison task and when answering questions about the relative position of shapes. Whereas several studies reported associations between activation in the hIPS, the manipulation of numbers (e.g., Cantlon, Brannon, Carter, & Pelphrey, 2006; Piazza, Pinel, Le Bihan, & Dehaene, 2007; Pinel, Piazza, Le Bihan, & Dehaene, 2004), and quantity estimation (Dehaene, Piazza, Pinel, & Cohen, 2003), in line with the findings of Prado, Noveck, and Van Der Henst (2010), earlier studies also demonstrated that activation in the IPS is not only related to the processing of numbers or magnitude information, but also to the processing of other types of stimuli (e.g., the processing of numerical and nonnumerical ordinality—Fias, Lammertyn, Caessens, & Orban, 2007; Kaufmann, Vogel, Starke, Kremser, & Schocke, 2009). Nevertheless, Prado et al.'s study goes beyond these earlier studies in that the participants were presented with a verbal description of spatial relations, instead of an array of simple symbols.

Prado, Chadha, and Booth (2011) conducted a meta-analysis to investigate the neural underpinnings of logical reasoning skills, including transitive reasoning. Prado et al. (2011) found that these reasoning tasks (including transitive arguments) elicit activation in a mostly left-lateralized brain system, which includes the left lateral and medial frontal cortices, the left parietal cortex, and the left basal ganglia. Nevertheless, within this brain network, there appeared to be several subsystems that were specific to a particular type of argument. Specifically, direct contrasts between studies revealed that activation of the posterior parietal cortex (in both the left and right hemispheres) and the right middle frontal gyrus was more consistently linked to transitive inferences (whether the materials were presented in a verbal or a nonverbal format) than to other types of arguments. Prado et al. (2011) concluded that transitive inferences utilize spatial mental representations, which are very similar to the mental number line.

Regarding behavioral evidence for a link between mathematics skills and transitive reasoning, Morsanyi, Kahl, and Rooney (2017), using verbal transitive inference problems, found that the mathematical skills of adolescents (as indicated by a variety of measures) were related to their transitive reasoning ability. Morsanyi, McCormack, and O'Mahony (2018) conducted a study with adults, which included both transitive and conditional reasoning tasks (both including verbal materials), basic and complex measures of mathematics skills (i.e., a number ordering task, an arithmetic fluency task, a number line task, and two reasoning tasks with numerical content), a measure of fluid intelligence, and a month ordering task. The two numerical reasoning tasks were the Cognitive Reflection Test-Long (CRT-L; Primi, Morsanyi, Chiesi, Donati, & Hamilton, 2016; an extended version of the CRT developed by Frederick, 2005) and the Probabilistic Reasoning Scale (PRS, Primi, Morsanyi, Donati, Galli, & Chiesi, 2017).

Although the transitive inference task was easy for adults, and performance was close to ceiling, transitive reasoning skills were related to performance on the number line task, mathematical reasoning skills, and fluid intelligence. However, transitive reasoning skills were unrelated to arithmetic fluency and conditional reasoning ability (these correlations were close to zero). The correlation between transitive reasoning skills and a combined measure of number and month ordering (where three numbers or three months were presented, and participants had to indicate whether these were in the correct or incorrect order) was nonsignificant, although this could possibly reflect a lack of statistical power ($r = 0.17$). The results of this study are in line with the neuroscience evidence regarding a link between transitive reasoning skills and number line performance. Additionally, the measure of fluid intelligence (Raven's Progressive Matrices) also requires reasoning on the basis of the spatial arrangement of items. Another notable finding of this study was that transitive reasoning skills were unrelated to conditional reasoning skills. That is, performance on two tasks that are commonly used to measure logical reasoning ability was unrelated. I will return to this issue in the next section.

Some studies also investigated the transitive reasoning skills of children with mathematics difficulties. It could be expected that individuals with dyscalculia might find transitive reasoning relatively difficult, because they also perform poorly on the number line task (e.g., Kucian et al., 2011; Morsanyi, van Bers, O'Connor, & McCormack, 2018; Sella, Berteletti, Brazzolotto, Luncageli, & Zorzi, 2013; Skagerlund & Träff, 2014). Nevertheless, this is not a trivial prediction, because at least some theorists would expect that dyscalculia is a domain-specific deficit, which only affects symbolic and nonsymbolic magnitude processing skills (e.g., Butterworth, Varma, & Laurillard, 2011). Nevertheless, others proposed that a single explanatory factor is not sufficient, because number processing and dyscalculia are based on a multicomponent network of skills that involve both domain-specific and domain-general components (Fias, Menon, & Szűcs, 2013).

Morsanyi et al. (2013) administered verbal transitive reasoning problems to three groups of 10-year-old children: children with dyscalculia, and children with average and high maths ability. The first two groups were matched on verbal and nonverbal IQ, verbal WM, and reading skills, whereas children in the high maths ability group had higher IQ and WM, and better reading skills than children in the other two groups. The transitive reasoning problems included both belief-laden and abstract/belief-neutral items (e.g., "*Dogs are stronger than lions*" and "*Tom is faster than John*").

Children with dyscalculia generally based their responses on the believability of the conclusion, whether it logically followed from the premises or not. Children with average and high maths ability were able to inhibit their beliefs and reason on the basis of logical structure, but children with high maths ability were better able to do this. Nevertheless, all groups performed poorly on the problems with belief-neutral items, possibly, because in the case of these problems it was more difficult to create a visual–spatial representation of the premise information.

Schwartz, Epinat-Duclos, Léone, Poisson, and Prado (2018) presented children between the ages of 8 and 12 years with stories that included transitive or nontransitive relations. Their sample included children with maths learning difficulties and typically developing (TD) children, and their brain activation patterns were recorded using fMRI. The children also answered some questions that either required transitive reasoning or they simply evaluated their memory of some details of the story.

The following is an example of a story including transitive relations:

"You would like some milk for your breakfast. You are going to milk cows with the farmer. *White cows give more milk than black cows. Black cows give more milk than brown cows.* You need to milk the cows giving the most milk. *Are you milking the brown cows or the white cows*?"

The following is an example of a story including nontransitive relations:

"The next morning, the farmer is baking pastries. The farmer is asking you to take them out of the oven now. *The chocolate cake is baking faster than the apple pie. The strawberry pie is baking faster than the cheesecake.* It is very hot in the kitchen. *Are you taking the pastries out of the oven now or later?*"

In terms of brain activation patterns, compared to nontransitive relations, listening to transitive relations was associated with enhanced activity in the IPS in TD children, but not in children with maths difficulties. Additionally, children with maths difficulties were also less accurate in their responses that required transitive inferences.

Overall, these findings suggest that, beyond a correlation between transitive reasoning and mathematics skills at the behavioral level, there are also shared neural networks that underlie these skills. These findings are not only important because they contribute to our understanding of the link between maths and logic, but because they also show that the IPS, which has been previously considered a domain-specific area of the brain specialized in numbers (Dehaene et al., 2003), also supports transitive inferences, even when the premises are presented verbally. On the other hand, these findings also show that rather than relying on abstract rules, logical reasoning (at least when we consider transitive inference) utilizes the basic process of mentally lining up items along a continuum, and making decisions about their relative position. In terms of theoretical approaches to developmental dyscalculia, these findings also suggest that it is not plausible to expect that the difficulties of individuals with dyscalculia would be restricted to the domain of numbers.

*Conditional reasoning and mathematics skills: is this relation different from the links between transitive inferences and mathematics skills?*

In the previous section, I provided an overview of the evidence that transitive reasoning skills relate to mathematics abilities, and, in particular, to processes that utilize the mental number line, and (more generally) spatial representations of relations between items. These links emerge early in development, and they are still present in the case of adolescents and adults. In contrast to transitive inferences, conditional inferences are more complex, and the ability to draw logical conclusions from conditional arguments develops slowly, and it remains error-prone even in the case of educated adults. Conditional reasoning performance is also very much dependent on the content of the tasks[b] (i.e., whether the content is familiar or abstract), and on the availability of counterexamples. Conditional reasoning performance has been linked to verbal WM resources, which are involved in the retrieval and inhibition of counterexamples from memory.

[b]Problem content also matters in the case of transitive inferences, but, typically, this effect is no longer present in the case of older adolescents and adults.

In terms of the neural underpinnings of conditional inference, there is much less evidence available than in the case of transitive inferences. Moreover, the available studies have only focused on the MP and MT arguments (e.g., Prado, Van Der Henst, & Noveck, 2010), and only one of these studies (Noveck, Goel, & Smith, 2004) included both MP and MT arguments together. Noveck et al. (2004), using fMRI, showed that MP was associated with activation in the left posterior parietal cortex (which is similar to the findings regarding transitive inferences), whereas MT engages the left inferior frontal gyrus (IFG), an area of the brain, which has been found to be engaged in rule-governed syntax processing in natural language (e.g., Friederici & Kotz, 2003; Grodzinsky & Santi, 2008). In their meta-analysis of neuroscience evidence regarding logical reasoning skills, Prado et al. (2011) contrasted the brain activation patterns associated with transitive and propositional reasoning (with the latter including conditional inferences). These analyses showed that activation of the left precentral gyrus was more consistently linked to propositional arguments than transitive inferences. The activation of this area might reflect the engagement of attentional and motor processes (Acuna, Eliassen, Donoghue, & Sanes, 2002).

Overall, the neuroscience evidence relating to conditional reasoning is sparse compared to the evidence relating to transitive inferences and gives less indication of why conditional reasoning might be linked to mathematics skills. This is not very surprising, given the complexity of conditional inferences, and that even behavioral studies suggest dissociations between people's ability to correctly reason about different types of conditional arguments, as well as differences in the strategies that they rely on. Indeed, Goel (2007) proposed that diverse results can be expected in neuroscience studies, because the neural bases of deduction are dependent on a variety of factors (e.g., the semantic content of the premises, the presence of conflicting information, and the degree of certainty in the conclusion). Additionally, Goel (2007) suggested that people might also differ in the degree to which they rely on visuospatial or rule-based representations during the reasoning process.

In the previous section, I described some of the results of Morsanyi, McCormack, and O'Mahony (2018), who investigated the relations between both transitive and conditional reasoning and a variety of mathematics tasks in an adult sample. In this study, the availability of counterexamples was also manipulated to investigate whether this affects the relation between conditional reasoning and maths performance. For example, in the case of the conditional statement "*if butter is heated, then it will melt*," it is difficult to retrieve counterexamples from memory. By contrast in the case of the statement "*if Anne falls from a tree, then she will be injured*," it is easy to think of counterexamples (e.g., she might wear protective clothing, or a mat is placed under the tree, or the tree is very small, etc.). Morsanyi, McCormack, and O'Mahony (2018), Morsanyi, van Bers, et al., 2018 found that performance on the different types of conditional arguments (i.e., MP, MT, DA, and AC) showed different trends. Specifically, correct performance on the MP argument was at ceiling and did not relate to performance on the other inferences. There was a strong positive correlation between correct performance on the DA and AC inferences, whereas performance on both of these inferences was negatively related to correct responding on the MT inference (see Newstead et al., 2004 for similar findings). The availability of counterexamples only affected performance on the DA and AC arguments.

Regarding the relationships between conditional reasoning and the other measures administered in the study (i.e., fluid intelligence, number and month ordering, arithmetic fluency, number line estimation, and performance on the CRT-L and PRS), performance on the MT inference was not significantly linked to any of these measures. Due to their strong association, performance on the DA and AC arguments was combined, and this measure was significantly and moderately related to arithmetic fluency, number and month ordering, and performance on the two reasoning tasks with numerical content (i.e., the CRT-L and PRS). However, the correlation between reasoning about the DA and AC arguments and transitive reasoning, fluid intelligence, and performance on the number line task was close to zero. The availability of counterexamples in the case of DA and AC arguments did not affect the correlation between these argument forms and the other measures.

Overall, the findings of this study showed a dissociation between transitive and conditional inferences. Apart from the notable finding that performance on the two types of logical reasoning measures was unrelated, it is also interesting that they differentially related to the measures of mathematics skills and fluid intelligence. In particular, whereas transitive reasoning skills were related to performance on the number line task and fluid intelligence, conditional reasoning performance (as indicated by performance on the AC and DA inferences) was related to arithmetic fluency. The only converging result between the two types of logical reasoning measures was that both showed relations with performance on the CRT-L. The finding that conditional reasoning (at least in the case of the invalid argument forms) related to arithmetic fluency is in line with our expectation that these inferences rely mostly on verbal working memory resources rather than spatial processing. Indeed, it appears that from late childhood, arithmetic performance mostly relies on verbal WM resources (e.g., Van de Weijer-Bergsma, Kroesbergen, & Van Luit, 2015). It is also notable that performance on the different types of conditional arguments was unrelated, with the exception of the invalid argument forms.

At the start of this chapter, I briefly referred to the Theory of Formal Discipline (TFD), the idea that training in a specific discipline could lead to improvements in other areas of knowledge. Inglis and colleagues have investigated this possibility in the case of mathematics learning in relation to logical reasoning skills (using conditional reasoning performance as an outcome measure) in a series of studies (Attridge & Inglis, 2013; Inglis & Simpson, 2008, 2009). In these studies, the reasoning skills of students studying post-compulsory mathematics were compared to the reasoning skills of arts students. Inglis and Simpson (2008) found better conditional reasoning performance among mathematics students than arts students, and Inglis and Simpson (2009) replicated these findings in a new sample of maths and arts students, who were matched on their level of intelligence. In another study, Attridge and Inglis (2013) investigated the changes in the reasoning skills of students studying maths and literature during the first year of their post-compulsory studies. The content of the problems was similar to the materials presented in Table 1 (i.e., the content was abstract with no reference to real-life knowledge). It is important to note that students studying mathematics were not taught any proof-based mathematics and were not any more familiar with formal approaches to conditional reasoning than the other student group.

Attridge and Inglis (2013) found that the conditional reasoning performance of literature students did not change between the start and the end of their first year of post-compulsory study. By contrast, maths students improved in their ability to accept the valid MP inference and reject the invalid AC and DA inferences. Additionally, they were also more likely to incorrectly reject MT inferences at the end of their first year of post-compulsory maths education. These findings are in line with Morsanyi, McCormack, and O'Mahony (2018); and Newstead et al. (2004) who also found similar patterns of conditional reasoning among adults with higher cognitive/mathematics ability. Although these results clearly show that mathematics study leads to, generally positive, changes in conditional reasoning performance, it is difficult to pin down what it is exactly about mathematical study that might drive these changes. Indeed, the curriculum for the first year of post-compulsory mathematics study includes various topics (i.e., algebra, geometry, calculus, trigonometry, probability, mathematical modeling, kinematics, and forces). One possibility is that students improve in their general ability to follow formal rules, and to work with abstract content (or symbols) which have no direct links to their everyday knowledge.

Overall, the studies on conditional reasoning and mathematics skills show that these two skills are related in adults, and the link between conditional reasoning skills and complex mathematics is especially well documented. Nevertheless, Morsanyi, McCormack, and O'Mahony (2018) also showed links between conditional reasoning and relatively basic mathematics tasks, including the number ordering task, and arithmetic fluency. It is notable that across studies, reasoning about the MT argument was either negatively related to mathematics skills or the two skills were unrelated. Together with the fact that neither transitive and conditional reasoning skills nor the various forms of conditional inference are generally related, these findings suggest that mathematics is not related to a generic logical reasoning ability, but to certain reasoning strategies that people employ. In fact, when people reason logically, they often rely on simple strategies (such as representing the items along a linear continuum or retrieving relevant examples from memory) to support the reasoning process.

## Summary and conclusions

In this chapter, I have provided an overview of the development of two types of logical inference (transitive and conditional reasoning), and their links to mathematics skills. I have also started with a set of questions that I aimed to answer. Regarding the question of what types of logical reasoning are related to mathematics, we can say that at least two types of logical reasoning (transitive and conditional inferences) are clearly linked to mathematics skills. I have focused on these two types of reasoning skills, because there is currently a lack of research evidence regarding the relations between other types of deductive inference and mathematical skills. I am only aware of a single paper that investigated the relations between reasoning about categorical syllogisms (e.g., *All fruits can be eaten. Hamburgers can be eaten. Therefore hamburgers are fruits.*) and mathematics skills in adolescents (Morsanyi et al., 2017). This study did not find any links between these two skills. Nevertheless, it is possible that the relatively low performance of adolescents on the reasoning problems masked a possible relation with mathematics skills. Thus this issue deserves further attention in future studies.

Another question that I posed at the start of this chapter was whether the links between maths and logic are stable across development. This appears to be the case with regard to transitive inferences. Indeed, a representation which is similar to the mental number line appears to underlie transitive reasoning performance not only in the case of animals and human infants but also in the case of children and adults. The only caveat regarding this finding is that transitive reasoning skills with abstract verbal content develop relatively late (around early adolescence), and there are some other types of transitive reasoning tasks that are difficult for young children. Thus, it is important to measure transitive inference skills with age-appropriate materials. Regarding conditional inferences, we have seen that performance on these tasks continues to develop into late adolescence, and it is also shaped by educational experiences. It is also likely that the way these problems are approached by people is dependent on their working memory capacity and level of intelligence. Nevertheless, at least in the case of adults, there is good evidence that conditional reasoning skills are linked to mathematics performance.

It is also of interest whether there are specific types of mathematics skills that are related to reasoning abilities, and whether the link between maths and logic is mediated by general cognitive factors, such as intelligence, working memory, or maybe some more basic cognitive processes. Regarding the involvement of working memory, there is strong evidence for the involvement of spatial WM in transitive reasoning, and verbal WM in conditional inferences. Additionally, measures of intelligence have been found to relate to both types of logical reasoning skill. Nevertheless, there are some studies that demonstrated relations between specific maths skills and logical reasoning even when the effect of intelligence was controlled. Specifically, Morsanyi, McCormack, and O'Mahony (2018) found that transitive inferences were related to number line estimation when the effect of fluid intelligence was controlled, although fluid intelligence was related to both transitive reasoning skills and number line estimation. Inglis and Simpson (2009) found differences in the reasoning abilities of post-compulsory maths and arts students, even though the two groups were matched on their level of intelligence. Attridge and Inglis (2013) showed that the changes in conditional reasoning performance associated with mathematical study relative to studying literature were still present when the effect of the students' level of intelligence at the start of the academic year was controlled. Finally, when comparing children with and without developmental dyscalculia, who were matched on intelligence, verbal WM, and reading ability, Morsanyi et al. (2013) found a difference in the children's transitive reasoning performance.

In this chapter, I have focused on working memory as a common underlying factor between reasoning and maths skills and did not consider any potential effect of other types of executive functions. In particular, several studies have linked both reasoning and mathematics performance to inhibition skills (i.e., the ability to suppress unwanted responses and the effect of distracting stimuli—see chapter by Borst in this volume for a detailed discussion). In general, inhibition skills are relevant for verbal reasoning problems when there is a conflict between the believability and the logical validity of conclusions (see, e.g., Morsanyi et al., 2013). In the case of conditional reasoning problems, it is assumed that correct performance on the MT inference requires the ability to suppress the retrieval of counterexamples from memory (e.g., De Neys et al., 2005). I have also presented an example earlier for a study (Markovits et al., 1995), where correct transitive reasoning required participants to overcome the effect of perceptual distraction. Even though inhibition has been assumed to be important for deductive reasoning skills, there are surprisingly few studies that investigated this effect by administering separate measures of inhibition and reasoning skills (although see Handley et al., 2004).

Inhibition is also considered to be important for general mathematics development (see Cragg & Gilmore, 2014 for a review), although the evidence regarding a link between inhibition and mathematics performance is inconsistent (see Bull & Lee, 2014). Gilmore, Keeble, Richardson, and Cragg (2015) conducted a detailed investigation into the relations between inhibition skills and overall mathematics achievement, as well as factual, procedural, and conceptual knowledge in children and adults. Gilmore et al. (2015) found that overall mathematics achievement was more strongly related to inhibition measured in numerical than in nonnumerical contexts. Additionally, there were some age-related differences in these findings, with inhibition skills relating to conceptual knowledge in older participants, but procedural skills in younger participants. Lee and Lee (2019) proposed that although it is a compelling idea that the ability to resist distraction is important for mathematics, research evidence might be weaker than expected for several reasons. These include age-related changes in executive function skills, issues with the sensitivity of commonly used inhibitory tasks, differences in the way inhibition skills and mathematical performance are assessed, and differences in the inhibitory demands of mathematics tasks that are used as outcome measures in these studies. Overall, although inhibition certainly plays a role in both mathematical and reasoning performance, there is stronger evidence for the role of context-specific inhibition processes than a generic inhibition ability. Nevertheless, future studies could investigate this question further by considering whether there is a link between people's susceptibility to context-specific manipulations of the inhibitory demand of mathematics and reasoning tasks.

In summary, studies have found a relationship between mathematics and logical reasoning skills in both children and adults, and these links might be present even when the effects of intelligence are controlled. Nevertheless, these relations are specific to particular types of inference. It is also important to consider that people do not seem to have a natural talent for logical reasoning. It appears that even highly educated adults with high level of cognitive ability only approximate formal logic, and they do this by relying on relatively simple strategies. The same basic strategies and building blocks (e.g., mentally lining up items along a continuum, and processing information about the ordinal relationships of items) seem to be very important for mathematics as well. An interesting future direction could be to check whether training in these basic skills could improve both reasoning and mathematics skills. Future studies could also investigate further the links between conditional reasoning and mathematics skills by involving child participants (using age-appropriate materials). Finally, it remains to be seen whether reasoning about categorical syllogisms also relates to mathematics skills.

## References

Acuna, B. D., Eliassen, J. C., Donoghue, J. P., & Sanes, J. N. (2002). Frontal and parietal lobe activation during transitive inference in humans. *Cerebral Cortex*, *12*, 1312–1321.

Attridge, N., & Inglis, M. (2013). Advanced mathematical study and the development of conditional reasoning skills. *PLoS One*, *8*(7).

Bryant, P. E., & Kopytynska, H. (1976). Spontaneous measurement by young children. *Nature*, *260*, 773.

Bryant, P. E., & Trabasso, T. (1971). Transitive inferences and memory in young children. *Nature*, *232*, 456–458.

Bull, R., & Lee, K. (2014). Executive functioning and mathematics achievement. *Child Development Perspectives*, *8*, 36–41.

Butterworth, B., Varma, S., & Laurillard, D. (2011). Dyscalculia: From brain to education. *Science*, *332*(6033), 1049–1053.

Cantlon, J. F., Brannon, E. M., Carter, E. J., & Pelphrey, K. A. (2006). Functional imaging of numerical processing in adults and 4-y-old children. *PLoS Biology*, *4*, e125.

Cragg, L., & Gilmore, C. (2014). Skills underlying mathematics: The role of executive function in the development of mathematics proficiency. *Trends in Neuroscience and Education*, *3*, 63–68.

Cummins, D. D. (1995). Naïve theories and causal deduction. *Memory & Cognition, 23*, 646–658.

De Neys, W., & Everaerts, D. (2008). Developmental trends in everyday conditional reasoning: The retrieval and inhibition interplay. *Journal of Experimental Child Psychology, 100*, 252–263.

De Neys, W., Schaeken, W., & d'Ydewalle, G. (2003). Working memory span and everyday conditional reasoning: A trend analysis. *Psychological Reports*, 296–311.

De Neys, W., Schaeken, W., & d'Ydewalle, G. (2005). Working memory and everyday conditional reasoning: Retrieval and inhibition of stored counterexamples. *Thinking & Reasoning, 11*, 349–381.

Dehaene, S., Bossini, S., & Giraux, P. (1993). The mental representation of parity and number magnitude. *Journal of Experimental Psychology: General, 122*, 371–396.

Dehaene, S., Piazza, M., Pinel, P., & Cohen, L. (2003). Three parietal circuits for number processing. *Cognitive Neuropsychology, 20*, 487–506.

Dias, M. G., & Harris, P. L. (1988). The effect of make-believe play on deductive reasoning. *British Journal of Developmental Psychology, 6*, 207–221.

Dolton, P. J., & Vignoles, A. (2002). The return on post–compulsory school mathematics study. *Economica, 69*, 113–142.

Evans, J. S. B., Barston, J. L., & Pollard, P. (1983). On the conflict between logic and belief in syllogistic reasoning. *Memory & Cognition, 11*, 295–306.

Fias, W., Lammertyn, J., Caessens, B., & Orban, G. A. (2007). Processing of abstract ordinal knowledge in the horizontal segment of the intraparietal sulcus. *Journal of Neuroscience, 27*, 8952–8956.

Fias, W., Menon, V., & Szűcs, D. (2013). Multiple components of developmental dyscalculia. *Trends in Neuroscience and Education, 2*(2), 43–47.

Friederici, A. D., & Kotz, S. A. (2003). The brain basis of syntactic processes: Functional imaging and lesion studies. *NeuroImage, 20*, S8–S17.

Gauffroy, C., & Barrouillet, P. (2009). Heuristic and analytic processes in mental models for conditionals: An integrative developmental theory. *Developmental Review, 29*(4), 249–282.

Gazes, R. P., Hampton, R. R., & Lourenco, S. F. (2017). Transitive inference of social dominance by human infants. *Developmental Science, 20*, e12367.

Gilmore, C., Keeble, S., Richardson, S., & Cragg, L. (2015). The role of cognitive inhibition in different components of arithmetic. *ZDM, 47*, 771–782.

Goel, V. (2007). Anatomy of deductive reasoning. *Trends in Cognitive Sciences, 11*, 435–441.

Goodwin, G. P., & Johnson-Laird, P. N. (2005). Reasoning about relations. *Psychological Review, 112*, 468–493.

Goodwin, G. P., & Johnson-Laird, P. N. (2008). Transitive and pseudo-transitive inferences. *Cognition, 108*, 320–352.

Grodzinsky, Y., & Santi, A. (2008). The battle for Broca's region. *Trends in Cognitive Sciences, 12*, 474–480.

Handley, S. J., Capon, A., Beveridge, M., Dennis, I., & Evans, J. S. B. (2004). Working memory, inhibitory control and the development of children's reasoning. *Thinking & Reasoning, 10*, 175–195.

Handley, S. J., Capon, A., Copp, C., & Harper, C. (2002). Conditional reasoning and the tower of Hanoi: The role of spatial and verbal working memory. *British Journal of Psychology, 93*, 501–518.

Handley, S. J., Newstead, S. E., & Trippas, D. (2011). Logic, beliefs, and instruction: A test of the default interventionist account of belief bias. *Journal of Experimental Psychology: Learning, Memory, and Cognition, 37*, 28–43.

Hogue, M. E., Beaugrand, J. P., & Laguë, P. C. (1996). Coherent use of information by hens observing their former dominant defeating or being defeated by a stranger. *Behavioural Processes, 38*, 241–252.

Inglis, M., & Attridge, N. (2017). *Does mathematical study develop logical thinking?: Testing the theory of formal discipline.* World Scientific.

Inglis, M., & Simpson, A. (2008). Conditional inference and advanced mathematical study. *Educational Studies in Mathematics, 67*, 187–204.

Inglis, M., & Simpson, A. (2009). Conditional inference and advanced mathematical study: Further evidence. *Educational Studies in Mathematics, 72*, 185–198.

Johnson-Laird, P. N. (1999). Deductive reasoning. *Annual Review of Psychology, 50*, 109–135.

Kallio, K. D. (1988). Developmental differences in the comprehension of simple and compound comparative relations. *Child Development, 59*, 397–410.

Kaufmann, L., Vogel, S. E., Starke, M., Kremser, C., & Schocke, M. (2009). Numerical and non-numerical ordinality processing in children with and without developmental dyscalculia: Evidence from fMRI. *Cognitive Development, 24*, 486–494.

Klaczynski, P. A., Schuneman, M. J., & Daniel, D. B. (2004). Theories of conditional reasoning: A developmental examination of competing hypotheses. *Developmental Psychology, 40*, 559–571.

Kucian, K., Grond, U., Rotzer, S., Henzi, B., Schönmann, C., Plangger, F., et al. (2011). Mental number line training in children with developmental dyscalculia. *NeuroImage, 57*, 782–795.

Lee, K., & Lee, H. W. (2019). Inhibition and mathematical performance: Poorly correlated, poorly measured, or poorly matched? *Child Development Perspectives, 13*, 28–33.

Markovits, H., & Barrouillet, P. (2002). The development of conditional reasoning: A mental model account. *Developmental Review, 22*, 5–36.

Markovits, H., Dumas, C., & Malfait, N. (1995). Understanding transitivity of a spatial relationship: A developmental analysis. *Journal of Experimental Child Psychology, 59*, 124–141.

Markovits, H., & Lortie-Forgues, H. (2011). Conditional reasoning with false premises facilitates the transition between familiar and abstract reasoning. *Child Development, 82*, 646–660.

Morsanyi, K., Devine, A., Nobes, A., & Szűcs, D. (2013). The link between logic, mathematics and imagination: Evidence from children with developmental dyscalculia and mathematically gifted children. *Developmental Science, 16*, 542–553.

Morsanyi, K., Kahl, T., & Rooney, R. (2017). The link between math and logic in adolescence: The effect of argument form. In M. E. Toplak, & J. Weller (Eds.), *Individual differences in judgment and decision making from a developmental context* (pp. 166–185). Hove: Psychology Press.

Morsanyi, K., McCormack, T., & O'Mahony, E. (2018). The link between deductive reasoning and mathematics. *Thinking & Reasoning, 24*, 234–257.

Morsanyi, K., van Bers, B. M., O'Connor, P. A., & McCormack, T. (2018). Developmental dyscalculia is characterized by order processing deficits: Evidence from numerical and non-numerical ordering tasks. *Developmental Neuropsychology, 43*, 595–621.

Moyer, R. S., & Landauer, T. K. (1967). Time required for judgements of numerical inequality. *Nature, 215*, 1519–1520.

Newstead, S. E., Handley, S. J., Harley, C., Wright, H., & Farrelly, D. (2004). Individual differences in deductive reasoning. *The Quarterly Journal of Experimental Psychology Section A, 57*, 33–60.

Newstead, S. E., Keeble, S., & Manktelow, K. I. (1985). Children's performance on set-inclusion and linear-ordering relationships. *Bulletin of the Psychonomic Society, 23*, 105–108.

Noveck, I. A., Goel, V., & Smith, K. W. (2004). Special section the neural basis of conditional reasoning with arbitrary content. *Cortex, 40*, 000–010.

Pears, R., & Bryant, P. (1990). Transitive inferences by young children about spatial position. *British Journal of Psychology, 81*, 497–510.

Piaget, J., & Inhelder, B. (1967). *The Child's Conception of Space, translated by F.J. Langdon and J.L.* Lunzer, New York: WW Norton.

Piaget, J., Inhelder, B., & Szeminska, A. (1960). *The Child's conception of geometry, translated by E.A. Lunzer*. Routledge and K. Paul.

Piazza, M., Pinel, P., Le Bihan, D., & Dehaene, S. (2007). A magnitude code common to numerosities and number symbols in human intraparietal cortex. *Neuron, 53*, 293–305.

Pinel, P., Piazza, M., Le Bihan, D., & Dehaene, S. (2004). Distributed and overlapping cerebral representations of number, size, and luminance during comparative judgments. *Neuron, 41*, 983–993.

Potts, G. R. (1974). Storing and retrieving information about ordered relationships. *Journal of Experimental Psychology, 103*, 431–439.

Prado, J., Chadha, A., & Booth, J. R. (2011). The brain network for deductive reasoning: A quantitative meta-analysis of 28 neuroimaging studies. *Journal of Cognitive Neuroscience, 23*, 3483–3497.

Prado, J., Noveck, I. A., & Van Der Henst, J. B. (2010). Overlapping and distinct neural representations of numbers and verbal transitive series. *Cerebral Cortex, 20*, 720–729.

Prado, J., Van Der Henst, J. B., & Noveck, I. A. (2008). Spatial associations in relational reasoning: Evidence for a SNARC-like effect. *Quarterly Journal of Experimental Psychology, 61*, 1143–1150.

Prado, J., Van Der Henst, J. B., & Noveck, I. A. (2010). Recomposing a fragmented literature: How conditional and relational arguments engage different neural systems for deductive reasoning. *NeuroImage, 51*, 1213–1221.

Primi, C., Morsanyi, K., Chiesi, F., Donati, M. A., & Hamilton, J. (2016). The development and testing of a new version of the cognitive reflection test applying item response theory (IRT). *Journal of Behavioral Decision Making, 29*, 453–469.

Primi, C., Morsanyi, K., Donati, M. A., Galli, S., & Chiesi, F. (2017). Measuring probabilistic reasoning: The construction of a new scale applying item response theory. *Journal of Behavioral Decision Making, 30*, 933–950.

Quinn, S., & Markovits, H. (1998). Conditional reasoning, causality, and the structure of semantic memory: Strength of association as a predictive factor for content effects. *Cognition, 68*, B93–B101.

Rabinowitz, F. M., & Howe, M. L. (1994). Development of the middle concept. *Journal of Experimental Child Psychology, 57*, 418–448.

Restle, F. (1970). Speed of adding and comparing numbers. *Journal of Experimental Psychology, 83*, 274–278.

Schroyens, W. J., Schaeken, W., & d'Ydewalle, G. (2001). The processing of negations in conditional reasoning: A meta-analytic case study in mental model and/or mental logic theory. *Thinking & Reasoning*, 7, 121–172.

Schwartz, F., Epinat-Duclos, J., Léone, J., Poisson, A., & Prado, J. (2018). Impaired neural processing of transitive relations in children with math learning difficulty. *NeuroImage: Clinical, 20*, 1255–1265.

Sella, F., Berteletti, I., Brazzolotto, M., Luncageli, D., & Zorzi, M. (2013). Number line estimation in children with developmental dyscalculia. *Learning Disabilities: A Contemporary Journal, 11*, 41–49.

Skagerlund, K., & Träff, U. (2014). Development of magnitude processing in children with developmental dyscalculia: Space, time, and number. *Frontiers in Psychology, 5*, 675.

Thompson, V. A. (1994). Interpretational factors in conditional reasoning. *Memory & Cognition, 22*, 742–758.

Van de Weijer-Bergsma, E., Kroesbergen, E. H., & Van Luit, J. E. (2015). Verbal and visual-spatial working memory and mathematical ability in different domains throughout primary school. *Memory & Cognition, 43*, 367–378.

Wright, B. C. (2001). Reconceptualizing the transitive inference ability: A framework for existing and future research. *Developmental Review, 21*(4), 375–422.

CHAPTER 4

# Cerebral underpinning of advanced mathematical activity

**Marie Amalric**
Department of Psychology, CAOs Lab, Carnegie Mellon University, Pittsburgh, PA, United States

## Introduction

Where have mathematical ideas originated from and how did they emerge? These questions, of great interest to intellectuals in antiquity, continue to spark attention of modern educators, philosophers, and scientists. Looking at mathematical history, it appears that number and space have been at the center of mathematics in many human civilizations around the world. Their oldest traces might even go back to Upper Paleolithic. Recent research in cognitive science suggests that the human brain is endowed from birth with a basic arithmetic system, shared with various animal species (Benson-Amram, Heinen, Dryer, & Holekamp, 2011; Cantlon, Merritt, & Brannon, 2016; Feigenson, 2011; Izard, Sann, Spelke, & Streri, 2009; Rugani, Fontanari, Simoni, Regolin, & Vallortigara, 2009). This system allows us to perceive small quantities, to approximately evaluate the number of objects in a set, and to perform simple operations such as addition. Intuitive numerical abilities also extend to the perception of ratios (Matthews & Chesney, 2015) and to the sense of probabilities (Xu & Garcia, 2008). Furthermore, humans, like other animal species, are endowed with spontaneous intuitions about spatial relations (Chiandetti & Vallortigara, 2007; Deipolyi, Santos, & Hauser, 2001; Lee, Sovrano, & Spelke, 2012), and also about shapes and their properties (Dehaene, Izard, Pica, & Spelke, 2006), angles in particular (Dillon, Duyck, Dehaene, Amalric, & Izard, 2019; Izard, Pica, Dehaene, Hinchey, & Spelke, 2011).

But mathematics in humans goes well beyond these approximate intuitions, in a way that seems unique among all animal species. At the scale of humanity, the development of mathematical concepts can be seen as pyramidal, i.e., starting from a base layer of knowledge, and progressively building successive layers that both integrate and refine the previous ones.

*Heterogeneous Contributions to Numerical Cognition*
https://doi.org/10.1016/B978-0-12-817414-2.00009-9

In Ancient Greece, mathematicians studied primarily abstract objects of arithmetic and geometry. They were interested in properties of rational number and integers, regular polygons, and so on. The famous Euclid "Elements" is still considered a foundational work and served as reference and starting point of new mathematical concepts for centuries. Arabic mathematicians drew the foundational principles of algebra from Greek, Indian, and Chinese mathematics, trying to fill the gap between arithmetic and geometry. When their works were transmitted to Europe, intellectuals progressively developed new concepts like complex numbers, and even new fields like infinitesimal calculus or analysis. Recent behavioral studies support the hypothesis of a pyramidal construction by evidencing, for example, a relation between number sense acuity and arithmetic achievement at schools (Gilmore, McCarthy, & Spelke, 2010; Libertus, 2015). A relation between ratio perception and success with symbolic fractions also seems to exist (Meng, Matthews, & Hubbard, 2018). In geometry, Dillon, Huang, and Spelke (2013) have suggested that nonsymbolic knowledge of geometry predicts the ability to use symbolic geometrical cues in 4-year-old children.

Yet, the question of the mechanisms by which more complicated mathematical concepts develop from these percepts remains unanswered. Two potential vectors of development, not necessarily mutually exclusive, have been hypothesized: (1) language and (2) visuospatial perception. On the one hand, for Hauser, Chomsky, and Fitch (2002), the same neural mechanisms operating over linguistic structures would be used for both language and mathematics. In particular in algebra, a widespread idea is that syntactical routines allowing to understand sentences such as "The girl kissed the boy who offered her roses" are directly used to interpret algebraic expressions such as $2\times(3\times(6+2)-5)$. On the other hand, visual experience of the world could heavily contribute to shape the human representation of mathematical concepts. For example, symmetry detection can drastically influence shape perception, as a square with sides at 45° to horizontal and vertical axes will be perceived as a diamond (Giaquinto, 2005). Moreover, De Toffoli and Giardino (2014) proposed that the visual representations enclosed in diagrams might not only help understand a mathematical object but could also promote the invention of new concepts.

Recent neuroimaging developments have offered new ways to investigate the long-debated issue of the origins of mathematics in the human brain. Neuroimaging techniques such as fMRI, EEG, NIRS, and EcoG have already elucidated many questions regarding numerical abilities. However, simple numerical knowledge seems hardly representative of the variety of domains that modern mathematics embraces. In the present chapter, I summarize some works that start to explore how the human brain represents advanced concepts from diverse domains such as analysis, algebra, geometry, or topology. This specific framework offers new perspectives on the long-debated questions of the relations between mathematics, language, and visual perception. For example, are language areas used by mathematicians when they do mathematics? Or does advanced mathematical reflection recycle core regions for number and space processing? Does the brain comprise a generic semantic system that stores mathematical knowledge alongside knowledge of history, geography, or famous people? Do neural mechanisms underlying mathematical reflection vary among domains? Is visual experience necessary to develop advanced mathematical reflection?

## Neural correlates of advanced math knowledge

### A math-responsive network in the brain

In three fMRI studies, we asked professional mathematicians to evaluate the truth value of advanced math and nonmath spoken statements (Amalric & Dehaene, 2016, 2018, 2019). In the first study, math statements were small problems respectively involving concepts of analysis, algebra, topology, and geometry (e.g., "Any matrix with coefficients in a principal ideal is equivalent to a companion matrix" is a false algebraic statement, while "A smooth function whose derivatives are all non-negative is analytic" is a true analysis statement). Nonmath statements similarly required to gather and combine relevant knowledge in order to answer (e.g., to answer to the statement "in France, in the Middle Ages, people were drinking alcohol made of potatoes," one has to bring together knowledge about Europe during the Middle Ages, and the discovery of the New World). We verified that all categories of statements were of matching difficulty. Participants were given a 4-s reflection period before answering whether the statement was true, false, or meaningless. Professional mathematicians only reached a 70% correct rate in this task. Thus in order to avoid potential confounding effects of difficulty, we presented simpler true or false statements in the following studies. To collect information about reaction time, we also removed the reflection period and gave participants a 2.5-s response period. In the second study, math statements spanned various solving strategies such as rote memory of algebraic and trigonometric facts (e.g., "$\cos^2 a + \sin^2 a = 1$"), algebraic calculation (e.g., "$(3y+1)(3y-1) = 9y^2 - 1$"), visualization of the unit circle as support to answer questions about trigonometry and complex numbers (e.g., "The angle between 1 and $1+i$ equals $\pi/4$"), and visualization of geometrical shapes and their relations (e.g., "The rotation of a line around a non-parallel axis generates a cone"). Nonmath statements were matching facts about arts, music, and literature (such as "Pointillism is a painting genre characterized by visible brush strokes"). This time, professional mathematicians answered almost perfectly, except on trigonometry statements. In the third study, we carefully matched very simple math and nonmath facts for syntax and included quantifiers, negation, or a combination of quantifiers and negation (e.g., "the sine function is periodical"; "[the] London buses are red"). The only quantifier we used was "some."

In all three studies, all math statements, regardless of their difficulty or domain, activated a reproducible set of bilateral intraparietal and ventrolateral temporal regions (Fig. 1). We call these regions a math-responsive network in the brain. Additional activations were observed in the dorsal prefrontal cortex depending on efforts deployed to understand and solve the problem (especially for the more difficult statements presented in the first experiment), as well as in the occipital cortex whenever solving problems required some visualization (especially for statements about trigonometry and complex numbers presented in the second experiment).

## Cortical dissociation between math and spoken language

All three abovementioned studies contained an independent syntax localizer that allowed us to perform single-subject analyses of the activity elicited by math vs nonmath statements in seven regions of interest (ROIs)—defined a priori as responsive to linguistic constituent structure (Pallier, Devauchelle, & Dehaene, 2011). Our first experiment showed that these language ROIs were transiently recruited by nonmath statements and only to a lesser extent by math statements. Moreover, the average percent signal change in these regions for each statement of experiments 2 and 3 proved to be linked with their syntactical complexity. In experiment 2, we observed a clear segregation between categories of statements (Fig. 2A) that significantly correlated with the number of grammatical morphemes. On the contrary, in experiment 3 where the syntax of math and nonmath statements was perfectly matched, no such difference was observed. Overall, language regions seemed to be recruited at a syntactical level but were entirely spared by the contrast of math vs nonmath reflection (Fig. 2B).

This contrast revealed that separate neural processes were involved in the semantic decision about the truth value of math vs nonmath statements (Fig. 1). Because our first experiment included a listening phase distinct of

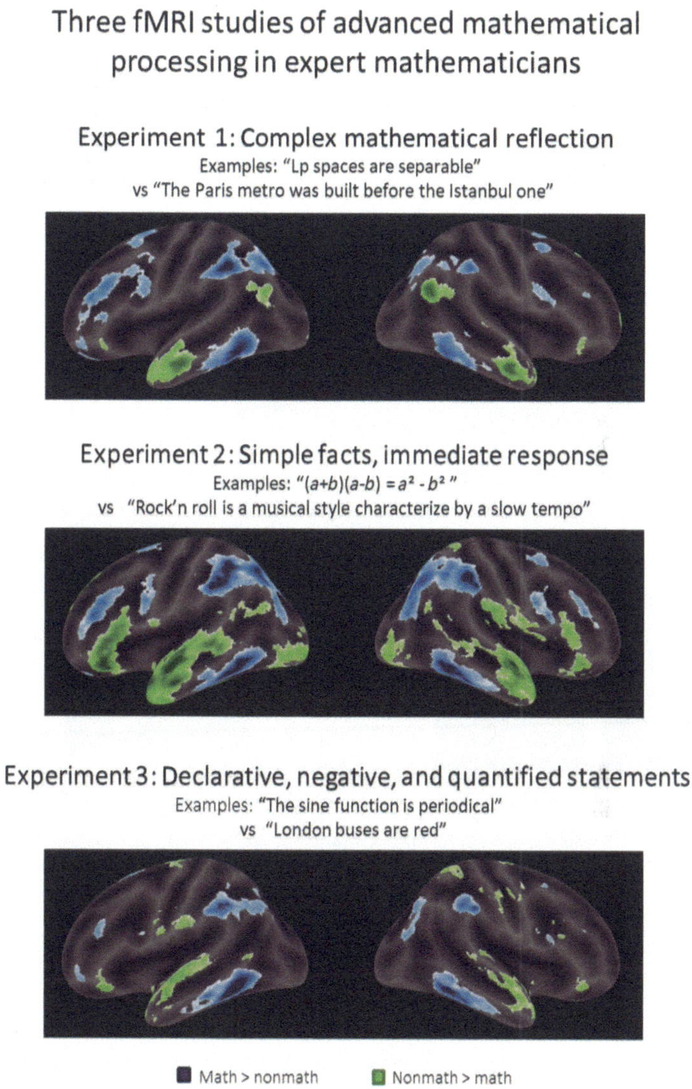


**Fig. 1** Brain activity in professional mathematicians evoked by complex mathematical and general semantic statements, simpler facts asking for an immediate response and elementary declarative, negative, and quantified statements. Brain maps showing areas of greater activation during reflection on mathematical statements vs nonmathematical statements (*blue*) or vice versa (*green*).

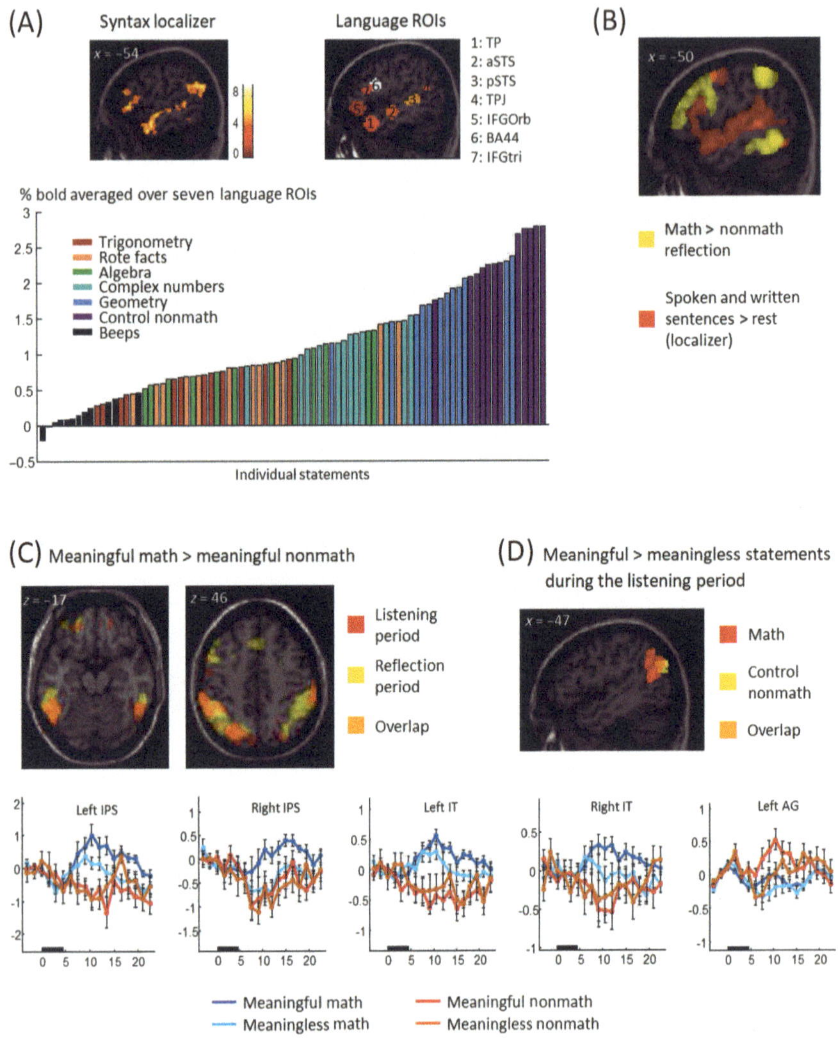


**Fig. 2** Syntax and semantics processing of math and nonmath statements. (A) Average activation for each statement presented in experiment 2 over TP, aSTS, pSTS, TPJ, IFGOrb, IFGtri, and BA44 defined by Pallier et al. (2011) and sorted in ascending order. Bar colors represent statement categories. Brain slices show the activation in the left hemisphere elicited by sentence processing in an independent syntax localizer, and the regions of interest. (B) Spatial relationship between the math and language networks. The contrasts of math vs nonmath reflection (in *yellow*) come from experiment 1 and the contrast of spoken and written sentences vs rest (in *red*) comes from an independent localizer. (C) Superposition of the contrasts of meaningful math > meaningful nonmath statements during their presentation and during the subsequent reflection period. (D) Activation to meaningful > meaningless math (in *red*) and to meaningful > meaningless nonmath (in *yellow*) during sentence presentation in the left angular gyrus. The intersection of both contrasts maps appears in *orange*. The time plots show the course of activation to meaningful and meaningless math and nonmath statements in the main four math-responsive regions and the left angular gyrus.

the reflection phase, as well as both meaningful and meaningless statements, we were also able to evaluate the neural correlates of semantic retrieving and meaning evaluation of math and nonmath statements. We observed an early activation of intraparietal and inferior temporal regions during the presentation of math statements (Fig. 2C). Importantly, only mathematicians exhibited this early activation, but not the controls—individuals with same academic standing but devoid of mathematical training beyond high-school, and present only in our first experiment—who nevertheless knew that statements were about math. We then observed a drop of activation in these intraparietal and inferior temporal regions when mathematicians identified meaningless math statements, while activation remained sustained during semantical decision on meaningful statements. On the contrary, nonmath statements elicited activation in bilateral temporal pole and anterior middle temporal gyri (similar to Fig. 1) that coincide with regions classically involved in the semantic processing of words or sentences (Binder, 2016). Partially overlapping transient activation to math and nonmath statements was nonetheless observed in the left angular gyrus during statements presentation (Fig. 2D). Altogether, these findings suggest that lexical semantics rapidly recruit regions that partly differ depending on the content, and that semantic decision is achieved using essentially the same regions that process the statements meaning. Our second and third experiments suggest that it is true even in the case of very simple math and nonmath statements that differ minimally in their surface form, and even in the presence of minimal logical operators such as quantifiers or negation.

## High-level mathematical knowledge recycles brain regions involved in numerical and visuospatial processing

Recent neuroimaging studies have revealed that parietal cortices activation jointly supports numerical and spatial processing (Harvey, Fracasso, Petridou, & Dumoulin, 2015; Hubbard, Piazza, Pinel, & Dehaene, 2005). In fact, the parietal cortices seem to be recruited by a wide range of spatial processes (Husain & Nachev, 2007) including mental projection in space (Gauthier & van Wassenhove, 2016) and spatial attention (Culham & Kanwisher, 2001; Silver, Ress, & Heeger, 2005; Simon, Mangin, Cohen, Le Bihan, & Dehaene, 2002). Parietal cortices also more generally support processes linked to human visuospatial working memory capacities (Klingberg, 2006; Ricciardi et al., 2006). Furthermore, the representation of visuospatial analogies calls upon a bilateral dorsal prefrontal and intraparietal network (Watson & Chatterjee, 2012). Moreover, among other regions, the left intraparietal sulcus has notably found to be associated with the visual perception of some geometrical characteristics of shapes, such as the presence of crossing, and the right intraparietal sulcus with the number of segments drawn to copy a given geometrical shape (Tzagarakis, Jerde, Lewis, Uğurbil, & Georgopoulos, 2009). Finally, a meta-analysis conducted by Zacks (2008) on all studies using the task of mental rotation introduced by Shepard and Metzler (1971) has demonstrated that such geometrical transformations recruit a large parietal region centered on the intraparietal sulcus and extended through the superior parietal lobule.

In addition, the intraparietal sulcus is historically the first region identified by neuroimaging studies that supports numerical processing. First, this sulcus contains neurons tuned to quantities in monkeys (Nieder, 2005). Neuroimaging studies of 6- and 7-month-old preverbal infants have also evidenced a numerical distance effect, following Weber's law, which appeared over right posterior sites in studies using electroencephalography (Libertus, Pruitt, Woldorff, & Brannon, 2008) or near-infrared spectroscopy (Hyde, Boas, Blair, & Carey, 2010). Adaption effect then progressively appears in the left intraparietal sulcus with age (Ansari & Dhital, 2006). Finally, bilateral activation in the intraparietal sulcus has systematically been reported in adults for all numerical tasks—comparison, estimation, calculation—whether approximate or exact, symbolic or nonsymbolic.

More recently, Shum et al. (2013) evidenced the bilateral involvement of inferior temporal sites in number processing, using electrophysiological signals recorded from surface and depth electrodes in epilepsy patients. These regions were initially called "visual number form areas" because of their strong response to Arabic digits more than other visual stimuli, and of their proximity to the "visual word form area" (Dehaene et al., 2010) and other category-specific regions of the ventral visual stream (Kanwisher & Dilks, 2013). These regions have now also been observed in fMRI (Grotheer, Herrmann, & Kovács, 2016; Grotheer, Jeska, & Grill-Spector, 2018; Yeo, Wilkey, & Price, 2017), and they do not activate only during the visual recognition of numbers, but also during calculation (Daitch et al., 2016; Pinheiro-Chagas, Daitch, Parvizi, & Dehaene, 2018).

In our first experiment, participants additionally performed simple subtractions and were exposed to written numbers and math expressions. In bilateral intraparietal sulci and inferotemporal regions, this allowed us to observe an almost perfect overlap of activation evoked by mathematical reflection with the activation evoked by calculation or mere viewing of numbers, though we verified that these overlapping activations could not be explained by a shared numerical component. In fact, we carefully avoided any direct mention of numbers in our high-level mathematical statements, and the results remained essentially unchanged after excluding all statements containing indirect references to numbers or to fractions. Furthermore, the overlap was confirmed by sensitive single-subjects representational similarity analyses. In bilateral intraparietal sulci and inferior temporal regions of interest, we found a higher degree of similarity between the activation patterns evoked by mathematical reflection and those evoked by calculation or the recognition of numbers and mathematical expressions, compared to the activation patterns evoked by nonmathematical reflection, sentence listening, face or words recognition. These results thus support the idea that formal mathematics build upon proto-mathematical intuitions of number and space.

## Does formal mathematics rely on visual experience?

The question then arises about the extent to which mathematical activity relies on visual experience. In our first study, after the fMRI exam, we asked participants to give ratings of "imageability" for each statement they heard. We found a positive correlation of their ratings with brain activation in two left occipital and inferior temporal sites, close to or even overlapping with the math-responsive network. In our second study, we showed that statements involving trigonometry and complex numbers that required visualizing the unit circle to answer in 2.5 s elicited activation in various occipital sites, including alongside the calcarine sulcus.

Cognitive science studies have long evidenced the role of visual processes in geometrical as well as arithmetic and algebraic activity. For example, Stoianov and Zorzi (2012), who managed to build a deep learning network developing classical representations of numerosity only from its exposure to visual arrays containing different numbers of objects, showed that numerical concepts can build upon the visual decomposition and integration of our environment. Numerical calculation also seems to be sensitive to visual cues of spacing and motion. For example, it becomes harder to perform arithmetic calculation when spacing does not match operator's precedence (Landy, Jones, & Goldstone, 2008). And solving simple algebraic equations can be impaired if the background is moving in a direction that is incongruent with the mental movement necessary to isolate the unknown variable on one side of the equal sign (Goldstone, Landy, & Son, 2010).

Related to geometrical activity, Howe and Purves (2005) have suggested that human conceptualization of angles and orientation could originate from natural-scene perception. Moreover, Giaquinto (2005) argued that our visual system spontaneously extracts symmetrical axes drawn through opposite vertices of a shape, and interpret it as the vertical axis, thus influencing shape perception drastically. Finally, visual representations enclosed in diagrams appear to be a major support of geometrical and topological reasoning. For example, Euclid made a systematic use of diagrams in the geometrical proofs of his "Elements" in a way that makes the underlying ideas spontaneously accessible to every reader (Hamami & Mumma, 2013).

However, is visual experience really mandatory to develop advanced mathematical representations? Actually, many great mathematicians in history were blind. How is it possible? Do they compensate through other modalities? To answer this question, we proposed to three blind mathematicians (one was congenitally blind and the others became blind, respectively, around the ages of 3 and 10) to perform our first and second experiments (Amalric, Denghien, & Dehaene, 2017). When comparing math and nonmath reflection, we observed a consistent and systematic activation of the math-responsive network that we had previously identified in sighted professional mathematicians (Fig. 3). In more detail, bilateral intraparietal, inferior temporal, and dorsal prefrontal sites were activated in both experiments in the group of sighted mathematicians as well as in each individual blind mathematician. The only difference we observed between blind and sighted mathematicians was an additional activation in the occipital cortex of blind mathematicians while they were thinking about math problems. Various explanations could account for these results such as the involvement of a certain form of visual imagery, a potential neuronal recycling of regions which function is especially relevant to math activity or mere cortical plasticity. Unfortunately, with only three subjects with various backgrounds, our experiment did not allow to distinguish among those three interpretations.

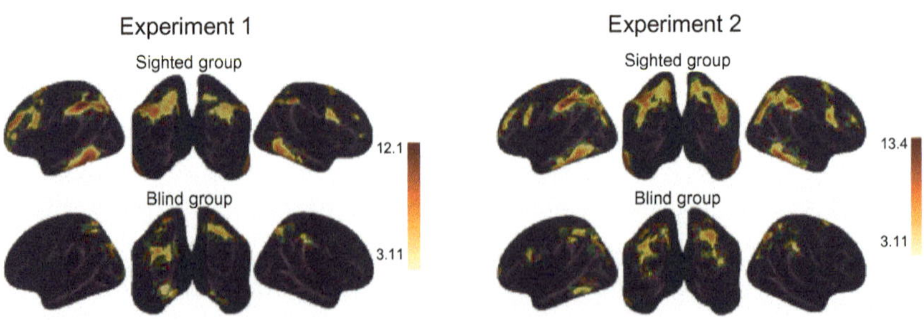


**Fig. 3** Maps of math > nonmath reflection elicited in our sighted participants (*top*) and our blind participants (*bottom*), during experiment 1 (*left*) and experiment 2 (*right*).

However, the results unambiguously support the hypothesis that visual experience is not necessary for the development of a normal cerebral representation of advanced mathematical concepts.

## Neural correlates of deductive reasoning

Mathematical activity as practiced in school and university curricula as well as in contemporary research could be described as defining and proving facts about mathematical objects. These proofs typically use logical reasoning. For that reason, a full understanding of the neural bases of mathematical activity seems impossible to achieve without investigating the neural underpinnings of abstract deductive reasoning.

For a few decades, fMRI studies of deductive reasoning have consistently used arguments constituted of two premises and one conclusion, successively displayed on a screen in order to distinguish between premise integration (i.e., the display of the second premise) and deduction (i.e., a task performed on the conclusion). Different types of reasoning have been examined, including syllogistic deduction—such as "all P are B; all B are D; all P are D" (Goel, 2004; Goel, Buchel, Frith, & Dolan, 2000; Reverberi et al., 2010; Rodriguez-Moreno & Hirsch, 2009); conditional reasoning—such as the famous Modus Ponens: "if A then B; A; then B" (Monti, Osherson, Martinez, & Parsons, 2007; Noveck, Goel, & Smith, 2004; Prado, Der Henst, & Noveck, 2010; Reverberi et al., 2010, 2007); or relational reasoning—such as "A is on top of B; B is on top of C; A is on top of C" (Goel & Dolan, 2001; Prado et al., 2010). To isolate abstract reasoning, experimental designs are usually complex, contrasting such arguments either simultaneously or separately with baseline arguments made of unrelated premises and conclusion (Goel, 2004; Goel et al., 2000; Noveck et al., 2004; Prado et al., 2010; Reverberi et al., 2010), abstract and concrete content (Goel et al., 2000; Goel & Dolan, 2001; Monti et al., 2007), logically valid and invalid arguments (Goel et al., 2000; Noveck et al., 2004; Rodriguez-Moreno & Hirsch, 2009), or different reasoning types (Goel, 2004; Goel & Dolan, 2001; Prado et al., 2010; Reverberi et al., 2010, 2007).

Probably because of all these differences in experimental design, no consensus has yet been reached on which brain areas underlie deductive reasoning. Goel (2007) has written that all studies "implicate some combination of occipital, parietal, temporal and frontal lobes, basal ganglia, and cerebellar regions in logical reasoning, and several implicate all of these regions." However, Coetzee and Monti (2018) and Monti and Osherson

(2012) overcame these differences and proposed the existence of two sets of brain areas involved in deductive reasoning regardless of content. The left rostrolateral and medial prefrontal cortices could constitute "core" regions underlying logical operations necessary to derive conclusion from premises. Second, a set of dorsolateral frontal and parietal regions have been hypothesized to provide additional support to mere deduction reasoning, allowing representing and maintaining the content-independent structure of arguments.

While there is simply not enough evidence yet in support of the dissociation between "core" and "cognitive support" regions in the processing of deductive reasoning, at least the similarity between this set of areas and the frontoparietal math-responsive network is striking, though one difference is worth mentioning: no activation in bilateral inferior temporal regions has yet been reported for deductive reasoning. These observations raise further questions as to the respective role of frontal, parietal, and inferior temporal areas in mathematical activity.

## Effects of math expertise on the brain

### Enhanced frontoparietal network

Despite the small number of neuroimaging studies concerned by mathematical expertise and giftedness, they offer converging evidence: mathematically gifted participants' abilities in math-related and fluid intelligence tasks are supported by an enhanced frontoparietal network and generally greater activation in the posterior parietal cortex. For example, math-gifted adolescents showed stronger activation in more bilateral and extensive brain regions of the frontoparietal network than their nongifted counterparts during fluid reasoning tasks (Desco et al., 2011). Greater activation to mental calculation was also observed in abacus experts in right premotor and parietal areas of the brain (Hanakawa, Honda, Okada, Fukuyama, & Shibasaki, 2003). Moreover, the frontoparietal network exhibited enhanced activation (O'Boyle et al., 2005) and enhanced functional connectivity (Prescott, Gavrilescu, Cunnington, O'Boyle, & Egan, 2010) during mental rotation in math-gifted adolescents. Recently, Navas-Sánchez et al. (2014) showed that the intrahemispheric connectivity in the frontoparietal tracts was positively correlated with math abilities in IQ-matched individuals. Finally, Jeon, Kuhl, and Friederici (2019) showed that the streamline density of the dorsal tract connecting the temporal lobes to the frontal lobes via the parietal cortex correlated with the degree of automaticity that expert mathematicians exhibit compared to nonexperts when computing complex arithmetic.

Various cognitive explanations could be compatible with these observations. In fact, mathematical giftedness has been associated with a wide range of abilities such as logical reasoning and fluid intelligence (Morsanyi, Devine, Nobes, & Szűcs, 2013), working memory (Dark & Benbow, 1991; Swanson, 2006), visuospatial processing (Sella, Sader, Lolliot, & Cohen Kadosh, 2016; Wei, Yuan, Chen, & Zhou, 2012), or pattern recognition (Leikin, Paz-Baruch, & Leikin, 2014), that all exhibit neural correlates within the frontoparietal network. However, without doing highly speculative backward inferences, current evidence is insufficient to untangle these possibilities.

## Partial reorganization of the ventral visual pathway

At the end of our first experiment, all participants were presented with images of numbers, math formulas, words, faces, bodies, houses, tools, and circular checkers. They had to press a button whenever they detected an image repetition (one-back task). Both professional mathematicians and controls exhibited a typical mosaic of ventral occipitotemporal preferences for each category of visual stimuli over all others (Fig. 4). Notably, we could recognize the right-hemispheric fusiform face area (FFA), bilateral parahippocampal place areas (PPA), bilateral extrastriate body areas (EBA), bilateral lateral occipital cortices for tools (LOC), and the left-hemispheric visual word form area (VWFA). Additionally, in both groups, we identified a strong number-related activation, as well as preferred responses to math formulas in bilateral regions of the inferior temporal gyrus (pITG).

It is very likely that these last two regions were present in control subjects because they had received high-school math education and could therefore recognize basic arithmetic and algebraic expressions. However, we observed that the volume of bilateral inferior temporal cortex activated by written math formulas was significantly enhanced in mathematicians relative to controls. We presume that mathematical expertise could lead to a bilateral enhancement of the visual representation of math symbols, in a similar way that reading expertise enhances the left ventral visual response to written letter strings (Dehaene et al., 2010). For numbers, mathematicians also exhibited more activation than controls for number versus nonsymbolic pictures at the peak of the left pITG (Fig. 4).

This expansion of math symbolic representations in the left pITG in expert mathematicians was accompanied by a significantly smaller response to faces in the right FFA (Fig. 4) and a significantly greater response to

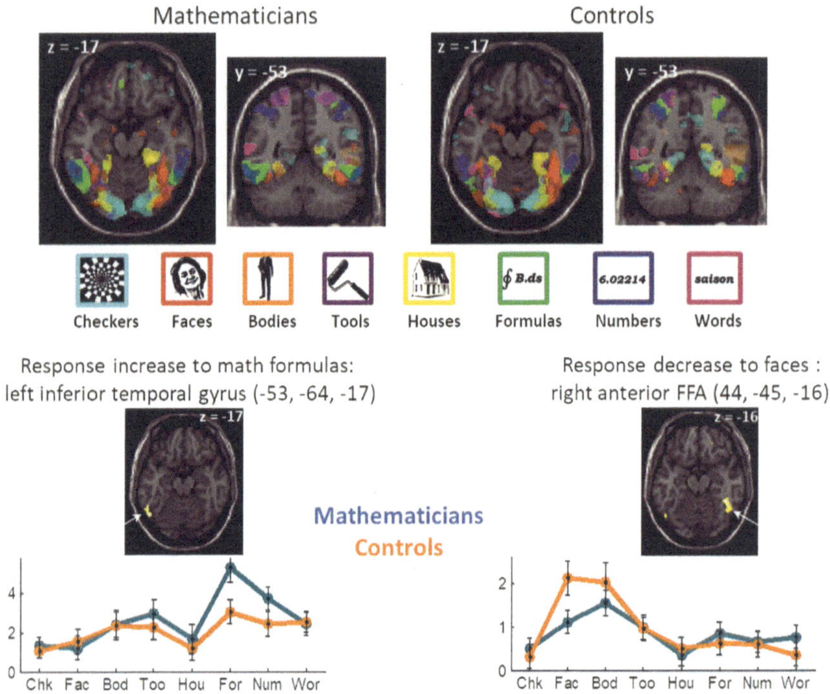


**Fig. 4** (*Top*) Mosaic of preferences for different visual categories in ventral visual cortex: activation for the contrast of a given category (represented by a specific color) minus all others. (*Bottom*) Larger response in mathematicians than in controls to formulas in left inferior temporal cortex and smaller response to faces in left lateral occipital cortex. Plots show the activation to each category relative to rest at the selected peak for mathematicians and controls. *Adapted with authorization from Amalric, M., Dehaene, S. (2016). Origins of the brain networks for advanced mathematics in expert mathematicians. Proceedings of the National Academy of Sciences of the United States of America 201603205. https://doi.org/10.1073/pnas.1603205113.*

tools in the left LOC. Although our experiment did not allow to draw any conclusion as to the behavioral impact or the causal link with math training for these differences, our results nonetheless suggest a partial reorganization of the ventral visual pathway in professional mathematicians.

## Discussion and perspectives

This chapter revealed that once advanced mathematical concepts are learnt, manipulating them seems to recruit a math-responsive network in the brain that dissociates from the classical syntactic and semantic networks. Hence, we can suppose that well-mastered mathematical concepts dispense with the most classical aspects of language. But mathematics is probably not independent of language in a broader sense, because words and written symbols are used to communicate about mathematical concepts, among mathematicians, but also in school. The mechanisms by which advanced mathematical concepts are learnt still remain unknown and our results do not necessarily exclude that linguistic symbols may play a role in the process by which the human basic mathematical intuitions are integrated with each other during learning.

A recent study—involving adults and children of various cultural backgrounds and amount of schooling in France and in the Amazon—has provided evidence suggesting that the human acquisition of geometrical rules relies on a "mathematical language," made of primitives and combinatorial rules that are geometric in nature (Amalric et al., 2017). Thanks to a following fMRI study, we have showed that the use of this "geometrical language" is completely independent of natural language (Wang et al., 2019). However, if the nonverbal implicit learning of certain mathematical laws dispenses with language, is it also the case when concepts are taught explicitly thanks to words and sentences? The current experimental evidence does not allow us to answer this question yet. Further investigation is thus necessary to understand the role of language in the general process of learning new mathematical concepts in naturalistic school context, and more specifically in the unfolding process of learning mathematical symbols and their meaning.

The second section of this chapter revealed that mathematical concepts can develop without visual experience. One possible hypothesis to account for this result is that math knowledge relies mostly on spatial processes. In fact, growing evidence suggests that performance in advanced mathematics is tightly linked to spatial abilities such as mental rotation, spatial working memory, or mental figure construction (Rohde & Thompson, 2007; Sella et al., 2016; Wei et al., 2012). Following this hypothesis, visual processes would simply be a preferred way to convey spatial information in sighted people but are not necessary. That naturally leads us to a deeper cognitive investigation of the role of visuospatial representations in mathematical thinking, and especially in the domains of geometry and topology that seem to more naturally rely on visuospatial representations. One class of objects relevant to geometrical thinking seems particularly suited for such an investigation: diagrams. Indeed, going back to Euclid's geometric practice, one of the most distinctive features of the "Elements" is its reliance on geometric diagrams. While the diagram-based geometric practice of Euclid's Elements has been extensively studied in several fields such as the history and philosophy of mathematics (Manders, 2008) or logic (Avigad, Dean, & Mumma, 2009), the cognitive role of diagrams in geometrical reasoning in humans is yet to be tackled by cognitive sciences.

Beyond the mere opposition between linguistic vs visuospatial skills, it is possible to reconcile these views by hypothesizing that the acquisition of mathematical principles relies on a completely abstract ability to identify the deep structure underlying a set of stimuli. We note here that the importance of identifying structures and their properties is particularly salient in mathematics. In particular, algebra is considered since the 20th century as the mathematical domain studying algebraic structures such as groups, fields, vector spaces, and so on (i.e., all types of structures composed of a set endowed with composition laws and satisfying certain axioms). In fact, the salience of structures is not only relevant to mathematicians, but also to all humans who spontaneously perceive the analogy existing between sequences of increasing numbers and increasing length (de Hevia & Spelke, 2010), or understand the properties of parallel lines in a plane (Izard, Pica, Spelke, & Dehaene, 2011), or even discover that integers are linearly organized (Dehaene, Izard, Spelke, & Pica, 2008), or understand that positions of objects on a surface can be represented by a Euclidian plane. In other words, to construct mathematical objects, we would start with our proto-mathematical percepts, find their commonalities, build a concept that encapsulates this commonality, apply a linguistic treatment to name this concept, and finally create a symbol that refers to the concept and encapsulates some of its properties. The concept now exists, along with other concepts, and the brain can search for their commonalities, which will be used to create a new concept, and so on so forth.

This chapter also points to the fact that mathematical expertise is accompanied by cerebral changes. Especially in the ventral visual pathway, these observations raise more questions than provide answers. In particular, how do the inferior-temporal regions involved in mathematics develop? Are they essentially tied to the learning of mathematical symbolism, like the "visual word form area" seems to develop when subjects learn how to read (Dehaene, Cohen, Morais, & Kolinsky, 2015)? To answer this question, we may need to study longitudinally the evolution of the organization of the ventral visual pathway throughout childhood while children progressively learn mathematical symbols in school. Moreover, similar to reading acquisition (Moulton et al., 2019), would we also observe an improvement of connectivity between these inferior-temporal regions and parietal regions? More generally, what is the relation between inferior-temporal and parietal regions in mathematical activity? The triple-code model postulated bidirectional exchanges between these regions, respectively, representing the visual number form of Arabic digits and the quantity as well as other aspects of number meaning. However, our results in professional mathematicians thinking about auditory statements devoid of numbers show that inferior temporal regions do not only recognize the visual form of numbers. As such, the role of the inferior temporal regions as well as the directionality of its connection with the intraparietal sulci needs to be further investigated.

Finally, we note that the third section of this chapter interrogates the cerebral basis of deductive reasoning and reveals that there is a partial overlap with the math-responsive network. However, our third experiment showed that minimal logical operators are not enough to trigger the activation of the math-responsive network. This may not be surprising since processing basic logical operators within a sentence is very different from applying deductive reasoning to a given situation. On the contrary, as one could argue that mathematical reflection partly relies on deductive reasoning, it is not farfetched to imagine that common processes underlie deductive reasoning and mathematical reflection. This raises a more general question of the function of these areas: though identified as a core network for mathematics, they also appear to be involved in a broad range of other cognitive processes. Beyond mathematics and domain-general logical, inferential, or relational reasoning, they also activate in various effortful problem-solving tasks (Duncan, 2010), and even when thinking about certain physics concepts such as "energy" or "wavelength" (Mason & Just, 2016). Nevertheless, our observation that this network does not activate during nonmath-related reflection of matched difficulty (Amalric & Dehaene, 2016) contradicts the hypothesis of a domain-general "multiple demand" system (Duncan, 2010). Identifying the commonality that explains the activation of similar neural substrates for arithmetic, geometry, logic, physics, IQ tests, etc. remains an open question that may, in turn, offer a new perspective on the question of the nature of mathematical knowledge (Kitcher, 1984).

## References

Amalric, M., & Dehaene, S. (2016). Origins of the brain networks for advanced mathematics in expert mathematicians. *Proceedings of the National Academy of Sciences of the United States of America, 201603205*. https://doi.org/10.1073/pnas.1603205113.

Amalric, M., & Dehaene, S. (2018). Cortical circuits for mathematical knowledge: Evidence for a major subdivision within the brain's semantic networks. *Philosophical Transactions of the Royal Society B, 373*, 20160515. https://doi.org/10.1098/rstb.2016.0515.

Amalric, M., & Dehaene, S. (2019). A distinct cortical network for mathematical knowledge in the human brain. *NeuroImage, 189*, 19–31. https://doi.org/10.1016/j.neuroimage.2019.01.001.

Amalric, M., Denghien, I., & Dehaene, S. (2017). On the role of visual experience in mathematical development: Evidence from blind mathematicians. *Developmental Cognitive Neuroscience*. https://doi.org/10.1016/j.dcn.2017.09.007.

Amalric, M., Wang, L., Pica, P., Figueira, S., Sigman, M., & Dehaene, S. (2017). The language of geometry: Fast comprehension of geometrical primitives and rules in human adults and preschoolers. *PLoS Computational Biology, 13*, e1005273. https://doi.org/10.1371/journal.pcbi.1005273.

Ansari, D., & Dhital, B. (2006). Age-related changes in the activation of the intraparietal sulcus during nonsymbolic magnitude processing: An event-related functional magnetic resonance imaging study. *Journal of Cognitive Neuroscience, 18*, 1820–1828. https://doi.org/10.1162/jocn.2006.18.11.1820.

Avigad, J., Dean, E., & Mumma, J. (2009). A formal system for euclid's elements. *Review of Symbolic Logic, 2*, 700–768. https://doi.org/10.1017/S1755020309990098.

Benson-Amram, S., Heinen, V. K., Dryer, S. L., & Holekamp, K. E. (2011). Numerical assessment and individual call discrimination by wild spotted hyaenas, Crocuta crocuta. *Animal Behaviour, 82*, 743–752. https://doi.org/10.1016/j.anbehav.2011.07.004.

Binder, J. R. (2016). In defense of abstract conceptual representations. *Psychonomic Bulletin & Review, 23*, 1096–1108. https://doi.org/10.3758/s13423-015-0909-1.

Cantlon, J. F., Merritt, D. J., & Brannon, E. M. (2016). Monkeys display classic signatures of human symbolic arithmetic. *Animal Cognition, 19*, 405–415. https://doi.org/10.1007/s10071-015-0942-5.

Chiandetti, C., & Vallortigara, G. (2007). Is there an innate geometric module? Effects of experience with angular geometric cues on spatial re-orientation based on the shape of the environment. *Animal Cognition, 11*, 139–146. https://doi.org/10.1007/s10071-007-0099-y.

Coetzee, J. P., & Monti, M. M. (2018). At the core of reasoning: Dissociating deductive and non-deductive load. *Human Brain Mapping, 39*, 1850–1861. https://doi.org/10.1002/hbm.23979.

Culham, J. C., & Kanwisher, N. G. (2001). Neuroimaging of cognitive functions in human parietal cortex. *Current Opinion in Neurobiology, 11*, 157–163. https://doi.org/10.1016/S0959-4388(00)00191-4.

Daitch, A. L., Foster, B. L., Schrouff, J., Rangarajan, V., Kaşikçi, I., Gattas, S., et al. (2016). Mapping human temporal and parietal neuronal population activity and functional coupling during mathematical cognition. *Proceedings of the National Academy of Sciences of the United States of America, 113*, E7277–E7286. https://doi.org/10.1073/pnas.1608434113.

Dark, V., & Benbow, C. (1991). Differential enhancement of working memory with mathematical versus verbal precocity. *Journal of Education & Psychology, 83*, 48–60.

de Hevia, M. D., & Spelke, E. S. (2010). Number-space mapping in human infants. *Psychological Science, 21*, 653–660. https://doi.org/10.1177/0956797610366091.

De Toffoli, S., & Giardino, V. (2014). Forms and roles of diagrams in knot theory. *Erkenntnis, 79*, 829–842. https://doi.org/10.1007/s10670-013-9568-7.

Dehaene, S., Cohen, L., Morais, J., & Kolinsky, R. (2015). Illiterate to literate: Behavioural and cerebral changes induced by reading acquisition. *Nature Reviews. Neuroscience, 16*, 234–244.

Dehaene, S., Izard, V., Pica, P., & Spelke, E. (2006). Core knowledge of geometry in an Amazonian indigene group. *Science, 311*, 381–384. https://doi.org/10.1126/science.1121739.

Dehaene, S., Izard, V., Spelke, E., & Pica, P. (2008). Log or linear? Distinct intuitions of the number scale in Western and Amazonian indigene cultures. *Science, 320*, 1217–1220. https://doi.org/10.1126/science.1156540.
Dehaene, S., Pegado, F., Braga, L. W., Ventura, P., Filho, G. N., Jobert, A., et al. (2010). How learning to read changes the cortical networks for vision and language. *Science, 330*, 1359–1364. https://doi.org/10.1126/science.1194140.
Deipolyi, A., Santos, L., & Hauser, M. D. (2001). The role of landmarks in cotton-top tamarin spatial foraging: Evidence for geometric and non-geometric features. *Animal Cognition, 4*, 99–108. https://doi.org/10.1007/s100710100103.
Desco, M., Navas-Sanchez, F. J., Sanchez-González, J., Reig, S., Robles, O., Franco, C., et al. (2011). Mathematically gifted adolescents use more extensive and more bilateral areas of the fronto-parietal network than controls during executive functioning and fluid reasoning tasks. *NeuroImage, 57*, 281–292. https://doi.org/10.1016/j.neuroimage.2011.03.063.
Dillon, M. R., Duyck, M., Dehaene, S., Amalric, M., & Izard, V. (2019). Geometric categories in cognition. *Journal of Experimental Psychology. Human Perception and Performance*. https://doi.org/10.1037/xhp0000663.
Dillon, M. R., Huang, Y., & Spelke, E. S. (2013). Core foundations of abstract geometry. *Proceedings of the National Academy of Sciences of the United States of America, 110*, 14191–14195. https://doi.org/10.1073/pnas.1312640110.
Duncan, J. (2010). The multiple-demand (MD) system of the primate brain: Mental programs for intelligent behaviour. *Trends in Cognitive Sciences, 14*, 172–179. https://doi.org/10.1016/j.tics.2010.01.004.
Feigenson, L. (2011). Predicting sights from sounds: 6-month-olds' intermodal numerical abilities. *Journal of Experimental Child Psychology, 110*, 347–361. https://doi.org/10.1016/j.jecp.2011.04.004.
Gauthier, B., & van Wassenhove, V. (2016). Time is not space: Core computations and domain-specific networks for mental travels. *The Journal of Neuroscience, 36*, 11891–11903. https://doi.org/10.1523/jneurosci.1400-16.2016.
Giaquinto, M. (2005). From symmetry perception to basic geometry. In *Visualization, explanation and reasoning styles in mathematics* (pp. 31–55). Springer.
Gilmore, C. K., McCarthy, S. E., & Spelke, E. S. (2010). Non-symbolic arithmetic abilities and mathematics achievement in the first year of formal schooling. *Cognition, 115*, 394–406. https://doi.org/10.1016/j.cognition.2010.02.002.
Goel, V. (2004). Differential involvement of left prefrontal cortexin inductive and deductive reasoning. *Cognition, 93*, B109–B121. https://doi.org/10.1016/j.cognition.2004.03.001.
Goel, V. (2007). Anatomy of deductive reasoning. *Trends in Cognitive Sciences, 11*, 435–441. https://doi.org/10.1016/j.tics.2007.09.003.
Goel, V., Buchel, C., Frith, C., & Dolan, R. J. (2000). Dissociation of mechanisms underlying syllogistic reasoning. *NeuroImage, 12*, 504–514. https://doi.org/10.1006/nimg.2000.0636.
Goel, V., & Dolan, R. J. (2001). Functional neuroanatomy of three-term relational reasoning. *Neuropsychologia, 39*, 901–909. https://doi.org/10.1016/S0028-3932(01)00024-0.
Goldstone, R. L., Landy, D. H., & Son, J. Y. (2010). The education of perception. *Topics in Cognitive Science, 2*, 265–284. https://doi.org/10.1111/j.1756-8765.2009.01055.x.
Grotheer, M., Herrmann, K.-H., & Kovács, G. (2016). Neuroimaging evidence of a bilateral representation for visually presented numbers. *The Journal of Neuroscience, 36*, 88–97. https://doi.org/10.1523/jneurosci.2129-15.2016.
Grotheer, M., Jeska, B., & Grill-Spector, K. (2018). A preference for mathematical processing outweighs the selectivity for Arabic numbers in the inferior temporal gyrus. *NeuroImage, 175*, 188–200. https://doi.org/10.1016/j.neuroimage.2018.03.064.
Hamami, Y., & Mumma, J. (2013). Prolegomena to a cognitive investigation of Euclidean diagrammatic reasoning. *Journal of Logic, Language and Information, 22*, 421–448. https://doi.org/10.1007/s10849-013-9182-8.

Hanakawa, T., Honda, M., Okada, T., Fukuyama, H., & Shibasaki, H. (2003). Neural correlates underlying mental calculation in abacus experts: A functional magnetic resonance imaging study. *NeuroImage, 19*, 296–307. https://doi.org/10.1016/S1053-8119(03)00050-8.

Harvey, B. M., Fracasso, A., Petridou, N., & Dumoulin, S. O. (2015). Topographic representations of object size and relationships with numerosity reveal generalized quantity processing in human parietal cortex. *Proceedings of the National Academy of Sciences of the United States of America, 112*, 13525–13530. https://doi.org/10.1073/pnas.1515414112.

Hauser, M. D., Chomsky, N., & Fitch, W. T. (2002). The faculty of language: What is it, who has it, and how did it evolve? *Science, 298*, 1569–1579. https://doi.org/10.1126/science.298.5598.1569.

Howe, C. Q., & Purves, D. (2005). Natural-scene geometry predicts the perception of angles and line orientation. *Proceedings of the National Academy of Sciences of the United States of America, 102*, 1228–1233. https://doi.org/10.1073/pnas.0409311102.

Hubbard, E. M., Piazza, M., Pinel, P., & Dehaene, S. (2005). Interactions between number and space in parietal cortex. *Nature Reviews. Neuroscience, 6*, 435–448. https://doi.org/10.1038/nrn1684.

Husain, M., & Nachev, P. (2007). Space and the parietal cortex. *Trends in Cognitive Sciences, 11*, 30–36. https://doi.org/10.1016/j.tics.2006.10.011.

Hyde, D. C., Boas, D. A., Blair, C., & Carey, S. (2010). Near-infrared spectroscopy shows right parietal specialization for number in pre-verbal infants. *NeuroImage, 53*, 647–652. https://doi.org/10.1016/j.neuroimage.2010.06.030.

Izard, V., Pica, P., Dehaene, S., Hinchey, D., & Spelke, E. (2011). Geometry as a Universal Mental Construction. In *Space, time and number in the brain* (pp. 319–332). Elsevier.

Izard, V., Pica, P., Spelke, E. S., & Dehaene, S. (2011). Flexible intuitions of Euclidean geometry in an Amazonian indigene group. *Proceedings of the National Academy of Sciences of the United States of America, 108*, 9782–9787. https://doi.org/10.1073/pnas.1016686108.

Izard, V., Sann, C., Spelke, E. S., & Streri, A. (2009). Newborn infants perceive abstract numbers. *Proceedings of the National Academy of Sciences of the United States of America, 106*, 10382–10385. https://doi.org/10.1073/pnas.0812142106.

Jeon, H.-A., Kuhl, U., & Friederici, A. D. (2019). Mathematical expertise modulates the architecture of dorsal and cortico-thalamic white matter tracts. *Scientific Reports, 9*, 1–11. https://doi.org/10.1038/s41598-019-43400-6.

Kanwisher, N., & Dilks, D. (2013). The functional organization of the ventral visual pathway in humans. In L. Chalupa, & J. Werner (Eds.), *The new visual neurosciences*.

Kitcher, P. (1984). *The nature of mathematical knowledge*. Oxford University Press.

Klingberg, T. (2006). Development of a superior frontal–intraparietal network for visuo-spatial working memory. *Neuropsychologia, Advances in Developmental Cognitive Neuroscience, 44*, 2171–2177. https://doi.org/10.1016/j.neuropsychologia.2005.11.019.

Landy, D. H., Jones, M. N., & Goldstone, R. L. (2008). How the appearance of an operator affects its formal precedence. In *Proceedings of the Thirtieth Annual Conference of the Cognitive Science Society* (pp. 2109–2114).

Lee, S. A., Sovrano, V. A., & Spelke, E. S. (2012). Navigation as a source of geometric knowledge: Young children's use of length, angle, distance, and direction in a reorientation task. *Cognition, 123*, 144–161. https://doi.org/10.1016/j.cognition.2011.12.015.

Leikin, R., Paz-Baruch, N., & Leikin, M. (2014). Cognitive characteristics of students with superior performance in mathematics. *Journal of Individual Differences, 35*, 119–129. https://doi.org/10.1027/1614-0001/a000140.

Libertus, M. E. (2015). The role of intuitive approximation skills for school math abilities. *Mind, Brain, and Education, 9*, 112–120. https://doi.org/10.1111/mbe.12072.

Libertus, M. E., Pruitt, L. B., Woldorff, M. G., & Brannon, E. M. (2008). Induced alpha-band oscillations reflect ratio-dependent number discrimination in the infant brain. *Journal of Cognitive Neuroscience, 21*, 2398–2406. https://doi.org/10.1162/jocn.2008.21162.

Manders, K. (2008). The Euclidean diagram. In P. Mancosu (Ed.), *The philosophy of mathematical practice* (pp. 80–133). Oxford University Press.

Mason, R. A., & Just, M. A. (2016). Neural representations of physics concepts. *Psychological Science*, *27*, 904–913. https://doi.org/10.1177/0956797616641941.

Matthews, P. G., & Chesney, D. L. (2015). Fractions as percepts? Exploring cross-format distance effects for fractional magnitudes. *Cognitive Psychology*, *78*, 28–56. https://doi.org/10.1016/j.cogpsych.2015.01.006.

Meng, R., Matthews, P., & Hubbard, E. (2018). Non-symbolic ratio sense supports symbolic fraction success. *Cognitive Science Society*.

Monti, M. M., & Osherson, D. N. (2012). Logic, language and the brain. *Brain Research*, *1428*, 33–42. https://doi.org/10.1016/j.brainres.2011.05.061.

Monti, M. M., Osherson, D. N., Martinez, M. J., & Parsons, L. M. (2007). Functional neuroanatomy of deductive inference: A language-independent distributed network. *NeuroImage*, *37*, 1005–1016. https://doi.org/10.1016/j.neuroimage.2007.04.069.

Morsanyi, K., Devine, A., Nobes, A., & Szűcs, D. (2013). The link between logic, mathematics and imagination: Evidence from children with developmental dyscalculia and mathematically gifted children. *Developmental Science*, *16*, 542–553. https://doi.org/10.1111/desc.12048.

Moulton, E., Bouhali, F., Monzalvo, K., Poupon, C., Zhang, H., Dehaene, S., et al. (2019). Connectivity between the visual word form area and the parietal lobe improves after the first year of reading instruction: A longitudinal MRI study in children. *Brain Structure & Function*, *224*, 1519–1536. https://doi.org/10.1007/s00429-019-01855-3.

Navas-Sánchez, F. J., Alemán-Gómez, Y., Sánchez-Gonzalez, J., Guzmán-De-Villoria, J. A., Franco, C., Robles, O., et al. (2014). White matter microstructure correlates of mathematical giftedness and intelligence quotient. *Human Brain Mapping*, *35*, 2619–2631. https://doi.org/10.1002/hbm.22355.

Nieder, A. (2005). Counting on neurons: The neurobiology of numerical competence. *Nature Reviews. Neuroscience*, *6*, 177–190. https://doi.org/10.1038/nrn1626.

Noveck, I. A., Goel, V., & Smith, K. W. (2004). The neural basis of conditional reasoning with arbitrary content. *Cortex*, *40*, 613–622. https://doi.org/10.1016/S0010-9452(08)70157-6.

O'Boyle, M. W., Cunnington, R., Silk, T. J., Vaughan, D., Jackson, G., Syngeniotis, A., et al. (2005). Mathematically gifted male adolescents activate a unique brain network during mental rotation. *Cognitive Brain Research*, *25*, 583–587. https://doi.org/10.1016/j.cogbrainres.2005.08.004.

Pallier, C., Devauchelle, A.-D., & Dehaene, S. (2011). Cortical representation of the constituent structure of sentences. *Proceedings of the National Academy of Sciences of the United States of America*, *108*, 2522–2527. https://doi.org/10.1073/pnas.1018711108.

Pinheiro-Chagas, P., Daitch, A., Parvizi, J., & Dehaene, S. (2018). Brain mechanisms of arithmetic: A crucial role for ventral temporal cortex. *Journal of Cognitive Neuroscience*, *30*, 1757–1772. https://doi.org/10.1162/jocn_a_01319.

Prado, J., Der Henst, J.-B. V., & Noveck, I. A. (2010). Recomposing a fragmented literature: How conditional and relational arguments engage different neural systems for deductive reasoning. *NeuroImage*, *51*, 1213–1221. https://doi.org/10.1016/j.neuroimage.2010.03.026.

Prescott, J., Gavrilescu, M., Cunnington, R., O'Boyle, M. W., & Egan, G. F. (2010). Enhanced brain connectivity in math-gifted adolescents: An fMRI study using mental rotation. *Cognitive Neuroscience*, *1*, 277–288. https://doi.org/10.1080/17588928.2010.506951.

Reverberi, C., Cherubini, P., Frackowiak, R. S. J., Caltagirone, C., Paulesu, E., & Macaluso, E. (2010). Conditional and syllogistic deductive tasks dissociate functionally during premise integration. *Human Brain Mapping*, *31*, 1430–1445. https://doi.org/10.1002/hbm.20947.

Reverberi, C., Cherubini, P., Rapisarda, A., Rigamonti, E., Caltagirone, C., Frackowiak, R. S. J., et al. (2007). Neural basis of generation of conclusions in elementary deduction. *NeuroImage*, *38*, 752–762. https://doi.org/10.1016/j.neuroimage.2007.07.060.

Ricciardi, E., Bonino, D., Gentili, C., Sani, L., Pietrini, P., & Vecchi, T. (2006). Neural correlates of spatial working memory in humans: A functional magnetic resonance imaging study comparing visual and tactile processes. *Neuroscience*, *139*, 339–349. https://doi.org/10.1016/j.neuroscience.2005.08.045.

Rodriguez-Moreno, D., & Hirsch, J. (2009). The dynamics of deductive reasoning: An fMRI investigation. *Neuropsychologia*, *47*, 949–961. https://doi.org/10.1016/j.neuropsychologia.2008.08.030.

Rohde, T. E., & Thompson, L. A. (2007). Predicting academic achievement with cognitive ability. *Intelligence*, *35*, 83–92. https://doi.org/10.1016/j.intell.2006.05.004.

Rugani, R., Fontanari, L., Simoni, E., Regolin, L., & Vallortigara, G. (2009). Arithmetic in newborn chicks. *Proceedings of the Royal Society of London - Series B: Biological Sciences*, *276*, 2451–2460. https://doi.org/10.1098/rspb.2009.0044.

Sella, F., Sader, E., Lolliot, S., & Cohen Kadosh, R. (2016). Basic and advanced numerical performances relate to mathematical expertise but are fully mediated by visuospatial skills. *Journal of Experimental Psychology. Learning, Memory, and Cognition*, *42*, 1458–1472. https://doi.org/10.1037/xlm0000249.

Shepard, R. N., & Metzler, J. (1971). Mental rotation of three-dimensional objects. *Science*, *171*, 701–703.

Shum, J., Hermes, D., Foster, B. L., Dastjerdi, M., Rangarajan, V., Winawer, J., et al. (2013). A brain area for visual numerals. *The Journal of Neuroscience*, *33*, 6709–6715. https://doi.org/10.1523/jneurosci.4558-12.2013.

Silver, M. A., Ress, D., & Heeger, D. J. (2005). Topographic maps of visual spatial attention in human parietal cortex. *Journal of Neurophysiology*, *94*, 1358–1371. https://doi.org/10.1152/jn.01316.2004.

Simon, O., Mangin, J.-F., Cohen, L., Le Bihan, D., & Dehaene, S. (2002). Topographical layout of hand, eye, calculation, and language-related areas in the human parietal lobe. *Neuron*, *33*, 475–487.

Stoianov, I., & Zorzi, M. (2012). Emergence of a "visual number sense" in hierarchical generative models. *Nature Neuroscience*, *15*, 194–196. https://doi.org/10.1038/nn.2996.

Swanson, H. L. (2006). Cognitive processes that underlie mathematical precociousness in young children. *Journal of Experimental Child Psychology*, *93*, 239–264. https://doi.org/10.1016/j.jecp.2005.09.006.

Tzagarakis, C., Jerde, T. A., Lewis, S. M., Uğurbil, K., & Georgopoulos, A. P. (2009). Cerebral cortical mechanisms of copying geometrical shapes: A multidimensional scaling analysis of fMRI patterns of activation. *Experimental Brain Research*, *194*, 369–380. https://doi.org/10.1007/s00221-009-1709-5.

Wang, L., Amalric, M., Pallier, C., Figueira, S., Sigman, M., & Dehaene, S. (2019). Representation of spatial sequences using nested rules in human prefrontal cortex. *Neuroimage*, *186*, 245–255. https://doi.org/10.1016/j.neuroimage.2018.10.061.

Watson, C. E., & Chatterjee, A. (2012). A bilateral frontoparietal network underlies visuospatial analogical reasoning. *NeuroImage*, *59*, 2831–2838. https://doi.org/10.1016/j.neuroimage.2011.09.030.

Wei, W., Yuan, H., Chen, C., & Zhou, X. (2012). Cognitive correlates of performance in advanced mathematics. *The British Journal of Educational Psychology*, *82*, 157–181. https://doi.org/10.1111/j.2044-8279.2011.02049.x.

Xu, F., & Garcia, V. (2008). Intuitive statistics by 8-month-old infants. *Proceedings of the National Academy of Sciences of the United States of America*, *105*, 5012–5015. https://doi.org/10.1073/pnas.0704450105.

Yeo, D. J., Wilkey, E. D., & Price, G. R. (2017). The search for the number form area: A functional neuroimaging meta-analysis. *Neuroscience and Biobehavioral Reviews*. https://doi.org/10.1016/j.neubiorev.2017.04.027.

Zacks, J. M. (2008). Neuroimaging studies of mental rotation: A meta-analysis and review. *Journal of Cognitive Neuroscience*, *20*, 1–19.

SECTION II

# Math education

CHAPTER 5

# Looking at the development of mathematical knowledge from the perspective of the framework theory approach to conceptual change: Lessons for mathematics education

**Stella Vosniadou**
College of Education, Psychology and Social Work, Flinders University, Bedford Park, SA, Australia

## Introduction

The chapter outlines an approach to the development of mathematical knowledge from the lenses of the *framework theory approach to conceptual change* and explains how it is applied to mathematics and especially to the development of rational number knowledge. The framework theory approach to conceptual change is not meant to be a complete theory of mathematical development. It is however an approach which sheds light to certain aspects of mathematical development and is especially relevant in explaining students' difficulties in understanding rational numbers. Unlike most theories, which are based on continuities between natural and rational number knowledge, the framework theory approach focuses on discontinuities. It argues that these discontinuities are responsible for the misconceptions students create during the learning of rational number and then proposes ways to address that can improve rational number instruction. In the pages that follow, we describe the framework theory approach as it applies to the development of rational number knowledge, outline the empirical evidence that supports it, and discuss some of its implications for mathematics education.

*Heterogeneous Contributions to Numerical Cognition*
https://doi.org/10.1016/B978-0-12-817414-2.00015-4

## Continuities between natural and rational number knowledge

Current approaches to the development of rational number knowledge emphasize the continuities between natural and rational numbers (Moss, 2005; Moss & Case, 1999; Siegler, 2016). For example, Moss and Case (1999) argue that children start with two distinct conceptual structures for natural and rational number. In the case of natural number, it is the numerical schema, which is primarily digital, verbal, and sequential, while in the case of rational number, it is the global quantitative schema, which is primarily spatial, analogic, and nonsequential. These distinct schemas gradually increase in complexity and are eventually mapped into each other creating new, integrated representations, which later become more explicit and flexible.

According to Siegler and his colleagues (Siegler, 2016; Siegler, Thompson, & Schneider, 2011), the core of numerical development is the understanding of magnitudes. According to Siegler, numerical development is a process of "progressively broadening the class of numbers that are understood to possess magnitudes and of learning the functions that connect that increasingly broad and varied set of numbers to their magnitudes" (p. 274). This progression is thought to reflect the gradual expansion of the mental number line to the right to include larger natural numbers, leftward to encompass negatives, and interstitially to include fractions and decimals (Siegler, 2016; Siegler & Lortie-Forgues, 2014).

Despite their differences, both Moss and Case (1999) and Siegler (2016) emphasize the importance of natural number knowledge in rational number learning. Both agree that the development of natural number knowledge precedes the development of rational number knowledge. Moss and Case consider natural number knowledge as a prerequisite for the development of the schema of splitting that allows for computations with percentages. Siegler and colleagues argue that an important mechanism underlying the development of rational number knowledge is reasoning by analogy to natural numbers. Their basic claim for instruction is that "drawing the explicit analogy that fractions are like whole numbers in having magnitudes that can be ordered and represented on number lines may be helpful" (Siegler et al., 2011, p. 291).

We agree with Siegler (2016) and acknowledge the important role of the number line in rational number instruction, a point that is also acknowledged by other mathematics education researchers (Kilpatrick, Swafford, & Findell, 2001). However, the proposal to build rational numbers on the idea that numbers have magnitudes and can be placed on number lines has certain limitations: It relies on the most abstract aspect of rational numbers that students will meet in their school career and it does not address the fundamental question of why rational numbers are considered numbers in a way that is meaningful for students. Most importantly, neither the Moss and Case (1999) nor the Siegler et al. (2011) approaches address the multiplicity of misconceptions that students form in the process of developing rational number knowledge. Numerous rational number-related misconceptions have been identified, such as in the comparison of decimals (e.g., "longer decimals are bigger and shorter decimals are smaller," Resnick et al., 1989), in the comparison of fractions (e.g., "the numerical value of a fraction increases when the numerator, denominator or both increase," Stafylidou & Vosniadou, 2004), and in arithmetical operations (e.g., "multiplication always makes bigger," "division always makes smaller," Fischbein, Deri, Nello, & Marino, 1985), to mention only a few. Students' overconfidence in their erroneous answers and the degree of resilience of their misconceptions have also been investigated (Durkin & Rittle-Johnson, 2015; Fischbein, 1987; Merenluoto & Lehtinen, 2004). Any theory of the development of rational number knowledge that does not address why and how students form such misconceptions and proposes instructional interventions to deal with them is incomplete.

The framework theory approach to conceptual change does not claim to be a complete theory of the development of mathematical knowledge. It is a limited approach that seeks to illuminate aspects of the development of this knowledge, focusing on students' difficulties and the creation of misconceptions. In contrast to the approaches that build on continuities between natural and rational number, Vosniadou, Vamvakoussi, and Skopeliti (2008) focus on their discontinuities and explain how these discontinuities are responsible for students' errors and misconceptions.

## The framework theory approach to conceptual change

The framework theory approach is one of several approaches to the problem of conceptual change. Conceptual change research investigates the learning of counterintuitive concepts that require substantial changes in prior knowledge. Conceptual change research is based on a host of empirical findings by science educators, which have shown that students' initial understandings of the physical world are very different from the scientific theories to which they are later exposed through systematic instruction and that the learning of science requires major reorganizations of prior knowledge, in other words, it requires *conceptual change* (Clement, 1982; Driver & Easley, 1978;

McCloskey, 1983). While there is wide agreement that young children form initial understandings of the physical and social world based on their early experiences, there is disagreement on exactly how to characterize these understandings. Some researchers have argued that initial understandings are best thought of as well formed but erroneous *intuitive theories* (McCloskey, 1983) and that these intuitive theories must be extinguished and replaced by the correct scientific ones. Posner, Strike, Hewson, and Gertzog (1982) argued that students need to undergo a radical conceptual change when it comes to learning counterintuitive concepts like force, heat, and energy. Following Kuhn's (1962) analysis of conceptual changes in the history of science, they proposed that conceptual change is something like a Gestalt-type restructuring that happens when students realize the shortcomings of their intuitive theories and the explanatory power of the scientific ones and proposed that cognitive conflict is the main instructional strategy to achieve such a conceptual change.

Other researchers, such as diSessa (1993), argued that students' initial understandings are not well-formed intuitive theories but "knowledge-in-pieces," consisting of a multiplicity of subconceptual elements known as p-prims. According to this approach, known as "knowledge-in-pieces," conceptual change consists of the gradual integration of these knowledge elements in cohesive conceptual structures such as scientific theories. The emphasis on integration and discrimination rather than on confrontation and cognitive conflict is the hallmark of the knowledge-in-pieces approach to instruction (Linn, 2013). However, the knowledge-in-pieces approach does little to explain students' relatively well-formed initial understandings, their resistance to instruction, and the construction of additional misconceptions in the process of learning.

The framework theory approach to conceptual change (Vosniadou, 1994, 2013; Vosniadou & Skopeliti, 2014) proposes a different explanation of the process of conceptual change. First, it agrees with the knowledge-in-pieces approach that initial understandings do not represent unitary intuitive theories but consist of clusters of ideas or beliefs based on everyday experience. Unlike diSessa (1993), however, the framework theory approach, based on evidence from cognitive developmental research, argues that these knowledge pieces become organized in loose but relatively coherent domain-specific *framework theories* from early on in childhood (Carey, 1985; Wellman & Gelman, 1998; Wiser & Smith, 2008). Framework theories are implicit skeletal structures based on interpretations of experience in the context of lay culture. Such systems lack the systematicity, cohesiveness, and explicitness of an explicit and socially shared scientific theory. They can be considered as theories, however, in that they are principle-based systems capable of producing explanation and prediction.

Some developmental psychologists have claimed that infants are born with biologically endowed systems of "core knowledge" for representing amongst others, objects, numbers, and space that shape the mental lives of children and adults (Spelke, 2004). Educational psychologists have made a distinction between biologically primary and secondary information (Geary, 2008). Framework theories are not considered to be necessarily biologically endowed. Nevertheless, they are meant to characterize the basic ontological and epistemological distinctions young children make as they create initial conceptual understandings out of the interpretations of their experiences within their physical and social world and in the context of lay culture (Vosniadou, 2013).

Unlike intuitive theories which are considered to be unitary, framework theories are conceptualized to consist of a number of constituent beliefs. As a result, conceptual change is not a sudden Gestalt-type restructuring of a faulty intuitive theory and its replacement with the correct scientific one, but a gradual constructive process, which follows a developmental progression. In fact, it is more accurate to talk about conceptual changes rather than conceptual change. During this constructive learning process, students can create various kinds of misconceptions. From the point of view of the framework theory approach, most of these misconceptions are hybrids—i.e., consist of a combination of scientific information with initial beliefs grounded in the framework theory. Vosniadou (2013) has distinguished between fragmented and synthetic (mis)conceptions. Fragmentation happens when scientific information is added on incompatible prior knowledge without awareness or concern for the inconsistencies that are being created. Synthetic models, although scientifically wrong, exhibit some concern for internal consistency and explanatory power (see Vosniadou, 2013; Vosniadou & Skopeliti, 2014 for a detailed discussion).

The rationale for applying the framework theory approach to conceptual change in the number domain was originally described in Vosniadou and Verschaffel (2004) as follows: In the domain of number, there is a great deal of evidence that children form an initial understanding of numbers as discrete quantities. While there does not yet appear to be sufficient evidence on the nature of children's initial understanding of number, existing research suggests that it is not innately based but rather the result of development (Carey, 2004; Mix, Levine, & Huttenlocher, 2002). The counting system created by humans to deal with practical needs is a powerful and pervasive cultural tool the acquisition of which by young children seems to influence the representation of number (Mix et al., 2002; Ni & Zhou, 2005). The acquisition of counting procedures, which is supported by lay culture and early mathematics education, enables children to build a concept of number as counting number (Gelman, 2000; Greer, 2004). Thus before they are exposed to formal rational number instruction, students have formed a rather coherent, explanatory framework for number based on features and properties of the natural numbers (Moss, 2005; Ni & Zhou, 2005; Smith, Solomon, & Carey, 2005; Vamvakoussi & Vosniadou, 2010). It is important to stress that it is not assumed that students are aware of these assumptions; rather, these are implicit in nature.

When nonnatural numbers are introduced in the curriculum, although they are also called "numbers," they clearly violate practically all the background assumptions underlying students' initial framework theory of number. The question arises, how do students assign meaning to these new mathematical objects? The framework theory approach to conceptual change is a constructivist approach. We argue that students draw heavily on their prior knowledge of natural numbers to make sense of nonnatural numbers. Not being aware that rational numbers do not obey some of the principles of natural numbers, they assimilate the information about fractions and decimals into their framework theory of number creating fragmented and synthetic conceptions. Consider, for example, the "longer is bigger" error in the case of decimals, which is very common, particularly for younger students. This error becomes less common with age while the "shorter is bigger" error becomes more prominent and remains present even in adulthood (Desmet, Grégoire, & Mussolin, 2010; Stacey, Helme, & Steinle, 2001). This error is often interpreted as an intrusion of knowledge about fractions. From our point of view, this is a synthetic conception, reflecting the assimilation of information about nonnatural numbers while retaining the idea that one can judge the size of a nonnatural number by the number of its digits.

The application of the framework theory approach to conceptual change in mathematics (Verschaffel & Vosniadou, 2004; Vosniadou & Verschaffel, 2004) has made possible not only to explain some of the systematic errors that children make when they are exposed to rational numbers but also to predict and discover new ones. The following section presents some of the empirical research on the development of rational number knowledge guided by the framework theory approach.

## Empirical research on the development of rational number knowledge guided by the framework theory approach

In line with the framework theory approach to conceptual change, Stafylidou and Vosniadou (2004) predicted that learning fractions will be slow and gradual, that the students will create misconceptions in the process, and that some of these misconceptions would be synthetic models located in the areas where fractions differ from natural numbers. For example, it was predicted that based on their experience with natural numbers, students would interpret fraction notation as consisting of two independent natural numbers when first exposed to it. Another example is the ordering of fractions and their relation to the unit. Here again, it was predicted that the students will have difficulty ordering fractions correctly and that they will form misconceptions in the areas in which fraction ordering differs from natural number ordering.

Two hundred students from fifth to tenth grade were asked to (a) write the smallest and biggest fraction that they could think of and explain their answers, and (b) compare and order fractions. As predicted, the development of fraction knowledge was gradual and exhibited a developmental progression. The great majority of these students (89%) were placed in three categories corresponding to an initial and two intermediate states of fraction understanding. The responses of the remaining students revealed fragmented conceptions. As predicted, in the first category, which consisted mainly of the younger students in the sample, the students had no understanding of fractions. They considered them to consist of two independent natural numbers and to follow the ordering of natural numbers. In the second, intermediate category, they started understanding that the numerator and denominator were related but considered fractions to be parts of a whole, always smaller than the unit. Only in the third category, which consisted mainly of the older students, the students were able to take into consideration the multiplicative relation between the numerator and denominator. Still most of these students believed that fractions are bounded by the smallest and biggest fraction.

Experiments by Vamvakoussi and Vosniadou (2004, 2010) investigated another area where the rational numbers differ from natural number, namely, the density of rational numbers. Within the natural number set, there is a finite number of numbers between any two numbers. On the contrary, rational numbers are dense. Within the rational number set, there are always infinitely many numbers between any two numbers. As predicted, Vamvakoussi and Vosniadou found that secondary school students are very likely to say that there is a finite number of intermediates between two given rational numbers, as is the case with natural numbers. Furthermore, as their knowledge of rational number developed, the students created intermediate levels, or otherwise synthetic models of understanding of the density property. For example, some students were more likely to answer that there were infinitely many numbers between two decimals than between two fractions or vice versa. They were also more likely to say that these infinite intermediate numbers have the same representational form as the interval points (i.e., only infinite decimals between decimals and only infinite many fractions between fractions). These results indicated that students view rational numbers as a collection of unrelated sets of decimals, fractions, and natural numbers, which behave differently with respect to their order. From the perspective of the framework theory, this is a synthetic conception of the rational number set. Other researchers reported similar findings with students from primary to tertiary education, although they were not able to establish that there are intermediate levels of understanding the density property (Giannakoulias, Souyoul, & Zachariades, 2007; Hannula, Pehkonen, Maijala, & Soro, 2006; Hartnett & Gelman, 1998; Merenluoto & Lehtinen, 2004; Smith et al., 2005).

More recent studies employed longitudinal designs and latent transition analysis to trace the development of rational number knowledge, with a focus on the conceptual knowledge of fractions in students from grades 3 to 5 (McMullen, Laakkonen, Hannula-Sormunen, & Lehtinen, 2015; McMullen & Van Hoof, 2020). The results showed that the students had substantial difficulties during their first years of formal learning of rational numbers and only a few demonstrated substantial conceptual changes in their conceptual understanding of fractions. Particularly knowledge about the density of rational numbers remained rare and tenuous for the great majority of students.

Similar results were obtained in a cross-sectional study of the development of conceptual knowledge of fractions in a sample of 516 grade 7–9 students from Indonesia (Haryanto, 2019). Eighty percent of these students had achieved only a part-whole understanding of fractions and were not capable of solving problems with improper fractions. Only 8% of these students had achieved some understanding of the unbounded infinity of fractions and only 3% indicated an understanding of density.

Studies investigating the precursors of rational number have shown that numerical magnitude knowledge, including whole number knowledge, are important predictors of the development of magnitude knowledge in fractions, although not of density knowledge (McMullen & Van Hoof, 2020; Rinne, Ye, & Jordan, 2017). These results have been erroneously interpreted to be inconsistent with the framework theory approach and with the natural number bias (Rinne et al., 2017). The framework theory approach claims that understanding rational numbers requires conceptual changes in the representation of numbers as whole numbers and that the errors students will make in their learning will be located in the areas where conceptually or in terms of procedures and operations rational numbers differ from whole numbers. These predications have been confirmed by the empirical evidence. The framework theory does not claim that students with superior conceptual and procedural knowledge of natural numbers will be disadvantaged in achieving a conceptual and procedural knowledge of fractions and decimals. On the contrary, based on the framework theory, one would expect that the students who are more mathematically advanced in their knowledge of whole numbers would be more likely to undergo the conceptual changes required to understand the more difficult and counterintuitive aspects of rational number and not the opposite.

In yet another series of study guided by the framework theory approach, Christou and Vosniadou (2009, 2012) investigated students' difficulties in the interpretation of literal symbols in algebra. A number of different methodologies, such as open and forced-choice questionnaires, and interviews were used. The results showed that students from seventh up to tenth grade tend to think that only natural numbers can be substituted for variables. For example, students answered that 2X stands only for multiples of 2, that a/b stands only for positive fractions, and that −b stands for negative integers. In agreement with our hypothesis, we also found that even the students who would accept nonnatural numbers as substitutes for the variables were not ready to accept any number as a possible substitute. In certain cases, students accepted decimal numbers—but not fractions—as possible substitutes for variables, and in other cases, they accepted fractions but not decimals. The finding that a literal symbol, denoting a (real) variable, stands only for specific types of numbers points to a synthetic conception of this mathematical notion. This finding was replicated by a study with Flemish secondary students, who tended to assign only natural numbers less frequently than their Greek peers but still had great difficulty to consistently assign any type of numbers to literal symbols (Van Dooren, Christou, & Vamvakoussi, 2010).

To sum up, the framework theory has provided a new perspective on rational number learning and has generated novel predictions regarding the interference of natural number knowledge in the context of rational numbers as well as in the context of algebra. A series of empirical studies have confirmed these predictions and showed that there are intermediate states in the transition from initial to more sophisticated understandings of complex mathematical concepts, as students' initial ideas interact with the new counterintuitive information coming from instruction to form synthetic conceptions.

## The coexistence of framework and scientific theories

For a long time, conceptual change researchers explicitly stated or implicitly assumed that when students learn a scientific theory, they use it to replace their initial understandings. However, a body of relatively recent research has started to be accumulated showing that this is not the case. On the contrary, initial understandings or framework theories coexist with scientific theories in a number of different knowledge domains. For example, psychological research has shown that both children and adults, across different cultures, frequently use materialistic and dualistic explanations for the mind, or creationist and evolutionary accounts of the origin of species. Shtulman and Valcarcel (2012) tested the coexistence hypothesis in 10 different domains of knowledge from the life sciences to the physical sciences to mathematics. Their results showed that undergraduates verified statements whose truth value differed across naïve and scientific theories significantly less accurately and more slowly than statements whose truth value was the same across both theories, suggesting that the coexistence of naïve and scientific theories extends across multiple domains of knowledge.

Similar results have been obtained in the domain of mathematics learning. In two experiments, DeWolf and Vosniadou (2015) investigated the coexistence hypothesis in timed fraction magnitude comparison tasks with skilled adults. In these experiments, the participants were presented with a fraction magnitude task in which half of the comparisons were consistent with natural number ordering and the other half were not. The results showed that both accuracy and response time depended on whether the fraction magnitudes were consistent or not with whole number ordering indicating that even skilled adults do not have direct access to the fraction magnitudes on the number line when the distance between the fraction pairs is very small. These results are consistent with the findings of other experiments which show that the whole number parts of which fractions are composed (the numerator and denominator) interfere with their processing (see Meert, Grégoire, & Noël, 2009; Obersteiner, Van Dooren, Van Hoof, & Verschaffel, 2013).

## The framework theory approach and the natural number bias

It could be argued that the framework theory approach with its emphasis on the adverse influence of natural number knowledge does not add anything new to what was already known in mathematics education. Prior research has shown the impact of initial understandings of number on the learning of a wide range of mathematical concepts (see Fischbein et al., 1985; Hartnett & Gelman, 1998; Smith et al., 2005). In fact, Ni and Zhou (2005) have argued that the adverse influence of natural numbers on rational number understanding is such a pervasive phenomenon that it should be attributed the status of a bias—"a robust tendency to use the single-unit counting scheme to interpret instructional data on fractions" (p. 28). The natural number bias has attracted the interest of researchers from diverse fields, such as mathematics education, educational psychology, cognitive and developmental psychology, and neuroscience. It has been studied using different methodologies such as paper-and-pencil tests, interviews, reaction time studies, and brain imaging studies with a great variety of different populations, from infants to adults, and from novices to expert mathematicians (DeWolf & Vosniadou, 2015; Kallai & Tzelgov, 2012; Meert et al., 2009; Obersteiner et al., 2013). What is therefore the contribution of the framework theory approach other than the focus on the adverse effects of the natural number bias?

In answering this question, it is important to first point out that the framework theory approach represents a larger attempt to explain conceptual development that applies to many domains of knowledge and not only to mathematics. From the point of view of the framework theory approach, the natural number bias is only one exemplar of a much larger phenomenon, which is present in all situations where a constructivist learning system must deal with the acquisition of new concepts incompatible with prior knowledge, in other words, all situations requiring conceptual changes.

Second, the framework theory is a developmental approach that attempts to explain how development happens and to predict students' difficulties and the creation of misconceptions in the process of learning. What is claimed is that framework theories create the knowledge base to which conceptually incompatible science and mathematics information is assimilated through knowledge building, constructive, mechanisms. During the learning process, students create intermediate, hybrid conceptions or misconceptions that can be fragmented or synthetic. The framework theory makes possible to generate predictions about the areas where students are likely to experience difficulties during the learning process as well as about the specific (mis)conceptions they may form. These predictions have been confirmed in many domains of knowledge, as, for example, in observational astronomy (Vosniadou & Brewer, 1992, 1994), mechanics (Ioannides & Vosniadou, 2001), biology (Carey, 1985; Inagaki & Hatano, 2008), theory of evolution (Evans, 2008), and theory of matter (Wiser & Smith, 2008). The theoretical framework has also been recently applied to the understanding of the change from visuospatial to analytic strategies in geometry (Kospentaris, Vosniadou, Kazi, & Thanou, 2016) and chemistry (Vlacholia et al., 2017).

This body of research provides specific information about students' conceptions that can be utilized to increase the effectiveness of mathematics education. In addition, the framework theory approach carries with it a number of principles for the design of instruction stemming from the conceptual change perspective on learning that are relevant for mathematics learning (Greer, 2004; Vosniadou et al., 2008; Vosniadou, Ioannides, Dimitrakopoulou, & Papademitriou, 2001; Vosniadou & Vamvakoussi, 2006). In the pages that follow, we illustrate briefly how such principles can be implemented in the design of curricula and instruction.

## Implications for instruction

As mentioned earlier, the importance of the framework theory approach to the design of mathematics instruction lies in that it makes possible to predict the areas where students are going to experience difficulties and are likely to create misconceptions. These are the areas where the new information coming from instruction might be inconsistent with the students' prior knowledge. In the case of rational number instruction, research has already provided a great deal of information about students' difficulties and misconceptions. This research can be utilized to design appropriate developmental, research-based curricula and instruction.

This is consistent with a recent trend in research in mathematics education to take a developmental approach to the design of curricula and instruction, which puts forward the notion of learning trajectories or progressions (Confrey & Maloney, 2015; Sztajn, Confrey, Wilson, & Edgington, 2012). Learning progressions are hypothetical, but empirically based and tested, models of students' transitions from less to more sophisticated understandings in a specific mathematical domain, under carefully designed instruction. Several learning progressions or trajectories have been articulated in the area of fractions (Arieli-Attali & Cayton-Hodges, 2014; Confrey & Maloney, 2015; Haryanto, 2019; Wright, 2014). Haryanto (2019) developed and validated a learning progression which distinguished between conceptual and procedural knowledge of fractions using a sample of 516 Indonesian students from grades 7 to 9. The progression specifies in great detail the order in which fraction conceptual knowledge and operations are acquired and provides information about the areas in which students experienced difficulties and the misconceptions they formed.

One way to deal with students' difficulties with rational numbers is to use dissonance-producing strategies, such as mild cognitive conflict, that alert students to the negative impact of their natural number knowledge on their conceptual and procedural understanding of fractions and decimals. This can be done in various ways. The students can be simply told of the situations where fractions and decimals are different from those used with natural numbers and be given examples. The use of contrasting cases that emphasize similarities and differences amongst examples has been associated with increased learning outcomes (Alfieri, Nokes-Malach, & Schunn, 2013). Teachers can identify common misconceptions with fractions and decimals and discuss them with their students. They can also create assessments containing problems with fractions and decimals in areas where misconceptions usually occur and then discuss students' mistakes explicitly pointing out the inconsistencies between fractions, decimals, and natural numbers and explain them. Students will gain a greater conceptual understanding of fractions and decimals when they understand why the procedures from natural numbers do not work for fractions rather than just memorizing the new procedures.

The instruction currently provided in the schools fails to provide children with explicit information about the inconsistencies between their natural number knowledge and rational number principles and operations. In a recent study, Van Dooren (2019) analyzed mathematics textbooks (including worksheets and teacher manuals) from grades 3 to 6 in five European countries to investigate the extent to which they explicitly addressed differences between natural and rational numbers in their representations, ordering, operations, and density. The results showed that explicit reference to differences between natural and rational numbers was very rare. On the contrary, there were many instances where similarities between natural and rational numbers were exploited.

An easy way to use dissonance-producing strategies with text is refutational text (e.g., Diakidoy, Kendeou, & Ioannides, 2003; Tippett, 2010). In a refutational text, students' initial conceptions, beliefs, or ideas are directly stated and immediately refuted as a means of introducing a counterintuitive concept or explanation (Hynd, 2001). Refutational texts not only provide arguments that falsify students' initial viewpoints but also present the rationale of the new, to be adopted, perspective. Refutations can also be included in lectures. In an experimental intervention, Christou (2012) exposed secondary students to a refutational lecture (Kowalski & Taylor, 2009) in order to address their difficulties with the interpretation of the phenomenal sign of algebraic expressions. The refutational lecture used the main principles of direct instruction. More specifically, the students were given the definition of the real variable in algebra and were explicitly told that literal symbols are used in algebra to stand for any real number, unless otherwise specified. Students' attention was called on to the differences between the (actual) sign of numbers in arithmetic, where the presence or absence of a minus sign indeed denotes a negative or positive value, and the (phenomenal) sign of algebraic expressions which does not necessarily denote an actual sign. Cognitive conflict was induced by providing examples and counterexamples of natural and nonnatural number substitutions to literal symbols in algebraic expressions, which either retained or changed their phenomenal sign. Students were also offered the strategy to always test the sign of an algebraic expression by substituting at least one negative number to the literal symbols. At the end of the intervention, all the above points were discussed with the students. The whole intervention lasted about 25 min. The results showed that the students who attended the refutational lecture made significantly fewer phenomenal sign errors compared to the students who did not attend the lecture. In addition, the learning profits were retained for at least 1 month after the intervention.

The use of dissonance-producing strategies such as refutation is one way to promote conceptual change. In order to be effective, these strategies must be followed with instruction that provides students with many opportunities for the deliberate practice of procedures and operations that are inconsistent with natural number principles. Another type of instruction is to build on prior knowledge. For example, many researchers, coming from different theoretical backgrounds, agree that the use of the number line can play an important role in rational number instruction (Kilpatrick et al., 2001; Siegler, 2016) or recommend the use of cross-domain analogies (Clement, 2008).

Vamvakoussi and Vosniadou (2012) invested on the cross-domain mapping between numbers and geometrical objects to foster students' understanding of the density of points and of numbers. Cross-domain mapping is considered a powerful mechanism for conceptual restructuring, because it allows for the transfer of inferences from a more familiar domain to the less familiar one (Gentner & Wolff, 2000; Vosniadou, 1989). The research on the density property presented above, and a pilot study reported in Vamvakoussi and Vosniadou (2012), had indicated that students are more prone to accept (a) that there are infinite points on a line segment than infinite numbers in an interval, and (b) that there are infinitely many intermediates (numbers or points) in an interval than that there is no successor at all. Vamvakoussi and Vosniadou used the line as a source domain and employed a *bridging analogy* (Brown & Clement, 1989; Clement, 2008) to decrease the gap between students' initial ideas and the sophisticated idea of points as a dense array. Specifically, the line was presented as an imaginary rubber band that never breaks, no matter how much it is stretched. It was hypothesized that the "rubber line" analogy has the potential to help students grasp the idea of density, particularly the "no successor" principle, because it evokes students' experiences with a real-world object, that is, the rubber band. It is consistent with students' experiences with physical representations of the number line and it is associated, via the imaginary property of being unbreakable, with a recursive process, which is an easier way of accessing the notion of infinity for students. Finally, the number line analogy produces a sequence of segments rather than a sequence of points that can be deemed discrete.

A text was designed that provided explicit information about the infinity of numbers in an interval, made explicit reference to the numbers-to-points correspondence, and used the "rubber line" bridging analogy to convey the idea that points (and numbers) can never be found one immediately after the other. The "rubber-line" text was experimentally tested against two other texts that contained the same explicit information. In addition, it presented examples of intermediate numbers or figures illustrating the examples. Six classes of 8th and 11th graders (one experimental class per grade), 149 students in total participated in the study. They received a pretest with density tasks in a arithmetical and a geometrical context, were administered the corresponding text, and then received a posttest containing all the pretest tasks as well as five additional tasks that examined students' abilities to deal with the no-successor aspect of density. All groups profited from the explicit information about the infinity of numbers presented in the text. However, the experimental group (8th as well as 11th graders) outperformed the other groups in the "no successor" items of the posttest and was more consistent in providing correct answers and justifications for them.

The interventions discussed above targeted specific areas and were short-termed. For the problem of the adverse effects of natural number knowledge in rational number learning to be tackled more effectively, a long-term perspective on the planning and design of instruction is necessary. We argue that instruction should target the problem at the earliest possible time, before the discrepancy between natural and rational number knowledge and experience gets too large. Rational number instruction is usually based on the part-whole aspect of fraction and elicits additive rather than multiplicative reasoning (Moss, 2005; Ni & Zhou, 2005). Alternative approaches to the introduction of fractions can include grounding instruction on ratio (e.g., Confrey, 1995), on fair sharing/equipartitioning (Confrey & Maloney, 2015; Nunes & Bryant, 1996), and on measurement (Davydov & Tsvetkovich, 1991). Sophian (2004, 2007) has suggested that if more attention were paid to the importance of units in counting at the first years of instruction, the gap between natural and rational numbers could be decreased.

No matter how early instruction on rational numbers starts, students will still have to learn to deal with the interference of their natural number knowledge. This is a case not only in mathematics but in many other subject matter areas where students need to deal with conceptual reorganizations and with the acquisition of new concepts that require taking new perspectives and reexamining well-entrenched beliefs. Instruction should aim not only on how to best present new concepts to students in order to make it easier for them to understand them, it should also aim towards strengthening student's abilities to manage and control their learning through the fostering of metacognition and the flexible use of appropriate learning strategies that promote cognitive flexibility and multiple perspective taking.

To sum up and conclude, it is possible to address students' difficulties with rational number more effectively than is currently done by utilizing the predictions and results of research grounded in the framework theory approach and by implementing principles for instruction stemming from conceptual change research in general. These instructional interventions support students to use, evaluate, compare, build, and discuss representations of mathematical constructs and understand their similarities and differences. They can explicitly address students' misconceptions and promote understanding. The results have shown that these teaching strategies can greatly foster the understanding of counterintuitive mathematical ideas (see also Greer, 2004; Vosniadou et al., 2008; Vosniadou & Vamvakoussi, 2006). Finally, we argued that it is worth investing on long-term approaches. One overarching principle for instruction stemming from conceptual change perspectives is that instruction should take a long-term developmental perspective by anticipating later expansions of meaning of mathematical ideas (Greer, 2004; Vosniadou & Vamvakoussi, 2006), and most importantly, by developing in students the skills and strategies needed for the effective, metacognitive control of their learning.

## References


Alfieri, L., Nokes-Malach, T. J., & Schunn, C. D. (2013). Learning through case comparisons: A meta-analytic review. *Educational Psychologist, 48*(2), 87–113.

Arieli-Attali, M., & Cayton-Hodges, G. (2014). Expanding the CBAL™ mathematics assessments to elementary grades: the development of a competency model and a rational number learning progression. *ETS Research Report Series*, (1), 1–41.

Brown, D. E., & Clement, J. (1989). Overcoming misconceptions via analogical reasoning: Abstract transfer versus explanatory model construction. *Instructional Science, 18*, 237–261. https://doi.org/10.1007/BF00118013.

Carey, S. (1985). *Conceptual change in childhood*. Cambridge, MA: MIT Press.

Carey, S. (2004). Bootstrapping and the origin of concepts. *Daedalus, 133*(1), 59–68. https://doi.org/10.1162/001152604772746701.

Christou, K. P. (2012). Helping students remedy the phenomenal sign bias: The case of a refutational lecture. In C. Prachalias (Ed.), *Proceedings of the 8th international conference on education* (pp. 643–648). Greece: Samos.

Christou, K. P., & Vosniadou, S. (2009). Misinterpreting the use of literal symbols in Algebra. In *Paper presented at the 33rd conference of the International Group for the Psychology of Mathematics Education, Thessaloniki, Greece.*

Christou, K. P., & Vosniadou, S. (2012). What kinds of numbers do students assign to literal symbols? Aspects of the transition from arithmetic to algebra. *Mathematical Thinking and Learning, 14*(1), 1–27. https://doi.org/10.1080/10986065.2012.625074.

Clement, J. (1982). Students' preconceptions in introductory mechanics. *American Journal of Physics, 50*(1), 66–70.

Clement, J. (2008). The role of explanatory models in teaching for conceptual change. In S. Vosniadou (Ed.), *The international handbook of research on conceptual change* (pp. 417–452). New York, US: Routledge.

Confrey, J. (1995). Student voice in examining "splitting" as an approach to ratio, proportions and fractions. In L. Meira, & D. Carraher (Eds.), *Vol. 1. Proceedings of the 19th international conference for the psychology of mathematics education* (pp. 3–29). Recife, Brazil: Universidade Federal de Pernambuco.

Confrey, J., & Maloney, A. (2015). Design research study of a curriculum and diagnostic assessment system for a learning trajectory on equipartitioning. *ZDM Mathematics Education, 47*, 919–932. https://doi.org/10.1007/s11858-015-0699-y.

Davydov, V. V., & Tsvetkovich, Z. H. (1991). The object sources of the concept of fraction. In V. V. Davydov, & L. P. Steffe (Eds.), *Soviet studies in mathematics education: Psychological abilities of primary school children in learning mathematics* (pp. 86–147). Reston, VA: National Council of Teachers of Mathematics.

Desmet, L., Grégoire, J., & Mussolin, C. (2010). Developmental changes in the comparison of decimal fractions. *Learning and Instruction, 20*, 521–532. https://doi.org/10.1016/j.learninstruc.2009.07.004.

DeWolf, M., & Vosniadou, S. (2015). The representation of fraction magnitudes and the whole number bias reconsidered. *Learning and Instruction, 37*, 39–49. https://doi.org/10.1016/j.learninstruc.2014.07.002.

Diakidoy, I. A. N., Kendeou, P., & Ioannides, C. (2003). Reading about energy: The effects of text structure in science learning and conceptual change. *Contemporary Educational Psychology*, *28*, 335–356. https://doi.org/10.1016/S0361-476X(02)00039-5.

diSessa, A. (1993). Towards an epistemology of physics. *Cognition and Instruction*, *10*(2–3), 105–225.

Driver, R., & Easley, J. (1978). Pupils and paradigms: A review of literature related to concept development in adolescent science students. *Studies in Science Education*, *5*, 61–84.

Durkin, K., & Rittle-Johnson, B. (2015). Diagnosing misconceptions: Revealing changing decimal fraction knowledge. *Learning and Instruction*, *37*, 21–29. https://doi.org/10.1016/j.learninstruc.2014.08.003.

Evans, M. E. (2008). Conceptual change and evolutionary biology. A developmental analysis. In S. Vosniadou (Ed.), *Handbook of research on conceptual change* (pp. 263–294). Mahwah, NJ: Lawrence Erlbaum Associates.

Fischbein, E. (1987). *Intuition in science and mathematics*. Dordrecht, The Netherlands: Reidel.

Fischbein, E., Deri, M., Nello, M., & Marino, M. (1985). The role of implicit models in solving problems in multiplication and division. *Journal of Research in Mathematics Education*, *16*, 3–17. https://doi.org/10.2307/748969.

Geary, D. C. (2008). An evolutionarily informed education science. *Educational Psychologist*, *43*, 179–195.

Gelman, R. (2000). The epigenesis of mathematical thinking. *Journal of Applied Developmental Psychology*, *21*, 27–37. https://doi.org/10.1016/S0193-3973(99)00048-9.

Gentner, D., & Wolff, P. (2000). Metaphor and knowledge change. In E. Dietrich, & A. Markman (Eds.), *Conceptual change in humans and machines* (pp. 295–342). Mahwah, NJ: Lawrence Erlbaum Associates.

Giannakoulias, E., Souyoul, A., & Zachariades, T. (2007). Students' thinking about fundamental real numbers properties. In D. Pitta-Pantazi, & G. Philippou (Eds.), *Proceedings of the fifth congress of the European society for research in mathematics education (ERME)* (pp. 416–425). Nicosia, Cyprus: Department of Education, University of Cyprus.

Greer, B. (2004). The growth of mathematics through conceptual restructuring. *Learning and Instruction*, *14*, 541–548. https://doi.org/10.1016/j.learninstruc.2004.06.018.

Hannula, M. S., Pehkonen, E., Maijala, H., & Soro, R. (2006). Levels of students' understanding on infinity. *Teaching Mathematics and Computer Science*, *4*(2), 317–337.

Hartnett, P., & Gelman, R. (1998). Early understandings of number: Paths or barriers to the construction of new understandings? *Learning and Instruction*, *8*(4), 341–374. https://doi.org/10.1016/S0959-4752(97)00026-1.

Haryanto, B. (2019). *Assessing learning progression in the domain of fractions*. Ph.D. dissertation College of Education, Psychology and Social Work, Flinders University.

Hynd, C. R. (2001). Refutational texts and the change process. *International Journal of Educational Research*, *35*, 699–714. https://doi.org/10.1016/S0883-0355(02)00010-1.

Inagaki, K., & Hatano, G. (2008). Conceptual change in naïve biology. In S. Vosniadou (Ed.), *International handbook of research on conceptual change* (pp. 240–262). Mahwah, NJ: Lawrence Erlbaum Associates.

Ioannides, C., & Vosniadou, S. (2001). The changing meanings of force. *Cognitive Science Quarterly*, *2*(1), 5–62.

Kallai, A. Y., & Tzelgov, J. (2012). When meaningful components interrupt the processing of the whole: The case of fractions. *Acta Psychologica*, *139*, 358–369. https://doi.org/10.1016/j.actpsy.2011.11.009.

Kilpatrick, J., Swafford, J., & Findell, B. (2001). *Adding it up. Helping children learn mathematics*. Washington, DC: National Academy Press.

Kospentaris, G., Vosniadou, S., Kazi, S., & Thanou, E. (2016). Visual and analytic strategies in geometry. *Frontline Learning Research*, *4*(1), 40–57.

Kowalski, P., & Taylor, A. K. (2009). The effect of refuting misconceptions in the introductory psychology class. *Teaching of Psychology*, *36*, 153–159. https://doi.org/10.1080/00986280902959986.

Kuhn, T. (1962). *The structure of scientific revolutions*. Chicago, IL: Chicago Press.

Linn, M. (2013). Teaching for conceptual change: distinguish or extinguish ideas. In S. Vosniadou (Ed.), *International handbook of research on conceptual change* (pp. 694–722). New York: Routledge.

McCloskey, M. (1983). Intuitive physics. *Scientific American, 248*(4), 122–130.

McMullen, J., Laakkonen, E., Hannula-Sormunen, M., & Lehtinen, E. (2015). Modeling the developmental trajectories of rational number concept(s). *Learning and Instruction, 37*, 14–20. https://doi.org/10.1016/j.learninstruc.2013.12.004.

McMullen, J., & Van Hoof, J. (2020). The role of rational number density knowledge in mathematical development. *Learning and Instruction, 65*, 101228. https://doi.org/10.1016/j.learninstruc.2019.101228.

Meert, G., Grégoire, J., & Noël, M.-P. (2009). Rational numbers: Componential versus holistic representation of fractions in a magnitude comparison task. *The Quarterly Journal of Experimental Psychology, 62*(8), 1598–1616. https://doi.org/10.1080/17470210802511162.

Merenluoto, K., & Lehtinen, E. (2004). Number concept and conceptual change: Towards a systemic model of the processes of change. *Learning and Instruction, 14*, 519–536. https://doi.org/10.1016/j.learninstruc.2004.06.016.

Mix, K. S., Levine, S. C., & Huttenlocher, J. (2002). Multiple cues for quantification in infancy: Is number one of them? *Psychological Bulletin, 128*(2), 278–294.

Moss, J. (2005). Pipes, tubes, and beakers: New approaches to teaching the rational-number system. In S. Donovan, & J. D. Bransford (Eds.), *How students learn: History, mathematics, and science in the classroom* (pp. 309–349). Washington, DC: National Academy Press.

Moss, J., & Case, R. (1999). Developing children's understanding of the rational numbers: A new model and an experimental curriculum. *Journal for Research in Mathematics Education, 30*, 122–147. https://doi.org/10.2307/749607.

Ni, Y. J., & Zhou, Y.-D. (2005). Teaching and learning fraction and rational numbers: The origins and implications of whole number bias. *Educational Psychologist, 40*(1), 27–52. https://doi.org/10.1207/s15326985ep4001_3.

Nunes, T., & Bryant, P. (1996). *Children doing mathematics*. Cambridge, MA: Blackwell.

Obersteiner, A., Van Dooren, W., Van Hoof, J., & Verschaffel, L. (2013). The natural number bias and magnitude representation in fraction comparison by expert mathematicians. *Learning and Instruction, 28*, 64–72. https://doi.org/10.1016/j.learninstruc.2013.05.003.

Posner, G. J., Strike, K. A., Hewson, P. W., & Gertzog, W. A. (1982). Accommodation of a scientific conception: Towards a theory of conceptual change. *Science Education, 66*(2), 211–227.

Resnick, L. B., Nesher, P., Leonard, F., Magone, M., Omanson, S., & Peled, I. (1989). Conceptual bases of arithmetic errors: The case of decimal fractions. *Journal for Research in Mathematics Education, 20*, 8–27. https://doi.org/10.2307/749095.

Rinne, L. F., Ye, A., & Jordan, N. C. (2017). Development of fraction comparison strategies: A latent transition analysis. *Developmental Psychology, 53*, 713–730.

Shtulman, A., & Valcarcel, J. (2012). Scientific knowledge suppresses but does not supplant earlier intuitions. *Cognition, 124*(2), 209–215.

Siegler, R. S. (2016). Magnitude knowledge: The common core of numerical development. *Developmental Science, 19*, 341–361. https://doi.org/10.1111/desc.12395.

Siegler, R. S., & Lortie-Forgues, H. (2014). An integrative theory of numerical development. *Child Development Perspectives, 8*, 144–150. https://doi.org/10.1111/cdep.12077.

Siegler, R. S., Thompson, C. A., & Schneider, M. (2011). An integrated theory of whole number and fractions development. *Cognitive Psychology, 62*(4), 273–296. https://doi.org/10.1016/j.cogpsych.2011.03.001.

Smith, C. L., Solomon, G. E. A., & Carey, S. (2005). Never getting to zero: Elementary school students' understanding of the infinite divisibility of number and matter. *Cognitive Psychology, 51*, 101–140. https://doi.org/10.1016/j.cogpsych.2005.03.001.

Sophian, C. (2004). A prospective developmental perspective on early mathematics instruction. In D. H. Clements, J. Sarama, & A.-M. DiBiase (Eds.), *Engaging young children in mathematics: Standards for early childhood mathematics* (pp. 253–256). Mahwah, NJ: Lawrence Erlbaum Associates.

Sophian, C. (2007). Rethinking the starting point for mathematics learning. In O. N. Saracho, & B. Spodek (Eds.), *Contemporary perspectives in early childhood education: Mathematics, science and technology in early childhood education* (pp. 21–44). Information Age Publishing.

Spelke, E. S. (2004). Core knowledge. In N. Kanwisher, & J. Duncan (Eds.), *Attention and performance, vol. 20: Functional neuroimaging of visual cognition*. Oxford: Oxford University Press.

Stacey, K., Helme, S., & Steinle,V. (2001). Confusions between decimals, fractions and negative numbers: A consequence of the mirror as a conceptual metaphor in three different ways. In *Paper presented at the 25th conference of the International Group for the Psychology of Mathematics Education, Utrecht*.

Stafylidou, S., & Vosniadou, S. (2004). Students' understanding of the numerical value of fractions: A conceptual change approach. *Learning and Instruction, 14*, 503–518. https://doi.org/10.1016/j.learninstruc.2004.06.015.

Sztajn, P., Confrey, J., Wilson, P. H., & Edgington, C. (2012). Learning trajectory based instruction: Toward a theory of teaching. *Educational Researcher, 41*(5), 147–156. https://doi.org/10.3102/0013189X12442801.

Tippett, C. D. (2010). Refutation text in science education: A review of two decades of research. *International Journal of Science and Mathematics Education, 8*(6), 951–970. https://doi.org/10.1007/s10763-010-9203-x.

Vamvakoussi, X., & Vosniadou, S. (2004). Understanding the structure of the set of rational numbers: A conceptual change approach. *Learning and Instruction, 14*, 453–467. https://doi.org/10.1016/j.learninstruc.2004.06.013.

Vamvakoussi, X., & Vosniadou, S. (2010). How many decimals are there between two fractions? Aspects of secondary school students' understanding of rational numbers and their notation. *Cognition and Instruction, 28*(2), 181–209. https://doi.org/10.1080/07370001003676603.

Vamvakoussi, X., & Vosniadou, S. (2012). Bridging the gap between the dense and the discrete: The number line and the "rubber line" bridging analogy. *Mathematical Thinking and Learning, 14*, 265–284. https://doi.org/10.1080/10986065.2012.717378.

Van Dooren, W. (2019). Tackling the natural number bias: A comparative textbook analysis. In *Paper presented in the biannual conference of the European Association for Research on Learning and Instruction, Aachen, Germany*.

Van Dooren, W., Christou, K. P., & Vamvakoussi, X. (2010). Greek and Flemish students' interpretation of the literal symbols as variables. In M. M. F. Pinto, & T. F. Kawasaki (Eds.), *Vol. 4. Mathematics in different settings – Proceedings of the 34th conference of the International Group for the Psychology in Mathematics Education* (pp. 257–264). Belo Horizonte, Brazil: PME.

Verschaffel, L., & Vosniadou, S. (Eds.). (2004). The conceptual change approach to mathematics learning and teaching. *Special Issue of Learning and Instruction, 14*.

Vlacholia, M., Vosniadou, S., Roussos, P., Salta, K., Kazi, S., Sigalas, M., et al. (2017). Changes in visual/spatial strategy use in organic chemistry with the development of expertise. *Chemistry Education Research and Practice, 18*, 763–773. https://doi.org/10.1039/c7rp00036g.

Vosniadou, S. (1989). Analogical reasoning as a mechanism in knowledge acquisition: A developmental perspective. In S. Vosniadou, & A. Ortony (Eds.), *Similarity and analogical reasoning* (pp. 413–436). Cambridge, MA: Cambridge University Press.

Vosniadou, S. (1994). Capturing and modelling the process of conceptual change. *Learning and Instruction, 4*, 45–69. https://doi.org/10.1016/0959-4752(94)90018-3.

Vosniadou, S. (2013). Conceptual change in learning and instruction: The framework theory approach. In S. Vosniadou (Ed.), *International handbook of research on conceptual change* (pp. 11–30). New York: Routledge. https://doi.org/10.4324/9780203154472.ch1. 2nd ed.

Vosniadou, S., & Brewer, W. F. (1992). Mental models of the earth. *Cognitive Psychology*, *24*, 535–585.

Vosniadou, S., & Brewer, W. F. (1994). Mental models of the day/night cycle. *Cognitive Science*, *18*, 123–183.

Vosniadou, S., Ioannides, C., Dimitrakopoulou, A., & Papademitriou, E. (2001). Designing learning environments to promote conceptual change in science. *Learning and Instruction*, *11*(4–5), 381–419. https://doi.org/10.1016/S0959-4752(00)00038-4.

Vosniadou, S., & Skopeliti, I. (2014). Conceptual change from the framework theory side of the fence. *Science and Education*, *23*, 1427–1445. https://doi.org/10.1007/s11191-013-9640-3.

Vosniadou, S., & Vamvakoussi, X. (2006). Examining mathematics learning from a conceptual change point of view: Implications for the design of learning environments. In L. Verschaffel, P. Dochy, M. Boekartz, & S. Vosniadou (Eds.), *Instructional psychology: Past, present and future trends. Sixteen essays in honour of Eric De Corte* Elsevier Press.

Vosniadou, S., Vamvakoussi, X., & Skopeliti, E. (2008). The framework theory approach to conceptual change. In S. Vosniadou (Ed.), *Handbook of research on conceptual change* (pp. 3–34). Mahwah, NJ: Lawrence Erlbaum Associates.

Vosniadou, S., & Verschaffel, L. (2004). Extending the conceptual change approach to mathematics learning and teaching. *Learning and Instruction*, *14*, 445–451. https://doi.org/10.1016/j.learninstruc.2004.06.014.

Wellman, H. M., & Gelman, S. A. (1998). Knowledge acquisition in foundational domains. In D. Kuhn, & R. Siegler (Eds.), *Cognition, perception and language. Volume 2 of the handbook of child psychology* (pp. 523–573). New York: Wiley. 5th ed.

Wiser, M., & Smith, C. L. (2008). Learning and teaching about matter in grades K-8: When should the atomic-molecular theory be introduced? In S. Vosniadou (Ed.), *International handbook of research on conceptual change* (pp. 205–239). New York: Routledge.

Wright, V. (2014). Towards a hypothetical learning trajectory for rational number. *Mathematics Education Research Journal*, *26*(3), 635–657.

CHAPTER 6

# Subtraction by addition: A remarkably natural and clever way to subtract?

**Lieven Verschaffel[a], Bert De Smedt[b], Stijn Van Der Auwera[a], and Joke Torbeyns[a]**
[a]Centre for Instructional Psychology and Technology, Katholieke Universiteit Leuven, Leuven, Belgium
[b]Parenting and Special Education Research Unit, Faculty of Psychology and Educational Sciences, KU Leuven, University of Leuven, Leuven, Belgium

## Two vignettes

*I still remember very well that my interest in mathematics started when I made, in Grade 1, my first great mathematical discovery. I discovered that there was a much easier way to do subtractions than the one I was taught by our teacher, namely by adding up from the smaller to the larger number (instead of subtracting the smaller from the larger number, which was the taught method). I had tried it out myself on several problems, and it always worked ....*

***Adrian Treffers, Dutch math educator***

*I learnt to subtract by addition at school, in the 1950s in Niedersachsen, Germany. At the start of secondary school, I had a mathematics teacher who once accused me of cheating. He did not believe that I made a difficult subtraction consisting of two 4-digit numbers by myself, because I had not written down any intermediate result. "This computation is too difficult to be done in the head!" – he said. "No, I can do it, and it is very easy", I said. So he gave me some other subtractions, and I computed them aloud in front of him by means of this method. We were both surprised: He was surprised because he did not know this clever method, and I was surprised that there are people who do it in a much more complicated way. Today I'm surprised that this method is still in discussion instead of being generally accepted.*

***Anonymous German math educator***

## Introduction

Both in the field of numerical cognition and mathematics education, strategy variety and flexibility are considered to be very important and valuable characteristics of arithmetic performance (Baroody & Dowker, 2003; Kilpatrick, Swafford, & Findell, 2001; Verschaffel, Greer, & De Corte, 2007;

*Heterogeneous Contributions to Numerical Cognition*
https://doi.org/10.1016/B978-0-12-817414-2.00005-1

Verschaffel, Luwel, Torbeyns, & Van Dooren, 2009). One subdomain of arithmetic that has attracted a lot of attention of researchers interested in strategy variety and flexibility is subtraction. Subtraction strategies can be analyzed and categorized in various ways. One way, which is the topic of this chapter, is whether the solution strategy is essentially subtractive or additive in nature. According to this criterion, two subtraction strategies can be distinguished: (1) direct subtraction (DS), in which the subtrahend is directly taken away from the minuend (e.g., $73-57=73-50-7=23-7=16$), and (2) subtraction-by-addition (SBA), in which one determines how much needs to be added to the subtrahend to arrive at the minuend (e.g., solving $73-57$ via $57+3=60$, $60+13=73$, so the answer is $3+13$ or 16).[a]

The SBA strategy is considered to be particularly efficient on subtractions with a large subtrahend (e.g., $81-79=?$), because in this case, the SBA solution process requires fewer, smaller, and/or easier calculation steps than the DS strategy (e.g., Torbeyns, De Smedt, Ghesquière, & Verschaffel, 2009a). In contrast, DS is generally assumed to be more efficient than SBA for subtractions wherein the subtrahend is small (e.g., $81-2=?$) on the basis of similar arguments. Despite its assumed efficiency, particularly for small-difference subtractions, current instruction—not only in Flanders, but also worldwide—hardly addresses the SBA strategy. Instead, DS is traditionally considered, taught, and practiced as the mainstream, if not the exclusive, strategy for solving subtractions (Selter, Prediger, Nührenbörger, & Hussmann, 2012; Torbeyns et al., 2009a).

SBA has fascinated researchers—cognitive psychologists as well as mathematics educators—for many decades and in a wide variety of subdomains of the mathematics curriculum (Baroody, Torbeyns, & Verschaffel, 2009; Fuson & Willis, 1988; Verschaffel, Bryant, & Torbeyns, 2012). It has been a focus of investigation of researchers interested in children's as well as adults' strategies for doing subtractions up to 20 as well as in their approaches to multidigit mental and written subtraction. SBA has attracted the interest of researchers active in the field of symbolic arithmetic as well as those investigating word problem solving.

[a]Using this criterion, De Corte and Verschaffel (1987) distinguish—besides direct subtraction and subtraction by addition—a third category of strategies, called indirect subtraction strategies, whereby the solver determines the difference between the minuend and the subtrahend by counting or computing how much has to be subtracted from the minuend to arrive at the subtrahend (e.g., solving $73-57$ via $73-13=60$, $60-3=57$, so the answer is $13+3$ or 16).

The reasons why SBA has attracted the attention of researchers are twofold. One reason, which could be called the "shortcut" reason and relates to the strategy's capacity to eliminate computational effort, fits into the interest of researchers—particularly cognitive psychologists—in understanding the flexibility of people's strategy choices in arithmetic tasks and cognitive tasks in general (e.g., Lemaire & Siegler, 1995; Siegler & Lemaire, 1997). The second reason relates to the conceptual underpinnings of the SBA strategy, especially the inverse relation between addition and subtraction, which is an essential part of understanding these arithmetic operations. Mathematics educators such as Greer (2012), Vergnaud (2012), and Wittmann (2010) argued that inversion, which conceptually underlies SBA, is a key component in the architecture of mathematical structures, from its central role in the arithmetic operations, to their algebraic generalizations, to their place within group theory, in the Fundamental Theorem of Calculus (i.e., the inverse relation between differentiation and integration), and beyond (Verschaffel et al., 2012).

In this chapter, we review and discuss the state of the art of this research, thereby paying particular attention to the contributions of our own research team. We start our review with the use of SBA for solving symbolic subtractions in the domains of mental subtraction up to 20, multidigit mental subtraction, and written algorithms for multidigit subtraction. Afterward, we review the research on SBA in word problem solving and on the conceptual underpinnings of SBA. For each of these themes, we summarize the major theoretical and empirical realizations and reflect upon remaining questions and challenges for future research. As such, this review aims to demonstrate how the combination of theories and methodologies of, on the one hand, cognitive and developmental psychology, and, on the other hand, mathematics education, fruitfully and successfully contributed to a better understanding of the development of varied and flexible strategy use in elementary arithmetic. The chapter also shows how this research can be used to design more adequate ways to support learners in acquiring a rich and flexible strategy repertoire.

## Mental subtraction up to 20

A first subdomain wherein researchers have investigated SBA, starting already several decades ago, is that of single-digit arithmetic. More specifically, they investigated the extent and the success with which people solve symbolically presented subtractions with minuends $<20$ via procedural or retrieval strategies that are based on the inverse operation of addition. People's strategies were investigated with a variety of data-gathering and data-analytic techniques such as verbal protocols, reaction time analyses, and comparing performance on these problems with their inverse counterparts (i.e., problems presented in addition format, $b+?=a$ instead of their canonical form $a-b=?$) (Ashcraft, 1982; Campbell, 2008; Campbell & Xue, 2001; Fuson, 1992; Geary, Frensch, & Wiley, 1993; Kirk & Ashcraft, 2001; LeFevre, DeStefano, Penner-Wilger, & Daley, 2006; Mauro, LeFevre, & Morris, 2003; Peters, De Smedt, Torbeyns, Ghesquière, & Verschaffel, 2010b, 2012; Robinson, 2001; Woods, Resnick, & Groen, 1975).

In one of the oldest studies in this domain, Woods et al. (1975) investigated second- and fourth-grade children's counting strategies when confronted with small and nontie single-digit subtractions (i.e., with a subtrahend $<10$ such as $9-2=?$). Several hypothetically possible counting models were evaluated by fitting observed reaction times to the linear regression predicted by each model. The solution strategies of most second graders and of all fourth graders were best fit by a model in which the child either counted down from the larger number or counted up from the smaller number, depending on which procedure required the fewest counting steps. According to this model, children solve a subtraction such as $9-2=?$ by counting down from the minuend (i.e., "9, 8 (−1), 7 (−2), so the answer is 7"), whereas they answer a subtraction such as $9-7=?$ by counting up from the subtrahend to the minuend (i.e., "7, 8 (+1), 9 (+2), so the answer is 2"). In sum, Woods et al.'s (1975) results suggested a developmental trend in which lower elementary school children move from a fixed model based on the routine use of a single counting strategy to a more flexible model characterized by the choice between two different strategies based on numerical judgment and estimation. Clearly, the counting-down-from-minuend and counting-up-from-subtrahend strategies can be considered as counting variants of, respectively, the DS and SBA approach introduced before.

Evidently, people do not continue to solve single-digit subtractions by means of counting. On the basis of experience and education, they increasingly solve these subtractions by means of strategies that are partly or completely based on the retrieval of known facts (Verschaffel et al., 2007). In an attempt to unravel adults' use of SBA in single-digit subtraction, Campbell (2008) compared adults' reaction times on small versus large single-digit subtractions (defined on the basis of the value of the minuend, i.e., $<$ or $=$ 10 versus $>10$) such as, respectively, $8-2=?$ versus $15-7=?$, in two different presentation formats, namely, the standard subtraction ($a-b=?$) and the addition ($b+?=a$) format. Campbell reasoned that if subtractions are solved by means of DS, they should be solved faster in the subtraction than in the addition format; if subtractions are solved by means of SBA, they should be solved faster in the addition than in the subtraction format. He found that large and nontie single-digit subtractions were solved faster in the addition than in the subtraction format and therefore concluded that these subtractions are often solved by retrieving the inverse addition fact. Although Campbell's interesting research paradigm allowed to demonstrate adults' use of SBA on single-digit subtractions as well as the impact of several numerical problem characteristics on adults' strategy choices, it was unable to reveal the role of the size of the subtrahend relative to the minuend on their strategy choices, as in Woods et al.'s (1975) study.

Combining the methodologies of the investigations of Woods et al. (1975) and Campbell (2008), Peters et al. (2010b, 2012) set up two similar studies—one with adults and one with third-to-sixth-grade children—about the use of SBA on large single-digit subtractions (i.e., minuend $>10$). They divided an extensive set of nontie large single-digit subtractions into four subtraction types, representing the relative size of the subtrahend through the combination of two characteristics: (1) whether the difference between the minuend and the subtrahend is small (as in $12-9=?$) or large (as in $12-3=?$) and (2) whether the numerical distance between the subtrahend and the difference is large (as in $12-3=?$ or $12-9=?$) or small (as in $12-7=?$ or $12-5=?$). Similar to the study of Woods et al. (1975), they fitted observed reaction times to the linear regression predicted by three different models: (1) always DS, (2) always SBA, and (3) flexible use of DS and SBA. As in Campbell (2008), they analyzed speed differences between the two formats for these four types of subtractions. This "double methodology" allowed them not only to find out if participants applied SBA on large single-digit subtractions, but also to test whether they solved these subtractions by flexibly switching between the DS and the SBA strategy depending on which of both strategies requires the fewest and smallest, or easiest, steps. The study with adult participants (Peters et al., 2010b) revealed that adults based their strategy choices on the relative size of the subtrahend: For both presentation formats ($a-b=?$ and $b+?=a$), they showed strategy use patterns that represent the use of DS when the subtrahend is relatively small and the use of SBA when the subtrahend is relatively large. However, a different result was found for the children (Peters et al., 2012): Their reaction time patterns suggested that they always used DS when subtractions are presented in the traditional subtraction format ($a-b=?$) and always applied SBA on subtractions presented in the addition format ($b+?=a$), independently of the relative size of the subtrahend. In other words, the children's strategy choices were completely determined by the operation cue (− or +) provided in the symbolic problem statement.

In sum, the available research on people's mental computation with symbolic single-digit subtractions suggests that second-grade and fourth-grade children use SBA to flexibly solve small single-digit subtractions by means of counting (Woods et al., 1975), that adults frequently solve large single-digit subtractions—but not small single-digit subtractions—by relying on the complementary addition facts (Campbell, 2008), and that adults' choice for SBA for large single-digit subtractions does not only happen frequently but also flexibly (Peters et al., 2010b). On the other hand, third-to-sixth-grade children did not show evidence of SBA on large single-digit subtractions, except when they were presented in the uncommon non-canonical $b+?=a$ format (Peters et al., 2012). So, the picture of solving single-digit subtractions by means of SBA seems to be mixed and complex. Apparently, the choice of SBA in this subdomain of arithmetic is affected by a variety of subject (e.g., age), task (e.g., small vs large single-digit subtractions), and context (e.g., participants' educational histories) factors.

## Multidigit mental subtraction

The curricular subdomain in which SBA has been investigated most intensively, is multidigit mental arithmetic (e.g., Barrouillet, Mignon, & Thevenot, 2008; Fuson, 1992; Robinson, 2001; Selter, 2001; Torbeyns et al., 2009a; Torbeyns, De Smedt, Ghesquière, & Verschaffel, 2009b). As for the subdomain of single-digit subtraction, many attempts have been made to make an inventory and categorization of the various ways in which such subtractions can be solved (e.g., Beishuizen, 1993; Blöte, Klein, & Beishuizen, 2000; Selter, 1998; Verschaffel et al., 2007). One categorization involves three main strategies based on the steps that are taken to actually execute the subtraction, distinguishing, for instance, between jumping, splitting, and varying strategies (Verschaffel et al., 2007). DS and SBA strategies can be further categorized on the basis of these three types. For instance, the subtraction $823-497=?$ can be solved via a jump (e.g., "$823-400=423$, $423-90=333$, and $333-7=326$"), a split (e.g., "$800-400=400, 20-90=-70, 3-7=-4$, so the answer is $400-70-4$ or 326"), or a varying version (e.g., $823-497=823-500+3=326$) of the DS strategy.

Our brief overview of the available research on SBA in multidigit mental arithmetic is split up into three parts, which relate to the three main methodologies being used, namely, (a) verbal strategy reports, (b) reaction-time experiments, and (c) choice/no-choice studies.

## Verbal strategy reports

In a first, and also the oldest, series of studies, researchers investigated elementary school children's strategy use by asking them, in the context of individual interviews, to mentally solve a set of multidigit subtractions and think aloud during the solution process or explain afterward how they solved the problem (verbal strategy report during or after solving the subtraction). For instance, Torbeyns et al. (2009a) investigated second- to fourth graders' DS and SBA use on symbolic two-digit subtractions in the number domain up to 100. All children had received traditional instruction in mental subtraction that focused on the mastery of the DS jump strategy (e.g., $823-467=823-400-60-7=423-60-7=363-7=356$). They all completed two tasks: a spontaneous strategy use task (SST) and a variability on demand task (VDT). The SST consisted of different types of two-digit subtractions, including small-difference subtractions (e.g., $41-39=?$) that are assumed to maximally trigger SBA. Children were instructed to mentally solve each subtraction as accurately and as fast as possible with their preferential strategy. The VDT consisted of the same types of two-digit subtractions as the SST, but now children were instructed to mentally solve each subtraction with as many different strategies as possible. The study resulted in three main findings. First, the analysis of children's strategies in the SST revealed that less than 10% of the second- and third graders and only 15% of the fourth graders spontaneously reported the SBA strategy at least once to answer the small-difference subtractions. Second, all children reported various strategies for solving the small-difference subtractions from the VDT, but only a minority of them reported SBA as a possible alternative strategy. Actually, only 5% of the second graders, 15% of the third graders, and 20% of the fourth graders mentioned the SBA strategy at least once as a possible alternative solution strategy. Third, despite the low overall occurrence of SBA in the three age groups, older children tended to report the SBA strategy more frequently than younger children. These results about the scarce self-reported SBA use among traditionally taught elementary school children were confirmed in several other studies using similar verbal report methods, both from our own research team (De Smedt, Torbeyns, Stassens, Ghesquière, & Verschaffel, 2010; Torbeyns et al., 2009b) and from various other teams (Blöte, Van der Burg, & Klein, 2001; Heinze, Marschick, & Lipowsky, 2009; Selter, 2001; Selter et al., 2012).

## Reaction-time experiments

In a second line of research, Peters, De Smedt, Torbeyns, Ghesquière, and Verschaffel (2010a, 2013) extended their reaction-time studies with adults and fourth-to-sixth-grade children in the number domain up to 20 to the number domain up to 100. As in their previous studies, they combined the methods of (a) presentation format manipulation ($a-b=?$ vs $b+?=a$) and (b) fitting linear regression models to people's reaction times. As in their studies with single-digit subtractions, they distinguished among four subtraction types: (a) subtractions with a small subtrahend and a large distance between subtrahend and difference (e.g., $83-4=?$), (b) subtractions with a large subtrahend and a large distance between subtrahend and difference (e.g., $77-68=?$), (c) subtractions with a small subtrahend and a small distance between subtrahend and difference (e.g., $92-44=?$), and (d) subtractions with a large subtrahend and a small distance between subtrahend and difference (e.g., $82-43=?$). Multilevel regression analyses showed that for both adults (Peters et al., 2010a) and children (Peters et al., 2013), the model in which the solver switched between DS and SBA based on the relative size of the subtrahend provided the best fit to the empirical reaction-time data. So, in contrast with the findings from the first series of studies in the domain of multidigit subtraction in which verbal reports were used to gather information about children's SBA strategy use, the results of these reaction-time studies led to the conclusions (1) that SBA belonged to the strategy repertoire of Flemish adults as well as upper elementary school children, and (2) that both groups flexibly applied the SBA strategy to solve two-digit subtractions by switching between the SBA and the DS strategy depending on the relative size of the subtrahend. These findings were replicated in other studies at our center, using different sets of subtractions (Peters et al., 2012) and different samples of children, including children with mathematical learning disabilities (Peters, De Smedt, Torbeyns, Verschaffel, & Ghesquière, 2014).

## Choice/no-choice studies

In a third and final line of research, Torbeyns and colleagues (Torbeyns, De Smedt, Peters, Ghesquière, & Verschaffel, 2011; Torbeyns, Ghesquière, & Verschaffel, 2009; Torbeyns, Peters, De Smedt, Ghesquière, & Verschaffel, 2018; Torbeyns, Verguts, De Smedt, & Verschaffel, 2018) applied Siegler and Lemaire's (1997); see also Lemaire & Siegler, 1995) choice/no-choice method to investigate adults' and upper elementary school children's SBA use on multidigit subtractions. In a first study with adults (Torbeyns et al., 2009), participants were offered three parallel series of multidigit subtractions up to 1000, which were divided into three subtraction types: (a) subtractions with a large subtrahend (e.g., 614 – 576 = ?), (b) subtractions with a medium subtrahend (821 – 475 = ?), and (c) subtractions with a small subtrahend (e.g., 734 – 47 = 687). These subtractions were individually administered in one choice and two no-choice conditions. In the choice condition, participants could choose between DS and SBA on each subtraction. In the no-choice DS condition, they had to solve the subtractions via a DS strategy. Likewise, in the no-choice SBA condition, all subtractions had to be solved via SBA. In each condition, participants were instructed to solve all subtractions as accurately and as fast as possible and to verbally report the applied strategy immediately after solving each subtraction. This first choice/no-choice study with adults provided convincing evidence for their remarkably frequent, efficient, and flexible use of SBA on multidigit subtractions (Torbeyns et al., 2009). First, the data from the choice condition revealed that overall SBA was spontaneously applied almost as frequently as DS. Second, the accuracy and speed data from the no-choice conditions demonstrated that both strategies were generally equally accurate and that SBA resulted in significantly shorter response times than DS. Finally, adults were found to flexibly apply SBA on multidigit subtractions in two different ways: (1) they took into account the numerical characteristics of the subtractions during the choice process, given that they more frequently selected SBA on subtractions with large subtrahends compared to subtractions with small subtrahends; (2) they fitted their strategy choices to their individual strategy speed competencies: The larger the speed difference between the no-choice SBA and the no-choice DS condition in favor of the SBA strategy, the more frequently they reported that latter strategy in the choice condition. Later studies with adults, involving an item set consisting of more extreme large (812 – 793 = ?) or small (713 – 18 = ?) subtrahends, yielded results that were even more in favor of the computational superiority of SBA compared to DS (Torbeyns et al., 2011).

A first subsequent study of these researchers following an identical design but involving sixth-grade children revealed (1) that most children also verbally reported SBA use in the choice condition; (2) that, in the two no-choice conditions, SBA was more efficient (in terms of accuracy and speed) than the DS strategy; and (3) that children flexibly applied SBA in the choice condition, taking into account both the numerical characteristics of the subtractions and their individual strategy efficiency (Torbeyns, Peters, et al., 2018). A recent replication study, with exactly the same design but including also younger elementary school children from Grades 4 and 5, showed that the above results about the frequent, efficient, and flexible use of the SBA strategy on symbolic multidigit subtractions among sixth graders also apply for Flemish children as young as Grade 4 (Torbeyns, Verguts, et al., 2018). Thus completely in line with our adult studies, the majority of upper elementary school children spontaneously and frequently applied the SBA strategy and executed it highly efficiently and flexibly, despite the exclusive focus on (the jump variant of) DS in their elementary mathematics classes.

## Conclusion

Taken together, the initial studies that relied on verbal reports and that were collected in conditions that can be described as choice conditions, suggested little, or no SBA use in elementary school children. The results of reaction-time and choice/no-choice studies were not in line with these verbal protocol data and indicated that not only adults but also elementary school children as young as 9–10 years old apply the untaught SBA strategy and do so remarkably frequently, efficiently, and flexibly. Reflecting upon these contrasting findings obtained by means of these distinct methodologies, Peters et al. (2010a, 2010b) point to the questionable validity of children's verbal reports. In their view, children may not always report the actual use of SBA, either because they do not find the appropriate words to articulate how they actually arrived at their answer when applying SBA or they may even deliberately hide that they actually used SBA because they think that the interviewer may consider it an inferior or even an unacceptable solution strategy. However, the reaction-time and choice/no-choice methodology are not without their problems either. For instance, the reaction-time analysis used in the abovementioned studies only allows for the assessment of a general strategy pattern for a large-item set and not for the identification of participants' strategy use on a singular item (Peters et al., 2010a, 2010b), and one important methodological problem of the choice/no-choice paradigm is that, in view of keeping the method practically feasible, the experimenter has to restrict the number of solution strategies to be allowed in the choice condition and to be used in the different no-choice conditions (Luwel, Torbeyns, Schillemans, Onghena, & Verschaffel, 2009).

Therefore it is highly recommendable to investigate SBA by combining various verbal and nonverbal research methods.

## Written multidigit subtraction

Although much less known, there is also literature on how SBA-based approaches can, and do, play a role in multidigit subtraction by means of written computational algorithms rather than through mental arithmetic. Historically, the algorithms for written computation of the four basic arithmetic operations have been taught and applied over many centuries for fast and accurate calculations with large numbers that are hard or impossible via retrieval or mental arithmetic. Even though their prominent place in the elementary mathematics curriculum has been questioned increasingly for various reasons (Verschaffel et al., 2007), they still constitute an important part of the elementary school curriculum in most countries today. When being applied properly by someone who masters and follows the procedure well, algorithms for written computation guarantee quick and effortless success for a large class of difficult problems. On the other hand, partly as a result of the way in which they typically are being introduced and practiced in schools, the computational algorithms have lost their connection with their conceptual underpinnings, resulting in many users' inability to explain how and why they work (Selter et al., 2012; Treffers, 1987; Verschaffel & De Corte, 1996).

Although elementary school curricula worldwide aim at the fluent mastery of only one standard algorithm for each arithmetic operation, there exist several algorithms for each operation (Selter et al., 2012; Verschaffel et al., 2007). Consequently, people from different cultures learn and apply different algorithms. The algorithms for written subtraction that are currently mostly taught and used worldwide are based on DS, implying that in each column of the subtraction, the solver has to directly subtract the lower digit from the upper digit (if necessary, after some kind of borrowing process). Fig. 1A and B represents two examples of algorithms that are fundamentally DS based.

There also exists an alternative type of subtraction algorithms—sometimes referred to as "Austrian subtraction" (Verschaffel et al., 2007)—which is essentially SBA based, in the sense that the steps included in the algorithm do not involve the execution of the subtraction operation. Rather, this algorithm involves, in each column, an additive step whereby the solver finds out how much needs to be added to a digit from the subtrahend to get at the

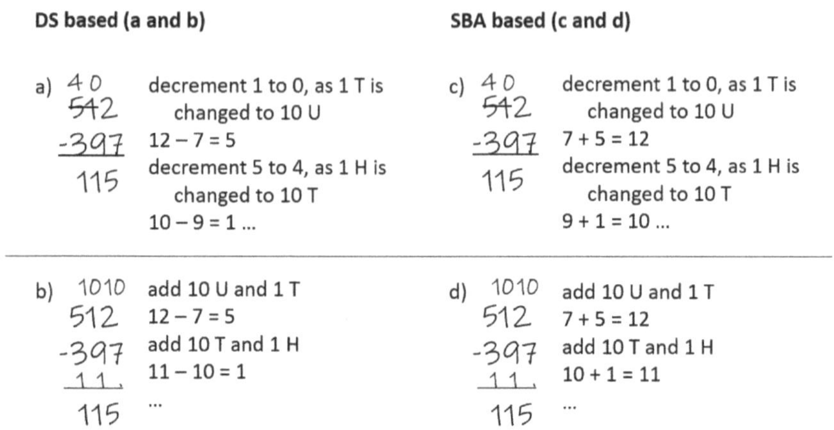


**Fig. 1** Examples of two DS-based (A and B) and two SBA-based (C and D) versions of the subtraction algorithm (based on Selter et al., 2012, p. 396).

corresponding digit from the minuend (if necessary, after some kind of carrying process). As explained in detail by Selter et al. (2012), there exist different variants of this SBA-based subtraction algorithm, depending on how the process of carrying takes place. Fig. 1C and D shows two examples of such SBA-based algorithms.

The discussion on which of these different subtraction algorithms should be favored dates back already a long time ago (e.g., Knight, Ruch, & Lutes, 1925). In some countries, such as Austria and Germany, this discussion has continued to attract the attention of mathematics educators, as evidenced by empirical studies by Mosel-Göbel (1988) and Fiori and Zuccheri (2005) and by a recent theoretical paper by Wittmann (2010). In Bavaria, for instance, the ongoing debate led to a change in the official elementary school curriculum from an SBA-based to a DS-based variant of the written subtraction algorithm in 2000 (A. Strohmaier, personal communication, 13-11-2018).

The scarce available empirical evidence does not allow to formulate convincing empirical arguments in favor of a particular subtraction algorithm. The abovementioned empirical investigations of Mosel-Göbel (1988) and Fiori and Zuccheri (2005) compared the procedural fluency and the conceptual understanding of the subtraction algorithm among groups of elementary school children who were taught different algorithmic methods, including an SBA-based one. Unfortunately, several methodological limitations of these studies make it impossible to draw valid conclusions about the educational value of the abovementioned subtraction algorithms. Summing up the available evidence, Selter et al. (2012, p. 397) concluded that "neither theoretical considerations nor results from empirical research give a clear picture which algorithm should be favored." However, these reflections and findings convincingly show that one of the SBA-based algorithms, namely, the "turning the counter" version (see Fig. 1B), is at least a coequal method to the traditionally taught one(s) based on DS.

## Word problem solving

Not only in the domain of symbolic arithmetic, but also in the domain of arithmetic word problem solving, researchers have studied the variety in and the efficiency and flexibility of the strategies people use to solve subtraction problems. Also in this line of research, SBA played a pivotal role. In the 1980s several research teams (e.g., Carpenter & Moser, 1984 ; De Corte & Verschaffel, 1987) set up longitudinal studies in which they interviewed a group of starting elementary school children several times during the school year. In each interview, children were confronted with a set of one-step addition and subtraction word problems with single-digit numbers, belonging to various semantic categories, including subtraction problems such as the following:

- Joe had eight marbles. He gave five marbles to Marie. How many marbles does Joe have now?
- Joe had three marbles. He got some marbles from Marie. Now Joe has eight marbles. How many marbles did Joe get from Marie?

Both problems are solvable by means of the subtraction 8 – 3 = ? and belong to the semantic category of change problems, as they involve a dynamic event involving three sets: a start, a change, and a result set (Carpenter & Moser, 1984; De Corte & Verschaffel, 1987; Riley, Greeno, & Heller, 1983; Verschaffel et al., 2007). The two problems also differ from each other in two important ways: (1) the change in the first problem involves a decrease, whereas the second problem describes an increase; (2) in the first problem, the result set is unknown, while in the second one, the change set is unknown. In terms of the well-known classification schema of Riley et al. (1983), these two problems are, respectively, called a Change 2 and a Change 3 problem.

The analysis of children's solution strategies, as evidenced by careful observations of their solution behavior and/or analyses of their verbal reports, revealed the following major findings. First, beginning elementary school children apply a wide variety of solution strategies, ranging from material counting strategies (involving blocks or fingers) and verbal counting strategies (based on upward- or downward-counting sequences) to mental solution strategies (based on known facts). Second, at each of these three levels of internalization (material, verbal, and mental), children use various strategies, which tend to reflect the semantic structure of the problem. For instance, the first abovementioned subtraction problem (Change 2) is mostly solved by a material, verbal, or mental variant of the DS strategy, whereas the majority of the children approach the second problem (Change 3) by means of a material, verbal, or mental variant of the SBA strategy. Third, even though children continue to prefer strategies that simulate the semantic problem structure throughout their development, older children tend to become more inclined to solve problems with a solution strategy that not necessarily fits the problem semantics, because the nature of the numbers in the problem asks for another type of strategy (Carpenter & Moser, 1984; De Corte & Verschaffel, 1987). Take, for instance, the problem "Joe had 91 marbles. He gave 89 marbles to Marie. How many marbles does Joe have now?," the semantics of which suggest a DS-based strategy, whereas from a computational point of view an SBA-based strategy seems more efficient.

Later on, other researchers have provided additional empirical findings that provide further evidence for the frequent use of SBA on subtraction word problems, especially on problem types with semantic features that nicely fit the additive nature of SBA (e.g., Brissiaud & Sander, 2010; Gvozdic & Sander, 2017; Peltenburg, van den Heuvel-Panhuizen, & Robitzsch, 2012; Van Lieshout, 1997).

## Conceptual underpinnings of subtraction by addition

So far, this review was completely oriented to the procedural side of SBA use. However, people's strategy use is linked to their understanding of mathematical principles (Baroody et al., 2009; Rittle-Johnson, Siegler, & Alibali, 2001; Verschaffel et al., 2012). These conceptual underpinnings of SBA have also been the object of a number of theoretical and empirical studies. Generally speaking, two different lines of research on these conceptual underpinnings can be distinguished: (1) studies on the subtraction-addition complement principle (i.e., if $a-b=c$, then $c+b=a$) and (2) studies on the different meanings of subtraction.

Only few studies have focused on children's understanding of the addition/subtraction complement principle in relation to the SBA strategy (Gilmore & Bryant, 2006; Rasmussen, Ho, & Bisanz, 2003; Torbeyns, Peters,

De Smedt, Ghesquière, & Verschaffel, 2016; Van den Heuvel-Panhuizen & Treffers, 2009). In an attempt to deepen our understanding of this relationship, Torbeyns et al. (2016) asked a group of third and fourth graders to solve two tasks in which verbal reports as well as accuracy and speed data were collected: a looking-back task and a baseline task. These two tasks differed only in the order of the subtractions and the instructions. In the looking-back task, subtractions were sometimes preceded by their complementary addition format (i.e., target subtractions preceded by their trigger, e.g., 48 + 45 = 93 presented on top of the screen when confronted with the subtraction 93 – 48 = ?) and children were told that they had to answer each item as accurately and fast as possible, and that sometimes the preceding item could help to answer the next one. In the baseline task, no helpful preceding problems were offered (i.e., no trigger-target subtraction pairs) and the instruction did only refer to the accuracy and speed of problem solving. The looking-back task included 10 such trigger-target problem pairs on the addition/subtraction complement principle. Children verbally reported looking back on about 40% of all target items in the looking-back task. The target items were also solved more accurately and faster than the same subtractions in the baseline task. While this study indicated that many 9–10-year-old children have an understanding of the complement principle, its design did not allow to measure the strength of relation between children's mastery of this principle, on the one hand, and their SBA strategy performance, on the other hand.

In addition to these studies focusing on the subtraction-addition complement principle, a second line of research addresses the conceptualization of the arithmetic operation of subtraction via both math-didactical and empirical analyses. The starting point of these analyses is the distinction between two models of subtraction that are put forward by math educators such as Freudenthal (1973) and Usiskin (2008) and that reflect a different intuitive conception of subtraction, namely, (1) the take-away model and (2) the comparison (or find-the-difference) model. The different conceptualizations of the subtraction operation on these two models lead to two different terms for the answer to the subtraction problem, respectively, the remainder (take-away) and the difference (comparison). In the take-away conception of subtraction, people think about an action, whereby a (smaller) subset is taken away from a (larger) set, and determine how much is left (= the remainder). The type of solution strategy that fits best with this conception of subtraction is DS. Alternatively, when subtraction is conceived as comparing or determining the difference between two sets, one thinks about how much is needed to reach from one quantity A to another quantity B (or, reversibly, from B to A). The answer to such a kind of subtraction is referred to as the difference. This conception of subtraction aligns much better to solutions based on an SBA approach (Gvozdic, 2020; Selter et al., 2012).

Both in Fischbein's (1987) theory of implicit models of arithmetic operations and in Lakoff and Nunez's (2000) conceptual metaphor approach, taking away is considered to be the most prominent, natural, and widespread intuitive conception of subtraction, because people encounter it frequently in their daily lives (Gvozdic, 2020; Selter et al., 2012). This prominence of the take-away conception is typically further enhanced by a strong and even exclusive focus on this model in preschool and elementary school mathematics education, leading to a very strong prompt of this conception—and thus to the accompanying DS-based solution strategies—when confronted with a minus sign in formal symbolically presented subtractions.

While the take-away model appears to be more natural and widespread than the determine-the-difference model, "a deeper understanding of mathematics requires dealing with aspects that appear to be of a theoretical nature, if they are being compared with a perceived reality. Thus, it is important to interpret mathematical objects and operations in different ways, depending on the requirements of the problem" (Selter et al., 2012, p. 390). Selter et al. (2012) made a plea for a much more prominent role of the alternative determine-the-difference model in teaching and learning subtraction from Grade 1, arguing that (a) seeing subtraction solely as taking away is too one-sided, (b) all mental strategies and written algorithms for subtraction can also be applied in the determine-the-difference model, and (c) the determine-the-difference model can be extended much more easily to subtracting negative numbers and algebraic manipulations. We are not aware of any empirical study that has explicitly linked people's understanding of these two meanings of subtraction to the frequency, efficiency, and flexibility with which they apply the SBA strategy in symbolic subtraction.

## Conclusion and discussion

In this chapter, we reviewed a long-standing and comprehensive line of research on the use of SBA when solving subtractions, in which our own research center has played an important role. We have documented this phenomenon in a variety of subdomains of the elementary arithmetic curriculum, ranging from single-digit mental subtraction, multidigit mental and written subtraction, to word problems in the domain of subtraction. We have also reviewed the available research that tries to link these findings about people's use of SBA-based strategies in these different curricular subdomains to their conceptual knowledge of the complementary relationship between addition and subtraction and to their intuitive conceptualization of the arithmetic operation of subtraction as "take-away" versus "determine-the-difference." In doing so, we have presented and discussed evidence coming from different research fields (cognitive and developmental psychology, mathematics education), a wide range of research methodologies (verbal report studies, reaction-time studies, studies applying the choice/no-choice paradigm, and theoretical reflections) and a large variety of participants (ranging from beginning elementary school children to adults). Taken as a whole, the reviewed research has yielded recurrent and convincing evidence for the frequent, efficient, and flexible use of SBA in the above curricular subdomains among both elementary school children and adults. Given that this strategy gets little or no instructional attention and that DS gets pivotal, if not exclusive, instructional attention, these are striking results. Hereafter, we reflect on some theoretical, methodological, and educational implications of these findings.

## Theoretical implications

The finding that SBA emerges in learners' strategy *repertoire*, in the absence of its explicit and systematic instruction, points to people's capacities to invent arithmetic strategies such as SBA. This fits with Siegler's (Shrager & Siegler, 1998; Siegler & Araya, 2005) general model of strategy change, which describes in detail how learners may discover and develop new strategies based on the increasingly fluent mastery of a strategy available in one's repertoire allowing them to devote the released attentional resources at looking for shortcuts and the exploration of alternative strategies (Shrager & Siegler, 1998; Siegler & Araya, 2005). While this model accounts well for the role of intensive practice of the subtraction strategy at hand (i.e., DS) in the discovery of a new one (i.e., SBA), it does not tell the full story. First, based on cognitively inspired instructional design theories (Verschaffel, Van Dooren, & Star, 2017), it is reasonable to assume that the precise nature of the subtraction items with which learners are confronted (e.g., the number of small-difference subtractions in the problem set, the extremity of the numbers used in these items, the number of times borrowing is needed, whether or not the way in which the different item types are alternated) will affect the speed and ease of learners' discovery of SBA. Second, general cognitive factors such as learners' executive functions (i.e., working memory, inhibition, flexibility) may also bear an effect on how quickly and smoothly the SBA strategy may become part of their strategy repertoire. While the role of these domain-general cognitive factors has already been documented amply for other arithmetic strategies (e.g., Imbo, Vandierendonck, & Vergauwe, 2017); to the best of our knowledge, their impact on the emergence of SBA has not been investigated yet. Third, there is the role of domain-specific cognitive factors, particularly learners' conceptual knowledge of subtraction. While researchers have yielded deep insight into the reciprocal relation between procedural and conceptual knowledge in the domain of arithmetic (e.g., between children's understanding of the commutativity principle and their use of larger-number-first addition strategies, e.g., Baroody & Gannon, 1984), empirical research into the development and role of the two subtraction-related principles discussed in the present review, namely, the addition-subtraction complement principle and the take-the-difference conception of subtraction, is almost nonexistent. Another potentially relevant aspect of learners' domain-specific content knowledge, which has received little or no attention of researchers so far, is their mastery of SBA in another subdomain of the mathematics curriculum (e.g., single-digit symbolic subtraction or subtraction word problems) and how their SBA-related knowledge in that subdomain may transfer to a new subdomain (e.g., multidigit symbolic mental subtraction). Finally, from a mathematics educational perspective, there is little doubt that the emergence of the SBA strategy will also—and maybe even primarily—be affected by learners' instructional environment, such as the intensity and quality of the explicit instruction in the SBA strategy or the classroom culture's implicit call for adaptive rather than routine expertise (Verschaffel et al., 2009). However, except for a few studies (e.g., De Smedt et al., 2010), relatively little is known about the role of these instructional factors on learners' use of SBA.

As far as the strategy parameter of *efficiency* is concerned, our review revealed the remarkable accuracy and speed of the SBA strategy as compared to DS, even when instructional attention goes exclusively to the latter one. This remarkable efficiency has been documented both in the subdomain of single-digit and multidigit mental symbolic subtraction, and both in adults and children and in high and low ability arithmeticians. Most remarkably, this efficiency was sometimes found to hold even for large-difference subtractions for which, from a rational task analytic perspective, DS seems more advantageous. Again, except for the role of the numerical features that have been systematically investigated in the studies reviewed, there are still many unclarities about the role of other task, subject, contextual, and educational factors on the efficiency of SBA-based strategies.

Third, when being confronted with single-digit and multidigit symbolic subtractions they have to solve mentally, people of different ages and achievement levels *flexibly* fit their SBA versus DS strategy to both item and individual strategy efficiency characteristics. So, not only do they select SBA most frequently on the subtractions for which this strategy is assumed most efficient, i.e., subtractions with a small difference between the minuend and subtrahend, learners also take into account their own individual strategy efficiency competencies when choosing between DS and SBA. This flexibility in strategy use has also been documented in a wide variety of cognitive tasks, even among young children (Verschaffel et al., 2009). However, as for the other strategy parameters, there is a need for additional research exploring the impact of task, subject, contextual, and educational factors on the flexibility with which people switch between SBA and DS strategies and investigating the impact of instruction on their strategy flexibility.

A final issue for future theoretical reflection and empirical research is *why* the untaught SBA strategy is so attractive and efficient among learners. Our best explanation at the moment is that the process of adding a number to a given number is inherently easier than the process of taking away a similar number from a given number. Although it is difficult to provide direct evidence for this claim, there are several elements from various sources in support of it. First, from a mathematical perspective, addition is (more) fundamental for the development of the natural numbers, the most basic number system (Kilpatrick et al., 2001). Natural numbers are constructed on the basis of the principle of recurrence, which means that every natural number is constructed on the basis of the previous number as the successor of $n$ (Ifrah, 1985). In other words, the system of natural numbers is created and defined in an additive order, almost automatically leading to the priority of addition to subtraction. Second, there is ample evidence in the psychological literature on early mathematics for young children's earlier mastery of and better performance on counting up compared to counting down, which can be considered as important precursors of the processes of adding and subtracting, respectively (Fuson, 1992; Verschaffel et al., 2007). Third, the logical and psychological primacy of addition is also reflected in children's curricula, textbooks, and teacher practices for elementary arithmetic. Worldwide, mathematics teachers first teach their pupils how to add numbers and only afterward start teaching subtraction (Kilpatrick et al., 2001). The earlier and more intensive practice of the addition operation in the classroom may further contribute to addition being more familiar and easier than subtraction. Together, all these elements may help to explain why learners like to solve subtractions by means of addition and why they are so good at it.

## Methodological implications

While the present overview already clearly highlights the need for using a multimethod approach when studying SBA combining and triangulating different research methods and techniques (e.g., reaction times, verbal protocols, choice/no-choice method), both between and within studies, the remaining research questions about the acquisition and development of the SBA strategy and its conceptual underpinnings as well as the contribution of learner characteristics and education on this acquisition and development ask for further broadening of the methodological scope.

It would be very interesting to investigate the development of SBA in the context of (a) microgenetic studies allowing a more scrutinized and fine-grained analysis of the emergence and initial growth of SBA; (b) longitudinal studies that trace the development of the SBA strategy in a large group of learners, in relation to various domain-general and domain-specific subject characteristics; and (c) intervention studies analyzing the impact of well-designed educational materials and approaches on the development of SBA strategy.

Furthermore, even though some valuable initial attempts have been made to measure the conceptual side of SBA use, there is a great need for reliable and valid tools for measuring knowledge of relevant mathematical concepts and principles (i.e., the addition-subtraction complementarity principle, the different meanings of subtraction), already from a young age on.

## Educational implications

From an educational perspective, the present study challenges the exclusive focus on DS in various subdomains of current elementary mathematics instruction worldwide. Our findings pointing to the competence of people of various ages and various mathematical achievement levels to self-invent SBA, and to efficiently and flexibly apply it on various kinds of subtraction tasks, question this extreme focus on DS in current mathematics education.

Even though the untaught SBA strategy's remarkable efficiency may be a first reason for giving it a more prominent place in the elementary mathematics curriculum, an even more important reason is, in our view, the role it may play in offering children the chance to build up a richer conception of subtraction (not only the take-away but also the find-the-difference conception), in sowing the seeds for a very powerful and pervasive mathematical idea of inversion, and in developing "a mathematical mind" in learners more broadly.

Besides the "why" question, mathematics educators also need to answer three other questions related to the teaching of SBA, namely, when, for whom, and how to teach it.

Concerning the "when" question, the available research evidence suggests that SBA can be introduced already early in children's mathematical development, namely, as a "clever" short-cut strategy for doing small single-digit subtractions by relying on known facts or counting. This research has revealed two task features that are particularly suitable for eliciting SBA in children: first, the numerical features of the subtraction (i.e., a small difference between the minuend and subtrahend, as in $9 - 8 = ?$), and, second, the semantic features of the situation in which the subtraction is embedded (i.e., a word problem describing an increase with a known start and result set, as in "In the morning Pete had 3 marbles. In the evening he had 8. How much did he won during the day?").

With respect to the "for whom" question, there is a debate among mathematics educators on whether teaching SBA as an alternative strategy for doing subtraction besides DS is a feasible and valuable goal for all children, including the mathematically weaker ones. Many practitioners argue that confronting these children with both strategies and teaching them to make flexible choices between them will do more harm than good to children with mathematical learning difficulties. The abovementioned research evidence that even children with mathematical learning difficulties are able to apply SBA efficiently and flexibly in multidigit subtraction (Peltenburg et al., 2012; Peters et al., 2014) supports the plea for not restricting the teaching of SBA to the mathematically stronger learners.

This brings us to the last question: how to design effective SBA instruction? Although we are aware of a few intervention studies that yield building blocks for such an instruction (De Smedt et al., 2010; Fuson & Willis, 1988), many questions about the design of this instruction remain, including the following ones: Is it better to introduce the SBA strategy only after children have developed good routine mastery in the corresponding DS strategy, or is it more efficient to introduce and practice both strategies simultaneously? When teaching for strategy flexibility, is it better to provide a clear-cut rule about when it is more efficient to apply DS and when SBA is more appropriate via direct instruction, or is it better to stimulate children to develop their own personal "meta-strategic knowledge" about when to use each strategy through more open guided-discovery or problem-based learning environments? And, finally, given the relevance of SBA in various subdomains of the elementary mathematics curriculum, how can instruction maximize the chance that learners' knowledge and skills in one subdomain will transfer to another one?

## References


Ashcraft, M. H. (1982). The development of mental arithmetic: A chronometric approach. *Developmental Review, 2*, 213–236.

Baroody, A. J., & Dowker, A. (Eds.). (2003). *The development of arithmetic concepts and skills: Constructing adaptive expertise* Lawrence Erlbaum Associates.

Baroody, A. J., & Gannon, K. E. (1984). The development of commutativity principle and economical addition strategies. *Cognition and Instruction, 1*, 321–329.

Baroody, A. J., Torbeyns, J., & Verschaffel, L. (2009). Young children's understanding and application of subtraction-related principles: Introduction [special issue]. *Mathematical Thinking and Learning, 11*, 2–9.

Barrouillet, P., Mignon, M., & Thevenot, C. (2008). Strategies in subtraction problem solving in children. *Journal of Experimental Child Psychology, 99*, 233–251.

Beishuizen, M. (1993). Mental strategies and materials or models for addition and subtraction up to 100 in Dutch second grades. *Journal for Research in Mathematics Education, 24*, 294–323.

Blöte, A. W., Klein, A. S., & Beishuizen, M. (2000). Mental computation and conceptual understanding. *Learning and Instruction, 10*, 221–247.

Blöte, A. W., Van der Burg, E., & Klein, A. S. (2001). Student's flexibility in solving two-digit addition and subtraction problems: Instruction effects. *Journal of Educational Psychology, 93*, 627–638.

Brissiaud, R., & Sander, E. (2010). Arithmetic word problem solving. A situation strategy first framework. *Developmental Science, 13*, 92–107.

Campbell, J. I. D. (2008). Subtraction by addition. *Memory and Cognition, 36*, 1094–1102.

Campbell, J. I. D., & Xue, Q. (2001). Cognitive arithmetic across cultures. *Journal of Experimental Psychology: General, 130*, 299–315.

Carpenter, T. P., & Moser, J. M. (1984). The acquisition of addition and subtraction concepts in grades one through three. *Journal for Research in Mathematics Education, 15*, 179–202.

De Corte, E., & Verschaffel, L. (1987). The effect of semantic structure on first graders' strategies for solving addition and subtraction word problems. *Journal for Research in Mathematics Education, 18*, 363–381.

De Smedt, B., Torbeyns, J., Stassens, N., Ghesquière, P., & Verschaffel, L. (2010). Frequency, efficiency and flexibility of indirect addition in two learning environments. *Learning and Instruction, 20*, 205–215.

Fiori, C., & Zuccheri, L. (2005). An experimental research on error patterns in written subtraction. *Educational Studies in Mathematics, 60*, 323–331.

Fischbein, E. (1987). *Intuition in science and mathematics. An educational approach*. Reidel.

Freudenthal, H. (1973). *Mathematics as an educational task*. Reidel.

Fuson, K. C. (1992). Research on whole number addition and subtraction. In D. A. Grouws (Ed.), *Handbook of research on mathematical teaching and learning* (pp. 243–275). MacMillan.

Fuson, K. C., & Willis, G. B. (1988). Subtracting by counting up: More evidence. *Journal for Research in Mathematics Education, 19*, 402–420.
Geary, D. C., Frensch, P. A., & Wiley, J. G. (1993). Simple and complex mental subtraction: Strategy choice and speed-of-processing differences in younger and older adults. *Psychology and Aging, 8*, 242–256.
Gilmore, C., & Bryant, P. (2006). Individual differences in children's understanding of inversion and arithmetical skill. *British Journal of Educational Psychology, 76*, 309–331.
Greer, B. (2012). Inversion in mathematical thinking and learning. *Educational Studies in Mathematics, 79*, 429–438.
Gvozdic, K. (2020). *How informal knowledge informs us about the teaching and learning of mathematics* (Unpublished doctoral dissertation). Université de Paris 8 Vincennes-Saint-Denis.
Gvozdic, K., & Sander, E. (2017). Solving additive word problems: Intuitive strategies make a difference. In B. C. Love, K. McRae, & V. M. Sloutsky (Eds.), *Proceedings of the 39th annual conference of the Cognitive Science Society*. London, UK: Cognitive Science Society.
Heinze, A., Marschick, F., & Lipowsky, F. (2009). Addition and subtraction of three-digit numbers: Adaptive strategy use and the influence of instruction in German third grade. *ZDM Mathematics Education, 41*, 591–604.
Ifrah, G. (1985). *From one to zero. A universal history of numbers.* Viking Penguin.
Imbo, I., Vandierendonck, A., & Vergauwe, E. (2017). The role of working memory in carrying and borrowing. *Psychological Research-Psychologische Forschung, 71*, 467–483.
Kilpatrick, J., Swafford, J., & Findell, B. (Eds.). (2001). *Adding it up. Helping children learn mathematics* National Academy Press.
Kirk, E. P., & Ashcraft, M. H. (2001). Telling stories: The perils and promise of using verbal reports to study math strategies. *Journal of Experimental Psychology: Learning, Memory, and Cognition, 27*, 157–175.
Knight, F. B., Ruch, G. M., & Lutes, O. S. (1925). How shall subtraction be taught? *Journal of Educational Research, 11*, 157–168.
Lakoff, G., & Nunez, R. E. (2000). *Where mathematics come from: How the embodied mind brings mathematics into being.* Basic Books.
LeFevre, J.-A., DeStefano, D., Penner-Wilger, M., & Daley, K. E. (2006). Selection of procedures in mental subtraction. *Canadian Journal of Experimental Psychology, 60*, 209–220.
Lemaire, P., & Siegler, R. S. (1995). Four aspects of strategic change: Contributions to children's learning of multiplication. *Journal of Experimental Psychology: General, 124*, 83–97.
Luwel, K., Torbeyns, J., Schillemans, V., Onghena, P., & Verschaffel, L. (2009). Strengths and weaknesses of the choice/no-choice method in research on strategy use. *European Psychologist, 14*, 351–362.
Mauro, D. G., LeFevre, J.-A., & Morris, J. (2003). Effects of problem format on division and multiplication performance: Evidence for mediation of division by multiplication. *Journal of Experimental Psychology: Learning, Memory, & Cognition, 29*, 163–170.
Mosel-Göbel, D. (1988). Algorithmusverständnis am Beispiel ausgewählter Verfahren der schriftlichen Subtraktion [Understanding algorithms using the example of selected methods of written subtraction]. *Sachunterricht und Mathematik in der Primarstufe, 12*, 554–559.
Peltenburg, M., van den Heuvel-Panhuizen, M., & Robitzsch, A. (2012). Special education students' use of indirect addition in solving subtraction problems up to 100. A proof of the didactical potential of an ignored procedure. *Educational Studies in Mathematics, 79*, 351–369.
Peters, G., De Smedt, B., Torbeyns, J., Ghesquière, P., & Verschaffel, L. (2010a). Adults' use of subtraction by addition. *Acta Psychologica, 133*, 323–329.
Peters, G., De Smedt, B., Torbeyns, J., Ghesquière, P., & Verschaffel, L. (2010b). Using addition to solve subtractions in the number domain up to 20. *Acta Psychologica, 133*, 163–169.

Peters, G., De Smedt, B., Torbeyns, J., Ghesquière, P., & Verschaffel, L. (2012). Children's use of subtraction by addition on large single-digit subtractions. *Educational Studies in Mathematics, 79*, 335–349.

Peters, G., De Smedt, B., Torbeyns, J., Ghesquière, P., & Verschaffel, L. (2013). Children's use of addition to solve two-digit subtraction problems. *British Journal of Psychology, 104*, 495–511.

Peters, G., De Smedt, B., Torbeyns, J., Verschaffel, L., & Ghesquière, P. (2014). Subtraction by addition in children with mathematical learning disabilities. *Learning and Instruction, 30*, 1–8.

Rasmussen, C., Ho, E., & Bisanz, J. (2003). Use of the mathematical principle of inversion in young children. *Journal of Experimental Child Psychology, 85*, 89–102.

Riley, M. S., Greeno, J. G., & Heller, J. I. (1983). Development of children's problem-solving ability in arithmetic. In H. P. Ginsburg (Ed.), *The development of mathematical thinking* (pp. 153–196). Academic Press.

Rittle-Johnson, B., Siegler, R. S., & Alibali, M. W. (2001). Developing conceptual understanding and procedural skills in mathematics: An iterative process. *Journal of Educational Psychology, 93*, 346–362.

Robinson, K. M. (2001). The validity of verbal reports in children's subtraction. *Journal of Educational Psychology, 93*, 211–222.

Selter, C. (1998). Building on children's mathematics. A teaching experiment in third grade. *Educational Studies in Mathematics, 36*, 1–27.

Selter, C. (2001). Addition and subtraction of three-digit numbers: German elementary children's success, methods and strategies. *Educational Studies in Mathematics, 47*, 145–173.

Selter, C., Prediger, S., Nührenbörger, M., & Hussmann, S. (2012). Taking away and determining the difference – A longitudinal perspective on two models of subtraction and the inverse relation to addition. *Educational Studies in Mathematics, 79*, 389–408.

Shrager, J., & Siegler, R. S. (1998). SCADS: A model of children's strategy choices and strategy discoveries. *Psychological Science, 9*, 405–410.

Siegler, R. S., & Araya, R. A. (2005). A computational model of conscious and unconscious strategy discovery. *Advances in Child Development and Behavior, 33*, 1–42.

Siegler, R. S., & Lemaire, P. (1997). Older and younger adults' strategy choices in multiplication: Testing predictions of ASCM using the choice/no-choice method. *Journal of Experimental Psychology: General, 126*, 71–92.

Torbeyns, J., De Smedt, B., Ghesquière, P., & Verschaffel, L. (2009a). Acquisition and use of shortcut strategies by traditionally schooled children. *Educational Studies in Mathematics, 71*, 1–17.

Torbeyns, J., De Smedt, B., Ghesquière, P., & Verschaffel, L. (2009b). Solving subtractions adaptively by means of indirect addition: Influence of task, subject, and instructional factors. *Mediterranean Journal for Research in Mathematics Education, 8*(2), 1–30.

Torbeyns, J., De Smedt, B., Peters, G., Ghesquière, P., & Verschaffel, L. (2011). Use of indirect addition in adults' mental subtraction in the number domain up to 1000. *British Journal of Psychology, 102*, 585–597.

Torbeyns, J., Ghesquière, P., & Verschaffel, L. (2009). Efficiency and flexibility of indirect addition in the domain of multi-digit subtraction. *Learning and Instruction, 19*, 1–12.

Torbeyns, J., Peters, G., De Smedt, B., Ghesquière, P., & Verschaffel, L. (2016). Children's conceptual knowledge of the addition/subtraction complement principle. *British Journal of Educational Psychology, 86*, 382–396.

Torbeyns, J., Peters, G., De Smedt, B., Ghesquière, P., & Verschaffel, L. (2018). SBA strategy use in children of varying mathematical achievement level: A choice/no-choice study. *Journal of Numerical Cognition, 4*, 2363–8761.

Torbeyns, J., Verguts, G., De Smedt, B., & Verschaffel, L. (2018). Efficient and flexible use of SBA in multi-digit subtraction in elementary school children. In E. Bergqvist, M. Österholm, C. Granberg, & L. Sumpter (Eds.), *Vol. 5. Proceedings of the 42nd conference of the International Group for the Psychology of Mathematics Education* (p. 175). Umeå, Sweden: PME.

Treffers, A. (1987). *Three dimensions: A model of goal and theory description in mathematics education. The Wiskobas project*. Reidel.

Usiskin, Z. (2008). The arithmetic curriculum and the real world. In D. De Bock, B. D. Sondergaard, B. G. Alfonso, & C. C. Litwin Cheng (Eds.), *Vol. 3. Proceedings of ICME-11- Topic Study Group 10: Research and development in the teaching and learning of number systems and arithmetic. 11th international congress on mathematics education, Monterrey, Mexico* (pp. 9–16).

Van den Heuvel-Panhuizen, M., & Treffers, A. (2009). Mathe-didactical reflections on young children's understanding and application of subtraction-related principles. *Mathematical Thinking and Learning, 11*, 102–112.

Van Lieshout, E. C. D. M. (1997). What can research on word and context problems tell us about effective strategies to solve subtraction problems? In M. Beishuizen, K. Gravemeijer, & E. C. D. M. Van Lieshout (Eds.), *The role of contexts and models in the development of mathematical strategies and procedures* (pp. 79–111). CDβ Press.

Vergnaud, G. (2012). Commentary. *Educational Studies in Mathematics, 79*, 439–446.

Verschaffel, L., Bryant, P., & Torbeyns, J. (2012). The inverse principle: Psychological, mathematical, and educational considerations: Introduction. *Educational Studies in Mathematics, 79*, 327–334.

Verschaffel, L., & De Corte, E. (1996). Number and arithmetic. In A. Bishop, K. Clements, C. Keitel, & C. Laborde (Eds.), *International handbook of mathematics education. Part I* (pp. 99–138). Kluwer.

Verschaffel, L., Greer, B., & De Corte, E. (2007). Whole number concepts and operations. In F. K. Lester (Ed.), *Second handbook of research in mathematics teaching and learning* (pp. 557–628). Macmillan.

Verschaffel, L., Luwel, K., Torbeyns, J., & Van Dooren, W. (2009). Conceptualizing, investigating, and enhancing adaptive expertise in elementary mathematics education. *European Journal of Psychology of Education, 24*, 335–359.

Verschaffel, L., Van Dooren, W., & Star, J. (2017). Applying cognitive psychology based instructional design principles in mathematics teaching and learning: Introduction. *ZDM Mathematics Education, 49*, 491–496.

Wittmann, E. C. (2010). Begründung des Ergänzungsverfahrens der schriftlichen Subtraktion aus der Funktionsweise von Zählern [Grounding the complement method for written subtraction in the way of functioning of numbers]. In C. Böttinger, K. Bräuning, M. Nührenbörger, R. Schwarzkopf, & E. Söbbeke (Eds.), *Mathematik im Denken der Kinder. Anregungen zur mathematikdidaktischen Reflexion [Mathematics in children's thinking. Suggestions for mathematics educational reflection]* (pp. 34–41). Klett-Kallmeyer.

Woods, S. S., Resnick, L. B., & Groen, G. J. (1975). An experimental test of five process models for subtraction. *Journal of Educational Psychology, 67*, 17–21.

# CHAPTER 7

# Probing the neural basis of rational numbers: The role of inhibitory control and magnitude representations

**Miriam Rosenberg-Lee**
Department of Psychology, Rutgers University, Newark, NJ, United States
Behavioral Neural Sciences Graduate Program, Rutgers University, Newark, NJ, United States

## Rational numbers beyond the classroom

Rational numbers are a notoriously challenging part of the elementary mathematics sequence (National Mathematics Advisory Panel, 2008). Less acknowledged is the role they play beyond the classroom or the job site (Handel, 2016; Matthes, Christoph, Janik, & Ruland, 2014), in areas like physical health and personal finance. For example, studies of health numeracy have found that items involving rational numbers are particularly challenging for patients—56% of diabetics could not correctly compute 18 g × 3.5 (servings) to determine the number of carbohydrates in a snack package (Cavanaugh et al., 2008)—making them strong predictors of health outcomes (Chakkalakal et al., 2017; Osborn et al., 2013). While millions lost their homes during the great recession, many did not. What differentiated these borrowers was not their incomes or credit histories, but their ability to compute with rational numbers (Gerardi, Goette, & Meier, 2013). These results indicate the far-reaching effects of rational number difficulties but also highlight the potential for insights from math cognition to have far-ranging benefits. But doing so requires understanding, both in terms of brains and behavior, what makes rational numbers so difficult for so many learners.

## Deconstructing "whole number bias"

Like other mathematical content domains, rational numbers introduce new notation—numerator, denominator, decimal point, percent symbol—that connotes new mathematical concepts. But in contrast to domains like geometry,

*Heterogeneous Contributions to Numerical Cognition*
https://doi.org/10.1016/B978-0-12-817414-2.00001-4

which formalize learners' previous experiences with shapes and angles, rational number understanding builds on, but can also contradict, learners' preexisting knowledge of the whole number system. Learners often incorrectly apply whole number principles and properties when processing rational numbers, a phenomenon that has been termed *whole number bias*[a] (Ni & Zhou, 2005).

While it may be tempting to attribute difficulties with rational numbers to simply overextending whole number properties, this stance oversimplifies the complicated relationship between whole and rational numbers. Table 1 identifies three categories of whole number properties that must be considered in order to fully understand difficulties that arise when learning about rational numbers: First are properties that never apply to rationals, such as, whole numbers have unique symbols and successors, while rational numbers do not (Category 1). Second are properties that always apply for whole numbers but only sometimes apply to rationals. For instance, more digits always connotes a larger quantity for whole numbers ($27 > 8$), but not so for all rational numbers ($0.27 < 0.8$) (Category 2). Finally, there are properties of whole numbers that also apply to rationals, although this fact may not be initially apparent by learners. In this category are deep properties of real numbers, such as all numbers represent magnitudes and can be uniquely ordered (Category 3). Indeed, initially learners may only have a superficial understanding that whole numbers have these properties, and mastering Category 3 for rationals may also induce conceptual change for whole numbers (De Smedt, this volume).

Crucially, these categories are not isolated but rather interact in challenging ways for learners. Take the case of symbolic representations: each whole number quantity can be specified uniquely by one (and only one) symbol,[b] but for rational numbers, there are infinite ways to denote a rational quantity. A "half" can be written as 1/2, 2/4, 5/10, 50%, 0.5, 0.500, ..., whereas there is only one symbolic expression for the quantity "two," namely, "2" (Braithwaite & Siegler, 2018). These multiple representations decouple previously learnt

[a]The terms *whole* and *natural number bias* have both been used to describe this phenomenon. Based on the classification of natural numbers as positive integers (1, 2, 3, ...) and whole numbers as positive integers and zero (0, 1, 2, 3, ...), and the observation that some properties which interfere with rational number processing are related to the role of zero, we use the term *whole number bias* here.

[b]Whole numbers are a subset of rational numbers. Therefore, within the rational number system there are multiple representations for whole number quantities ($2 = 4/2 = 200\% = 2.00 = \ldots$). However, within the whole number system, symbols are unique, thus the only whole number symbol that represents the quantity "two," is "2."

**Table 1** Examples of properties of whole numbers and their status for rational numbers.

| Category | Property | Whole numbers | Rational numbers |
|---|---|---|---|
| 1 | Unique symbol | ✔ | ✘ |
| | Finite number in interval | ✔ | ✘ |
| | Unique successor | ✔ | ✘ |
| 2 | More digits → larger number | ✔ | ? |
| | Larger numerals → larger number | ✔ | ? |
| | Multiply → larger number | ✔ | ? |
| 3 | Have magnitude | ✔ | ✔ |
| | Uniquely ordered | ✔ | ✔ |

Category 1 properties never apply to rational numbers. Category 2 properties sometimes apply to rational numbers and sometimes do not. Category 3 properties always apply to rational numbers.

associations between number symbols and their magnitude, which results in Category 2 properties, like more digits can represent a smaller rational number (Huber, Klein, Willmes, Nuerk, & Moeller, 2014; Varma & Karl, 2013). Finally, multiple representations may obscure the fact that rational numbers represent quantities and can be uniquely ordered just like whole numbers. Thus 1/2 is less 52%, but more than 0.2. In fact, in this last case, "whole number bias" would actually be helpful, as these properties of whole numbers should be extended to rational numbers. Yet, many individuals struggle to understand that rational numbers refer to numerical quantities.

In this chapter, I argue that deep understanding of rational numbers requires mastering properties from all three of the categories summarized in Table 1. To do so, learners must *expand* and *refine* their understanding of numbers. In mastering Categories 1 and 2, learners must overcome their prior whole number knowledge, a task made more difficult because this knowledge is often acquired implicitly. I propose that the key cognitive capacity that supports refining numerical representations is inhibitory control, the ability to resolve interference from competing information. In mastering Category 3 properties, learners must expand their representation of whole numbers to include rational numbers. I propose that the key cognitive capacity that supports expansion is developing a magnitude representation of rational number quantities.

In this chapter, I begin by highlighting the role implicit understanding may play in rational number understanding and then review behavioral studies examining the role of inhibitory control in rational number processing using individual differences methods. Next, I consider magnitude representation of rational numbers in both symbolic and nonsymbolic contexts, highlighting the challenges in creating stimuli to independently assess whole and rational number influences. Next, the handful of neuroimaging studies which have examined rational number processing are summarized and situated with respect to symbolic and nonsymbolic whole number imaging results. Finally, I offer suggestions and recommendations for future work to leverage our growing understanding of rational numbers to develop brain-informed instruction for this vital, but challenging, class of numbers.

## Implicit understanding of number systems

A growing body of research examines how mathematical content can be learned implicitly, through statistical regularities, rather than from explicit instruction. While implicit understanding may be a hallmark of expertise (Weber, 2001), it can also result in overgeneralization and misconceptions (Chang, Koedinger, & Lovett, 2003). For example, when learning to count, young children often assume that they must proceed from left to right, absorbing this nonessential feature because all examples they have seen follow that order (Gelman & Gallistel, 1978). Another example, which illustrates the role instruction and context can play in implicit learning, comes from the phenomenon of misinterpreting the equal sign as a directive to complete an operation. In elementary school, instruction emphasizing the operational meaning further reinforces this misinterpretation (Baroody & Ginsburg, 1983). Yet, in middle school varying the prior context (i.e., $4+8+5+4=$___ vs $4+8+5=4+$___) can lead either to the operational misinterpretation or to the correct interpretation as a symbol of equivalence (McNeil & Alibali, 2005). These latter results suggest both that instruction can lead to systematic misunderstanding, but also that context can implicitly activate appropriate (or inappropriate) representations in mathematical domains.

A similar phenomenon may be happening in rational number instruction. For example, early emphasis on mastering the count sequence may inadvertently render understanding the concept of fractional numbers more difficult. Similarly, students may have already noticed that numbers with more digits are larger than numbers with fewer digits, well before learning the place value system in early elementary school (National Governors Association Center for Best Practices and Council of Chief State School Officers, 2010). Although I know of no formal research on this topic, informal discussions with educators confirm that in typical instruction, the introduction of rational numbers does not explicitly highlight properties of these numbers that contradict students' prior implicit knowledge. Illustratively, one math reference work for middle-grade teachers (Kaplan and Great Source Education Group, 2003) uses only consistent examples in the sections on comparing and ordering decimals (sections 018-020). Even the "Math Alert," which highlights common mistakes, focuses on the fact that lower numbers can be better in timed sporting events, and no mention is made of the possibility that more digits can connote a smaller number.

While many sources have noted the challenge for learners presented by differences between rational and whole numbers properties (Carey, 2011; Obersteiner, Dresler, Bieck, & Moeller, 2019), less acknowledged is the distinction between Category 1 properties, where whole number understanding is never appropriate, and Category 2 where, in some instances, correct responses are compatible with whole number properties. For example: 0.87 is larger than 0.2, a conclusion which could be erroneously arrived at by noting that 87 is larger than 2. That is, one's prior knowledge of whole numbers would actually lead to the correct answer in many instances. While the distinction between Category 1 and Category 2 is rarely stated explicitly, it forms the basis of most experimental designs in this domain, which specifically balance the number of problems that are congruent or incongruent with whole number knowledge (Obersteiner, Van Dooren, Van Hoof, & Verschaffel, 2013; Vamvakoussi, Van Dooren, & Verschaffel, 2012; Varma & Karl, 2013).

From the implicit, or statistical learning, point of view, these designs present an additional hurdle: prior context may be influencing the performance in the current situation. Negative priming studies illustrate the effects of context in moving between whole and rational representations. Studies in these domains employ a cue trial, which is thought to activate a particular representation, and then a probe trial is presented to assess the effects of the cue on subsequent performance. In the case of decimals, Roell and colleagues (Roell, Viarouge, Houde, & Borst, 2017, 2019) have found that comparing decimal pairs incongruent with whole number knowledge reduces performance on a subsequent congruent trial, relative to a neutral cue trial.

Beyond the laboratory, the majority of one's everyday numerical experiences involve whole numbers, and modern cultural practices further reduce real-world opportunities to process rational numbers with features incongruent with whole number information. The adoption of the decimal system for money, with two and only two significant digits—as recently as 1971 in the United Kingdom (Lee, 2010)—essentially removes the possibility for this confusion in the currency system. Likewise, metrification usually replaces fractional systems (Vera, 2012), and the flexibility of the International System of Units (SI) means decimal quantities can easily be replaced by whole numbers (0.6 kg = 600 g). One notable exception, the Library of Congress Classification system, still uses both whole and decimal numbers, a consistent source of confusion for university library users (Murphy, Long, & MacDonald, 2013). Future work is needed to assess the ways that both shorter (contextual) and longer term (cultural) exposure to whole numbers (and compatible rational numbers) influences rational number processing.

## Behavioral evidence for the role of inhibitory control in rational number processing

Highlighting the role of implicit knowledge in understanding number systems motivates the necessity of inhibitory control in mastering rational numbers. Information that is learned implicitly and hence activated automatically may be especially difficult to override. This contention meshes with the robust body of literature demonstrating performance deficits on rational number tasks when correct responses conflict with whole number knowledge, in children (Meert, Gregoire, & Noel, 2009, 2010) and adults (Bonato, Fabbri, Umiltà, & Zorzi, 2007; DeWolf & Vosniadou, 2015; Huber et al., 2014; Varma & Karl, 2013). Yet, to establish that these effects are driven by the need for inhibitory control requires more than reporting impairment on incongruent cases. Individual differences designs leverage natural variation in inhibitory control across participants to test the need for this capacity in rational number processing. Far fewer studies have employed this approach, as the inclusion of nonmathematical tasks introduces several potential pitfalls. In this section, I review individual differences studies examining the connection between inhibitory control and rational number processing, with reference to two crucial methodological issues: how to (1) measure inhibitory control and (2) account for the contributions of other executive functions.

### Measuring inhibitory control

The individual differences design requires the selection of measures to capture the cognitive construct of inhibitory control. Unfortunately, at the most basic level, the field lacks consensus about how best to measure inhibitory control (Lee & Lee, 2019), or even how to organize the myriad measures of inhibition. Several theoretical accounts do not treat the construct as unitary, instead dividing it into two or more forms (Friedman & Miyake, 2004; Kipp, 2005). Here, following Avgerinou and Tolmie (2019), I distinguish response inhibition (the inhibition of prepotent responses) from semantic inhibition (the inhibition of prepotent knowledge). Under this view, tasks like Go/No Go or Stop-Signal that set up a particular response which must be inhibited on some trials constitute response inhibition, whereas tasks which involve inhibiting responses which are favored by prior knowledge, like color-word and numerical Stroop, are considered measures of semantic inhibition. This division contrasts with the organization of Friedman and Miyake (2004), which considered both Stop-Signal and color-word Stroop as examples of prepotent response inhibition, which they contrasted with resistance to distractor interference. The division between behavioral and cognitive inhibition (Harnishfeger, 1995) is more akin to the ideas presented here, as this view distinguishes between inhibiting actions (i.e., motor responses, impulse control) from the active suppression of the contents of working memory (Kipp, 2005). However, the latter case often includes within cognitive inhibition, episodic memory paradigms, such as directed forgetting (Harnishfeger & Pope, 1996), where the information that must be inhibited was only recently and briefly presented. By contrast, in semantic inhibition the information that must be inhibited has been acquired over multiple, prolonged exposures and is thus stored in long-term memory (see Borst, this volume).

In the Go/No Go task, a classic measure of response inhibition, participants are presented with a stream of stimuli, where one stimulus is presented often (80% of the time) and the other rarely (20% of the time) and participants must respond to the frequent stimuli but withhold responses to the rare stimuli, setting up a propensity to respond. In the Animal Stroop task (Merkley, Thompson, & Scerif, 2015), a more recently developed measure of semantic inhibition, participants must identify the larger of two animals in real life, while the physical size of the presented animals is manipulated (e.g., a mouse drawn larger than an elephant). Avgerinou and Tolmie (2019) used these tasks to examine the contributions of both forms of inhibition to the processing of intuitive and counterintuitive fractions and decimal problems, in 8–10-year old children. They also assessed the effects of cognitive load, by manipulating how easily relevant information could be gleaned from a short-word problem. As expected, participants were slower and less accurate when the fraction and decimal quantities contradicted prior whole number knowledge (larger denominator indicating smaller fraction, more digits indicate smaller decimal). Although they did not provide first-order correlations, they do report regression analyses, showing response inhibition predicting counterintuitive performance under moderate cognitive load, and both response and semantic inhibition predicting under high-load conditions. These results confirm the need for inhibitory control in processing rational numbers, especially in more naturalistic settings where learners must seek information within a text display.

A common semantic inhibition measure in studies of numerical and mathematical cognition is the numerical Stroop task (Henik & Tzelgov, 1982), where single-digit numbers are compared, while the physical size of the numerals is manipulated to be either congruent (3 vs 7) or incongruent (3 vs 7) with numerical quantity. Recently, Gómez, Jiménez, Bobadilla, Reyes, and Dartnell (2015) used this task to assess relations between inhibitory control and rational number understanding, as measured by a fraction comparison task. They found that better performance on the numerical Stroop task was related to better performance on the fraction comparison task, especially in the incongruent cases. However, they also reported that math skills (as measured by in-school testing) explained this relationship. The authors interpret their results to imply that math abilities support both inhibitory control and rational number knowledge, and hence there is no direct link between inhibition and fraction understanding.

An alternative interpretation of these results is possible. First, note that the numerical Stroop task is a domain-specific measure of inhibitory control, as it measures inhibition between physical size and numerical magnitude. Thus those who excel at this task could have better understanding of numerical magnitude, which would support both fraction comparison and mathematical skills (Schneider et al., 2017; Ye et al., 2016). Without a domain general measure, one cannot assess the pure contribution of inhibitory control to fraction comparison. Second, these results could be equally consistent with the proposal that inhibitory control contributes to math skills, not just directly, but also by supporting the acquisition of fraction magnitude knowledge, which in turn supports other skills, like rational number arithmetic (Coulanges et al., 2021). Longitudinal studies in children support this conceptual trajectory, as students first acquire fraction magnitude understanding before mastering fraction arithmetic skills (Kainulainen, McMullen, & Lehtinen, 2017; Van Hoof, Degrande, Ceulemans, Verschaffel, & Van Dooren, 2018). Unfortunately, no studies to date have considered the effect of inhibitory control in the progression toward more sophisticated rational number understanding.

Another classic measure of inhibitory control is the Flanker task (Eriksen & Eriksen, 1974), where participants must respond according to the middle stimuli, ignoring the flanking stimuli. In modern formulations, a centrally presented stimulus points to the left or right and is surrounded by either congruent (e.g., <<<<<) or incongruent (e.g., >><>>) stimuli. While this task fits generally under the category of response inhibition in the current scheme, as the stimuli itself sets up the inhibition context, it can also be considered a form of visual distractor inhibition task (Friedman & Miyake, 2004; Lee & Lee, 2019). Matthews, Lewis, and Hubbard (2016) used this task to control for the potential role of inhibitory control in the relationship of nonsymbolic ratio comparison with symbolic fraction understanding and general math skills. They report that their key result—nonsymbolic fraction acuity predicts performance on symbolic measures—was robust to the inclusion of inhibitory control. In fact, zero-order correlations revealed that Flanker task performance was not significantly related to any of their outcome measures (number line estimation of rational numbers, symbolic fraction knowledge, or students' college entrance exam score). These null results suggest that visual distractor inhibition measures, like the Flanker task, may not be capturing the relevant aspect of inhibition required by rational number processing. More broadly, individual differences studies suggest that measures of semantic inhibition carry more predictive power than response inhibition measures.

## Accounting for other executive functions

Studies on the specific role for inhibitory control in rational number processing play out within the broader context of research examining relations between executive functions (EF) and academic achievement. Here, I take the tripartite view of executive functioning as comprising (1) working memory, the ability to maintain and manipulate information; (2) inhibitory control, the ability to resolve interference from competing information; and (3) cognitive flexibility, the ability to switch between cognitive modes (Diamond, 2013). Among the three, the majority of work has focused on the relationship of math outcomes with working memory (Peng, Namkung, Barnes, & Sun, 2016), while fewer studies have examined on the role of inhibitory control and cognitive flexibility, also called shifting (Friso-van den Bos, van der Ven, Kroesbergen, & van Luit, 2013). In the realm of rational numbers, the focus thus far has been on working memory, either in terms of its direct relations (Fuchs et al., 2014; Hansen et al., 2015) or as a control measure when relating fraction understanding to later academic achievement (Siegler et al., 2012).

Robust relations are found between all the executive constructs and mathematical outcomes (Friso-van den Bos et al., 2013; Peng et al., 2016), and the separability of these constructs has been challenged (Miyake et al., 2000), especially in children (Best & Miller, 2010; Lerner & Lonigan, 2014). Thus any specific role for inhibitory control must be established over and above working memory capacity and to a lesser extent cognitive flexibility, yet most studies do not measure more than one executive function. In our own work, my colleagues and I have addressed these issues by employing tasks which capture semantic, rather than response inhibition, and collected multiple executive function measures. Specifically, we have used the color-word Stroop task (Stroop, 1992) where participants must report the font color of a written color word, which can either be congruent ("green" in green ink) or incongruent ("green" in red ink). Visual spatial working memory was assessed using an adaptive computerized version of the Corsi block-tapping task. Finally, we assessed cognitive flexibility with a task-switching measure where participants were first cued to report either the color or shape of a subsequent stimulus. In the subsequent probe phase, a colored shape is presented (e.g., red triangle) and participants then reported the appropriate dimension based on the preceding cue.

In terms of measuring of rational number understanding, we have focused our efforts on decimal numbers. We targeted the Category 2 property of *more digits, larger number*, based on the robustness of the effect (Huber et al., 2014; Varma & Karl, 2013). We found that performance on incongruent problems was correlated with color-word Stroop performance, as was working memory but not the cognitive flexibility task (Coulanges et al., 2021). Moreover, there were no correlations with problems which were consistent with prior whole number knowledge. With respect to relations with math achievement, all three executive function measures predicted performance on the Calculation subtest of the Woodcock-Johnson (Woodcock, McGrew, & Mather, 2001), but only inhibitory control was an independent predictor. Finally, we found that performance on the inconsistent decimal trials mediated the relationship between inhibitory control and math achievement, suggesting that greater inhibitory control may support understanding of counterintuitive rational numbers concepts, which may further support future math content learning.

In summary, the small set of studies using individual differences designs suggest that inhibitory control (especially, as measured by domain general semantic inhibition tasks) contributes to successful counterintuitive rational number processing. However, work remains to establish the independence of this relationship relative to other components of executive functioning and to track the developmental trajectory of this contribution. Another open question is the effects of whole number understanding on rational number learning: Would strong whole number knowledge require more inhibitory control to overcome whole number bias? While no studies to date have examined this question using an independent measure of inhibitory control, Van Hoof, Verschaffel, and Van Dooren (2017) found that better performance in whole number symbolic comparison and number line estimation were related to incongruent rational number understanding. One hypothesis from this line of reasoning is that superficial understanding of whole numbers (Category 2 properties) may require more inhibitory control to overcome, whereas grasping deeper features whole numbers, like magnitude coding (Category 3 properties), may set up a foundation for understanding these properties in rational numbers.

## Magnitude representations of symbolic rational number quantities

The evidence presented thus far suggests that successful rational number processing draws on inhibitory control capacities to suppress interference from whole number knowledge. However, it must be acknowledged that while interference effects are pervasive, in proficient adults the effects are modest, especially at the highest ability levels. For example, a study of expert mathematicians found accuracy was over 94% on all categories of fraction comparison (Obersteiner et al., 2013). Only reaction times revealed the effects of whole number congruency, with common numerator problems solved more slowly than common denominator problems. And even this difference only amounted to 13% longer reaction times (1796 ms vs 2048 ms).

If inhibitory control were the only factor in successful rational number processing, we might expect that the proficiency would be marked by a reduction in performance costs for incongruent problems. That is, learners would simply improve in their capacity to execute the opposite of the whole number response. An alternative characterization proposes that mastering rational numbers involves conceptual change, that is, developing new knowledge structures, which enable ideas not possible in the original structure (Werner, 1957). Carey (2011) has noted that rational number learning has the hallmarks of conceptual change, in that it is very difficult to learn, and is marked by discontinuity in knowledge, noting "[w]hether children can properly order fractions and decimals, how they justify their ordering, how they explain the role of each numeral in a fraction expressed x/y, whether they agree there are numbers between 0 and 1 and whether they believe that repeated division by 2 will ever yield 0 are all interrelated" (p. 119). This assertion contrasts with longitudinal studies which suggest that magnitude-based ordering of rational numbers typically precedes arithmetic understanding, while understanding the density of rational numbers—there are infinite rational numbers between any two numbers—is the last of these concepts to develop (Kainulainen et al., 2017; Van Hoof et al., 2018).

Regardless of whether rational number understanding builds gradually (see Vosniadou, this volume) or reflects a discontinuous change (Carey, 2011), growing evidence supports the contention that deep understanding of rational numbers involves building new knowledge structures. In this chapter, I focus on the structure of rational numbers based on their magnitudes for three reasons. First, doing so demonstrates an expansion of magnitude-based representation from whole numbers, thus mastering a crucial Category 3 property. Second, individual differences in this capacity have been linked to general math achievement, suggesting it is a foundational capacity which promotes further mathematics learning (Matthews et al., 2016; Siegler, Thompson, & Schneider, 2011). Finally, from a practical perspective, magnitude-based representation can be operationalized in terms of the *distance effect* and therefore can be probed rigorously in behavioral and neuroimaging experiments.

The behavioral hallmark of magnitude processing—the distance effect—describes the phenomenon of higher accuracy and faster reaction times for numerical comparisons that are far apart (2 vs 9) relative to close (8 vs 9) comparisons (Moyer & Landauer, 1967). Among whole numbers, distance effects are found in both symbolic and nonsymbolic comparison task, and crucially, performance on these tasks has been linked to mathematics achievement (Schneider et al., 2017). Thus, there has been considerable interest in establishing that the processing of symbolic rational numbers is distance-driven.

Like other multicomponent numbers (Nuerk, Moeller, Klein, Willmes, & Fischer, 2011), rational numbers afford multiple distance dimensions. Here, I define *rational distance* as the actual difference between two quantities, which contrasts with *whole distance,* the distance between potentially interfering whole number quantities within a rational number. In the case of fractions, at least two whole dimensions can be identified: (1) distance between the numerators, and (2) distance between the denominators. Fig. 1 plots the relationship between rational and whole distance for all the single-digit

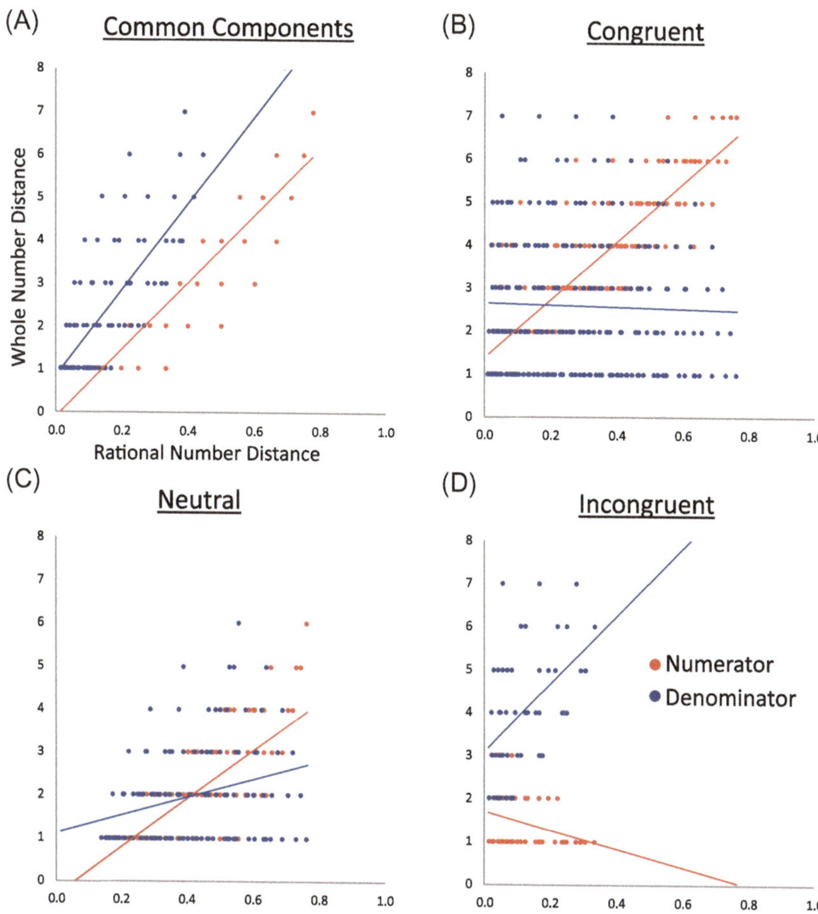


**Fig. 1** Whole and rational number distance for all single-digit proper fractions. (A) *Common components*: same numerator problems (2/8 vs 2/3, rational distance = 0.42, denominator distance = 3, *blue*) and same denominator problems (3/7 vs 6/7, rational distance = 0.43, numerator distance = 3, *red*). (B) *Congruent*: larger numerator and larger denominator both indicate the larger fraction (1/7 vs 5/9 rational distance = 0.41, numerator distance = 4, denominator distance = 2). (C) *Neutral*: larger numerator and smaller denominator indicate the larger fraction (4/9 vs 6/7 rational distance = 0.41, numerator distance = 2, denominator distance = 2). (D) *Incongruent*: smaller numerator and smaller denominator both indicate the larger fraction (4/9 vs 3/4 rational distance = 0.31, numerator distance = 1, denominator distance = 5).

proper fraction comparison problems. As can be seen in the plot of common components problems (common numerators or common denominators, Fig. 1A), problems with comparable rational distance can vary on their whole distance. For example, 4/7 versus 4/5 and 1/3 versus 1/9 have a rational difference of 0.222 and 0.229 yet have denominator distance of 2 and

6, respectively. Thus if rational number magnitude is driving performance, these two problems should have comparable response times, whereas they should differ if whole distance is driving responses.

An early study, which tackled this question in adults, examined distance effects relative to a standard value (1/5, 0.2) for unit fractions (1/1, 1/2, 1/3, etc.) and in separate experiments, problems with common denominators (Bonato et al., 2007). Across four experiments, they found problems were solved componentially, or in our parlance, with respect to whole distance. Crucially, however, their experimental design never required a magnitude computation to reach the correct answer. For example, for unit fraction comparison, reversing whole number responses (i.e., selecting the smaller denominator in 1/2 vs 1/5) would always lead to the correct answer. Moreover, while the study reported for unit fractions that denominator distance better fit response times than rational distance, there was no acknowledgement that rational and whole distance correlates strongly for unit fractions. In fact, for all single-digit common numerator problems, there is a large correlation between these rational and denominator distance ($r(82) = 0.746$), and similarly among unit fractions ($r(26) = 0.742$). Even within the eight stimuli used in Bonato et al. (2007), this correlation is in the large range ($r(6) = 0.598$).

Bonato et al. (2007) also considered the role of expertise, by contrasting psychology students with engineering, physics, and computer science students. However, their samples were small, ($n = \sim 10$ in each group) and gender was confounded with expertise. Regardless, they found a main of effect of expertise, but no difference in the reliance on whole distance. More recently, Obersteiner et al. (2013) examined a larger sample, with even greater expertise, mathematicians ($n = 44$, 25% female, PhD students, postdocs and professors). They sampled stimuli from the set of 1–2 digit numerators and denominators fractions, in blocks with common components (either same denominator, e.g., 22/49 vs 18/49, or same numerator, e.g., 4/17 vs 4/39) or without common components, which could be congruent with whole number (larger fraction has larger numerator and larger denominator, e.g., 17/21 vs 9/14), incongruent (larger fraction has smaller numerator and smaller denominator, e.g., 11/25 vs 8/13), or neutral (larger fraction has larger numerator and smaller denominator, e.g., 7/16 vs 5/29). They found evidence that numerator, denominator, and rational distance all drove performance, in at least one of the conditions. Notably, the effects were strongest for rational distance for the problems without common components. The use of two-digit numerators and denominators affords more flexibility in stimulus selections. Yet, the pervasive correlations found in single-digit problems (Fig. 1) remain in this stimulus set. For example, among neutral problems, there are large correlations of rational distance with numerator ($r(16)=0.549$) and denominator distance ($r(16)=0.781$), and all three distance measures predicted reaction times among the mathematicians. Without orthogonalizing rational and whole distance, it is impossible to disambiguate the effects of each dimension and therefore to definitively assess the extent of magnitude-based representations of rational number.

Another factor complicating the interpretation of distance sensitivity in fraction comparison is the effects of strategy variation, which are especially prevalent in fraction comparisons (Obersteiner & Tumpek, 2016). As noted by Siegler (1999), efficient strategy choice is influenced by problem type, among other factors. For example, the incongruent problem 14/23 versus 5/6 has large numerator and denominator distances favoring the incorrect answer (9 and 17, respectively), yet if one compares the gaps between numerator and denominator (9 vs 1), it becomes clear that larger fraction (5/6) is also much closer to 1.

Strategies, like benchmarking (comparing to known values like 1/2) and comparing gaps, introduce new numerical dimensions, which in turn drive distance effects. For example, among the incongruent stimuli of Obersteiner et al. (2013), there is no meaningful correlations of rational distance with numerator ($r(16)=0.034$) or denominator distance $r(16)=0.169$). Yet, there is a large correlation with gap distance ($r(16)=0.539$). In fact, gap distance will always lead to the correct response on incongruent problems, that is, the largest fraction always has the smaller gap (e.g., 8/11 vs 12/23, gap of 3 vs 11). In contrast, it can sometimes lead to the wrong answer for congruent problems (e.g., 17/39 vs 8/25, gap = 22 vs 17, yet, 17/39 is larger than 8/25). Although Obersteiner et al. (2013) did not explicitly test for the effects of gap distance, they did find unexpectedly slower responses on congruent relative to incongruent problems, and for incongruent problems, neither rational, nor numerator, nor denominator distance explained fraction reaction times among expert mathematicians. These results are all consistent with the possibility that gap distance drove performance for the mathematicians and the reported behavioral correlations with rational and whole distance may instead reflect stimulus correlations with gap distance. Gap-based strategies are a particularly fruitful area for future research because they may be a marker of more sophisticated relational understanding of fractions than componential strategies among learners who have yet to develop full-fledged magnitude-based representation of fractions.

To summarize these studies of fraction comparison in proficient adults, experimental evidence supports both whole and rational distance effects. However, insufficient attention has been paid to confounding factors within stimuli sets or to the role of strategy variation. Further compounding the effects of strategy variation are underappreciated contextual effects, which may bias participants toward or against certain representations. Presenting stimuli in blocks by type versus interspersing them may change the need to activate or suppress various magnitude representations, as in Bonato et al. (2007). In conclusion, for fractions it can be difficult, if not impossible, to design experiments that can definitively establish the independent effects of whole and rational distance.

In contrast, decimal comparison is much less studied and while some studies have examined magnitude-based processing (Kallai & Tzelgov, 2014; Zhang, Chen, Lin, & Szucs, 2014), to my knowledge, none have explicitly contrasted whole versus rational distance effects in this notation. Decimals are a ripe area for future research as the stimuli space affords greater opportunities for orthogonalizing distance dimensions. I have already highlighted the salient interference effect found in decimal numbers between the number of digits and numerical magnitude. Two explanations are given for this effect in the literature: *the string length congruity effect* proposes that participants independently compare individual place values and the number of digits in a string (Huber et al., 2014). In contrast, *the semantic interference effect* proposes that not just the number of digits, but the value of those digits influence comparison processes (Varma & Karl, 2013). For example, problems like 0.2 versus 0.12 and 0.9 versus 0.82 are both incongruent with whole number knowledge and had have the same rational distance of 0.08. Yet, if whole number knowledge is being processed (i.e., 2 vs 12, 9 vs 82), the first is much closer comparison than the second (10 vs 73), which may be further pushing participants toward the whole number response. Among congruent problems, comparisons like 0.1 versus 0.23 and 0.8 versus 0.93 should be equally difficult based on rational distance of 0.13, but based on whole distance the first is a much closer comparison than the second (22 vs 85).

Explicitly testing whether whole or rational distance is driving behavior requires carefully constructing the stimuli space. Fig. 2 plots all the combinations of tenths (0.1–0.9) versus hundredths (0.10–0.99) comparison problems with respect to whole and rational distance, organized by whole number congruency. As in fractions, there is a natural bias in the full space: among congruent problems, whole and rational distance are positively

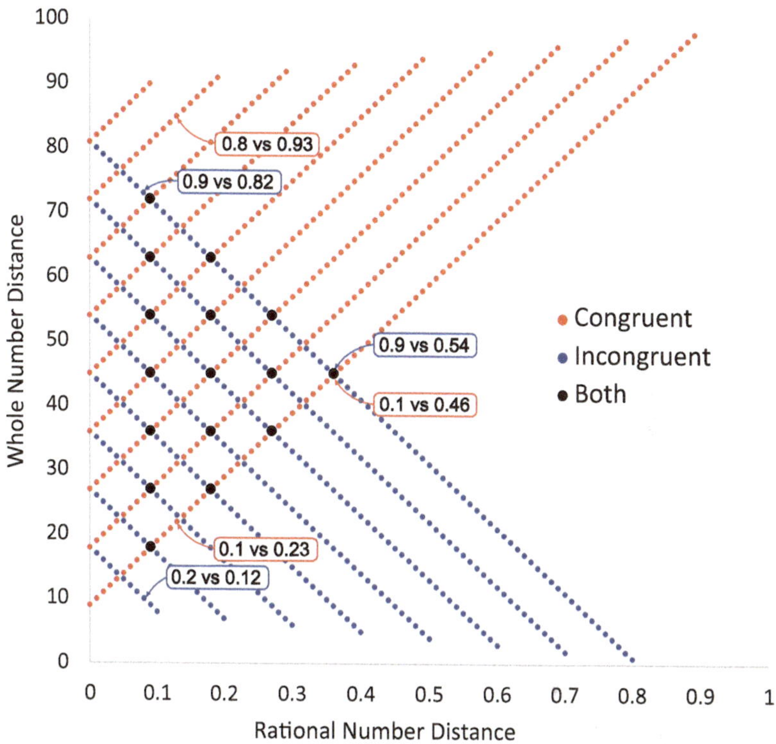


**Fig. 2** Whole and rational number distance for all the combinations of tenths (0.1–0.9) vs hundredths (0.10–0.99) decimal comparison problems. In *red*: examples of congruent problems with identical rational distance (0.13), but differing whole distance (12 vs 85). In *blue*: examples of incongruent problems with identical rational distance (0.08), but differing whole distance (10 vs 73). In *black*: stimulus set with identical rational (e.g., 0.36) and whole (e.g., 45) distances for both congruent and incongruent problems. Among the *black* stimuli, whole and rational distance are matched across congruent and incongruent comparisons, and the distance dimensions are orthogonal.

correlated ($r(403) = 0.583$), while the correlation is negative among incongruent problems ($r(322) = -0.584$). In the latter case, that means that, on average, harder incongruent problems with smaller rational distances also have larger whole distances, further tipping the scales toward the whole number response.

My colleagues and I have controlled for rational distance across congruency conditions, in part by explicitly sampling near (rational distance: 0.16) and far problems (rational distance: 0.64) (Coulanges et al., 2021). However, by selecting problems with far rational distance, this approach confounds whole number distance with congruency (whole distance: congruent = 70.6; incongruent = 23.1). Thus to fully assess the impact of whole and rational distance on decimal processing, one must devise a stimulus set which orthogonalizes these properties and balances them over the condition types. Fortunately, within the full stimuli set is a region of overlapping whole and rational distance (Fig. 2), enabling just such an approach. In ongoing work, we have specifically selected problems with identical whole and rational distance for both congruent and incongruent stimuli (Pincus et al., 2020). This design can assess whether participants' responses are sensitive to whole distance while solving decimal comparison problems, and whether congruency effects are independent of the confounds of whole distance found in previous studies. Further, this design enables probing brain responses to whole versus rational distance, within the same individuals in a completely orthogonal design.

## Nonsymbolic rational numbers and inhibitory control

The goal of studies of symbolic magnitude representation is to characterize knowledge states at various levels of instruction. In contrast, for nonsymbolic representations, work has focused on identifying precursor capacities that might predict later learning or be useful targets for intervention. To this end, researchers have focused on nonsymbolic, visuospatial representations of rational numbers, noting that individuals who perform better on these tasks also have better symbolic fraction understanding and math skills. For example, Siegler et al. (2011) found that better accuracy in placing fractions on a number line is related to higher math achievement. Another method of assessing magnitude understanding is comparison tasks, for example comparing two pairs of line lengths or sets of colored dots to determine the larger proportion. Using this task, Matthews et al. (2016) found that greater nonsymbolic comparison ability, which they term the ratio processing system (RPS), predicts symbolic understanding of rational numbers and performance on a college entrance exam.

These correlational studies provide tantalizing hints that instruction which emphasizes nonsymbolic representations could lead to improvements in rational number understanding. But to develop such curricula requires first selecting among the multiple possible nonsymbolic representations. Area models, the traditional approach to fraction instruction, involves dividing two-dimensional shapes into equal segments, like circles into wedges. This approach has been criticized because it often leads to misconceptions, like fractions can only be quantities less than one (Obersteiner et al., 2019). In contrast, the measurement approach, which divides objects along one axis, avoids some of these pitfalls and growing evidence supports the educational effectiveness of linking these nonsymbolic representations to symbolic fractions (Fuchs, Malone, Schumacher, Namkung, & Wang, 2017).

Here, I highlight another key distinction: continuous versus discrete representations. Both circles and line lengths can be presented continuously or discretized, while dots can only be discrete (Fig. 3A). While receiving less attention from the educational research community, a growing literature in cognitive development finds superior discrimination of continuous formats relative to discrete stimuli (Boyer, Levine, & Huttenlocher, 2008). For example, Jeong, Levine, and Huttenlocher (2007) found that children as young as six can select the larger proportion from two annuli with continuous sections. However, when those sections were divided into segments, performance dropped, specifically for stimuli where the larger proportion had fewer segments (see Fig. 3B). Further, while children's performance on continuous comparisons improved gradually from 6 to 10 years, performance on these "counting misleading" problems remained consistently poor over this same age range.

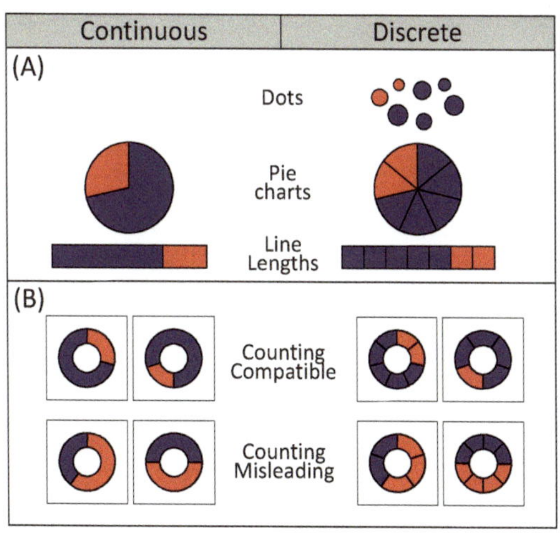


**Fig. 3** Nonsymbolic stimuli format. (A) Sets of dots are always discrete stimuli, but pie charts and line lengths can be continuous, or segmented into discrete stimuli. (B) Sample stimuli from Jeong et al. (2007), adapted in Abreu-Mendoza, Coulanges, Ali, Powell, and Rosenberg-Lee (2020), contrast counting compatible stimuli (larger ratio has a larger number of discrete segments) with counting misleading stimuli (larger ratio has smaller number of discrete segments). These stimuli would be congruent (2/7 vs 1/5) and incongruent (3/5 vs 4/8) as symbolic fraction comparisons.

Difficulties with counting misleading discrete comparisons may be analogous to the congruency effects found in symbolic fraction comparison studies. In fact, the same stimuli that are incongruent in fraction comparison are counting misleading in segmented proportional reasoning (e.g., 3/5 vs 4/8, Fig. 3B). Given the relations between inhibitory control and incongruent symbolic rational number comparison (Avgerinou & Tolmie, 2019; Gómez et al., 2015), my colleagues and I reasoned that inhibitory control should also play a role in processing these misleading problems. In a study of second grade children (ages 7–8), we collected a computerized version of the Spinners task (Jeong et al., 2007) and the Hearts and Flowers task, a reliable measure of response inhibition for children as young as 3 years of age (Wright & Diamond, 2014). As expected, children with better inhibitory control also performed better on the counting misleading discrete trials (Abreu-Mendoza et al., 2020). Future work, in older children and adults, should examine relations between symbolic and nonsymbolic fraction performance and inhibitory control, to confirm this hypothesized connection between incongruent symbolic comparisons and their counting misleading nonsymbolic counterparts.

Another line of work in proportional reasoning examines improving discrete performance by priming participants with continuous information (Boyer & Levine, 2015; Hurst & Cordes, 2018). In our sample of second graders, we contrasted two groups of students: one saw continuous stimuli immediately before a discretized version of the same problems, while the other saw intermixed discretized stimuli (i.e., the segments of the same color were not adjacent). We hypothesized that seeing the continuous stimuli would prime proportional processing while the intermixed stimuli would activate counting strategies. Indeed, we found that immediately prior exposure to continuous information improved performance, specifically on the counting misleading trials (Abreu-Mendoza et al., 2020). Interestingly, among the continuously primed participants, the relationship with inhibitory control was reduced relative to the discretely primed participants. Although the interaction was not significant, these results are consistent with the proposal that subtle changes in context can bias comparison tasks towards or against the appropriate magnitude representations and thus impact the need for inhibitory control.

## Brain imaging of rational number processing

The key contention of this chapter is that success with rational numbers requires refining and expanding one's representation of number. Support for the refining hypothesis would be provided by neural activity during incongruent rational number processing in brain regions identified with inhibitory control. For the expanding hypothesis, we would expect to see activity—especially magnitude-coding activity—for rational numbers in brain regions previously identified in whole number magnitude processing. Thus, this final section interrogates the small set of neuroimaging studies of rational number for evidence in support of these two predictions. Figs. 4 and 5 (Tables 2 and 3) present the reported parietal peaks from these studies

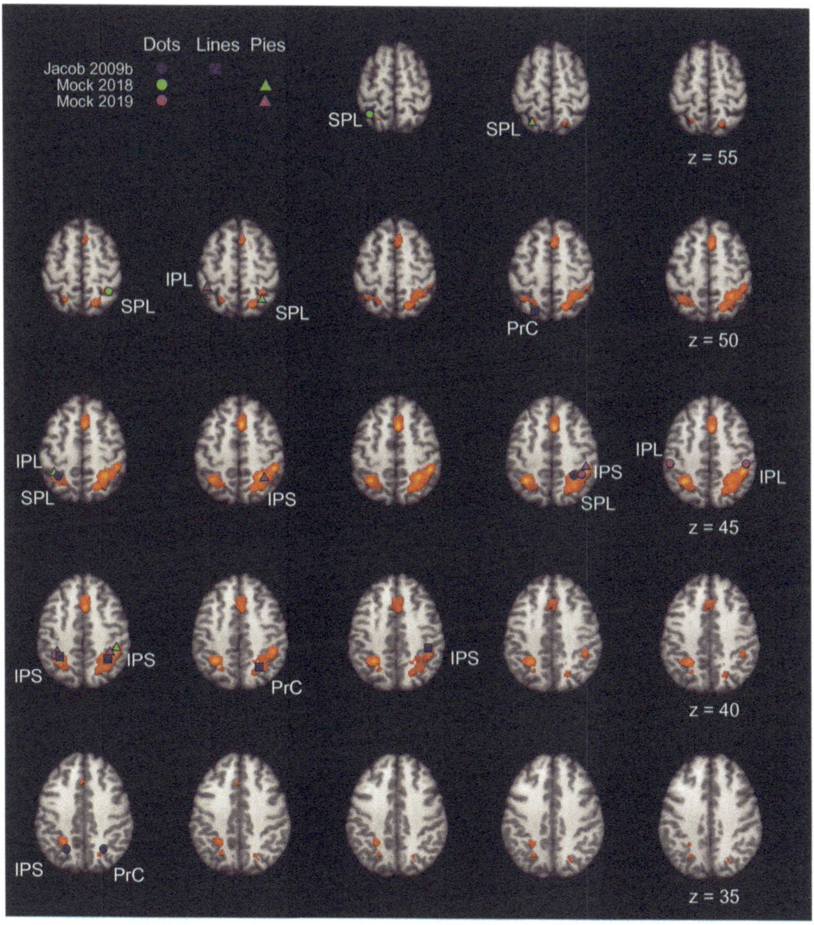


**Fig. 4** Nonsymbolic parietal peaks for studies of rational numbers. ROIs: nonsymbolic peaks from Table 2; Functional clusters: nonsymbolic map (Sokolowski et al., 2017); Underlay brain: colin 1 mm$^3$, Talairach coordinates. IPS regions identified from the Juelich Histological Atlas, all other regions from the Talairach Daemon. *IPL*, inferior parietal lobule; *IPS*, intraparietal sulcus; *PrC*, precuneus; *SPL*, superior parietal lobule.

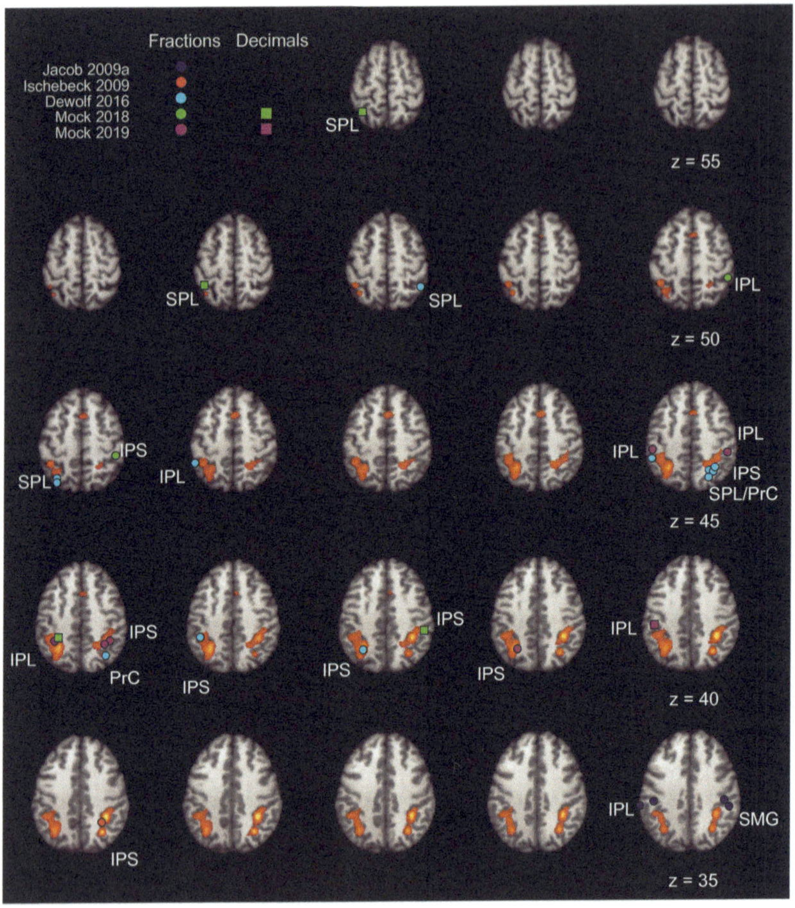


**Fig. 5** Symbolic parietal peaks for studies of rational numbers. ROIs: symbolic peaks from Table 3; Functional clusters: symbolic map (Sokolowski et al., 2017); Underlay brain: colin 1 mm$^3$, Talairach Coordinates. IPS regions identified from the Juelich Histological Atlas, all other regions from the Talairach Daemon. *IPL*, inferior parietal lobule; *IPS*, intraparietal sulcus; *PrC*, precuneus; *SMG*, supramarginal gyrus; *SPL*, superior parietal lobule.

superimposed on recently published meta-analytic maps of nonsymbolic and symbolic whole number studies (Sokolowski, Fias, Mousa, & Ansari, 2017). Fig. 6 depicts the peaks from the only study to report the contrast of incongruent versus congruent fractions, overlaid on a "meta-analysis" of inhibition studies. The meta-analysis map was generated using Neurosynth.org, with the search term "inhibition," which identified 601 studies, using the forward inference test, at a false discovery rate of $P < .01$ (Yarkoni, Poldrack, Nichols,

**Table 2** Nonsymbolic parietal peaks for studies of rational numbers (plotted in Fig. 4).

| | | MNI coordinates | | | Talairach coordinates | | |
|---|---|---|---|---|---|---|---|
| **Study** | **Contrast** | x | y | z | x | y | z |
| Jacob and Nieder (2009b) | Dots (habituation) | 24 | −63 | 42 | 25 | −59 | 39 |
| | | 27 | −51 | 51 | 28 | −47 | 46 |
| | | −21 | −63 | 42 | −21 | −60 | 39 |
| | | −30 | −54 | 54 | −30 | −51 | 49 |
| | Lines (habituation) | −30 | −51 | 48 | −30 | −48 | 44 |
| | | −30 | −54 | 63 | −30 | −50 | 57 |
| | | −18 | −72 | 57 | −18 | −68 | 51 |
| | | 24 | −63 | 48 | 25 | −59 | 43 |
| | | 30 | −54 | 48 | 31 | −50 | 44 |
| | | 39 | −39 | 45 | 40 | −36 | 42 |
| Mock et al. (2018) | Dots (distance effect) | 33 | −52 | 60 | 34 | −48 | 54 |
| | | −30 | −57 | 63 | −30 | −53 | 57 |
| | Pie charts (distance effect) | 41 | −40 | 48 | 42 | −37 | 44 |
| | | 28 | −60 | 60 | 29 | −56 | 53 |
| | | −22 | −65 | 63 | −22 | −61 | 56 |
| | | −35 | −50 | 53 | −35 | −47 | 49 |
| Mock et al. (2019) | Dots > baseline (univariate contrast) | 46 | −35 | 48 | 46 | −32 | 45 |
| | | 36 | −50 | 50 | 37 | −46 | 46 |
| | | −47 | −37 | 48 | −47 | −34 | 45 |
| | | −32 | −50 | 48 | −32 | −47 | 44 |
| | Pie charts > baseline (univariate contrast) | 33 | −45 | 48 | 34 | −42 | 44 |
| | | 41 | −40 | 50 | 42 | −36 | 46 |
| | | 31 | −55 | 53 | 32 | −51 | 48 |
| | | −37 | −45 | 58 | −38 | −42 | 53 |
| | | −32 | −45 | 48 | −32 | −42 | 44 |

MNI coordinates were converted to Talairach coordinates (http://sprout022.sprout.yale.edu/mni2tal/mni2tal.html).

Van Essen, & Wager, 2011). The forward inference map, now called the uniformity test, represents the probability of finding activation in a voxel, given the search term was included in a study. This more liberal map was chosen, to maximally detect voxels identified with "inhibition," rather than the more conservative, reverse inference map (now called association test), which depicts the probability of a study mentioning the search term, given there is activity in that region.

**Table 3** Symbolic parietal peaks for studies of rational numbers (plotted in Fig. 5).

| | | MNI coordinates | | | Talairach coordinates | | |
|---|---|---|---|---|---|---|---|
| **Study** | **Contrast** | **x** | **y** | **z** | **x** | **y** | **z** |
| Jacob and Nieder (2009b)[a] | Fraction numerals (distance effect) | −62 | −38 | 37 | −61 | −35 | 35 |
| | | 51 | −38 | 37 | 51 | −35 | 35 |
| | Fraction words (distance effect) | 43 | −30 | 37 | 43 | −28 | 35 |
| | | −45 | −31 | 37 | −44 | −29 | 35 |
| Ischebeck et al. (2009) | Fraction numerals (distance effect) | 27 | −54 | 42 | 28 | −51 | 39 |
| DeWolf et al. (2016) | Fraction > integers (univariate contrast) | −28 | −62 | 46 | −28 | −59 | 42 |
| | | −30 | −74 | 54 | −30 | −69 | 49 |
| | | −40 | −46 | 46 | −40 | −43 | 43 |
| | | 30 | −60 | 50 | 31 | −56 | 45 |
| | | −30 | −70 | 54 | −30 | −66 | 49 |
| | | −48 | −50 | 52 | −48 | −47 | 48 |
| | | −48 | −48 | 48 | −48 | −45 | 45 |
| | | 32 | −68 | 48 | 33 | −64 | 44 |
| | | 28 | −64 | 50 | 29 | −60 | 45 |
| | | 24 | −68 | 48 | 25 | −60 | 45 |
| | | 42 | −54 | 58 | 43 | −50 | 52 |
| | | 22 | −72 | 50 | 23 | −68 | 45 |
| Mock et al. (2018) | Fraction numerals (distance effect) | 43 | −42 | 53 | 44 | −38 | 49 |
| | | 46 | −45 | 55 | 47 | −41 | 50 |
| | Decimals (distance effect) | 48 | −40 | 45 | 48 | −37 | 42 |
| | | −37 | −52 | 58 | −37 | −48 | 53 |
| | | −30 | −57 | 63 | −30 | −53 | 57 |
| | | −27 | −45 | 48 | −27 | −42 | 44 |
| | | −60 | −45 | 30 | −59 | −43 | 30[b] |
| Mock et al. (2019) | Fractions > baseline (univariate contrast) | −47 | −37 | 48 | −47 | −34 | 45 |
| | | −32 | −50 | 48 | −32 | −47 | 44 |
| | | −25 | −62 | 45 | −25 | −59 | 41 |
| | | 46 | −40 | 48 | 47 | −37 | 45 |
| | | 36 | −50 | 48 | 37 | −47 | 44 |
| | | 28 | −55 | 48 | 29 | −51 | 44 |
| | Decimals > baseline (univariate contrast) | −45 | −32 | 43 | −44 | −30 | 40 |

[a] Peaks from Jacob and Nieder (2009a) are not reported in the paper, but are plotted with z-coordinate values in Fig. 2C (Fraction numerals) and Fig. 3B (Fraction words). These figures were used to approximate MNI co-ordinates. MNI coordinates were converted to Talairach coordinates (http://sprout022.sprout.yale.edu/mni2tal/mni2tal.html).

[b] Not plotted as outside the range of the z coordinates of the other ROIs.

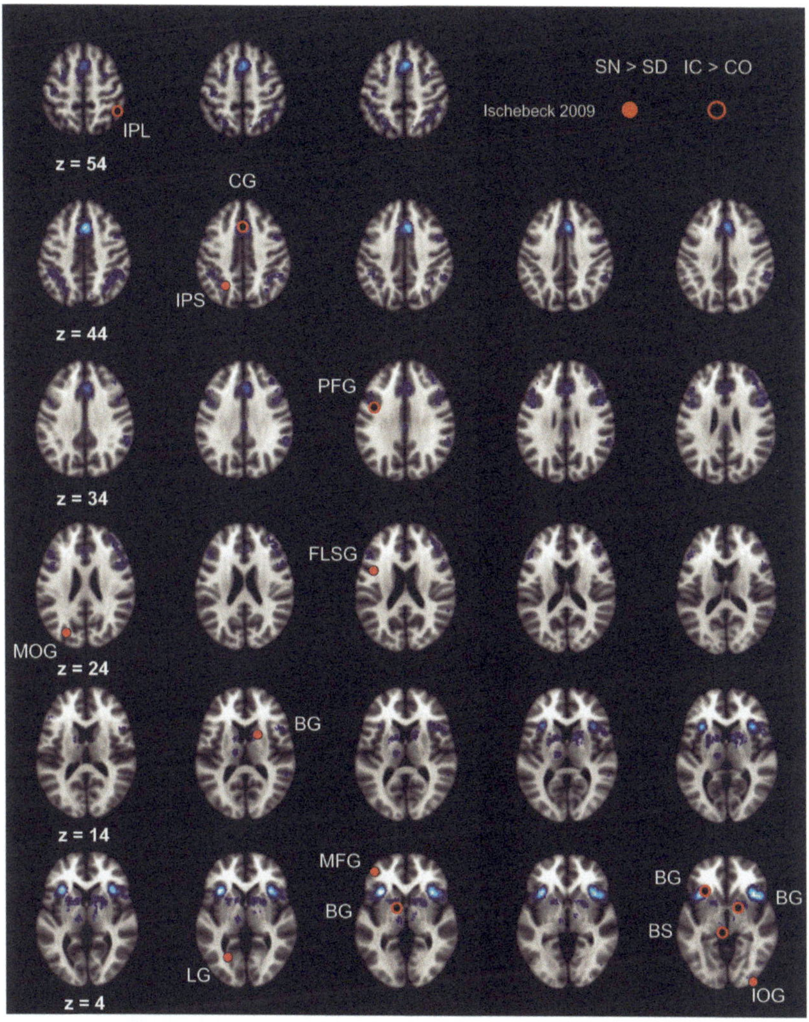


**Fig. 6** Incongruency with whole number knowledge, whole brain peaks for studies of rational numbers. ROIs: incongruency contrasts of Ischebeck et al. (2009), peaks from Table 4; Functional clusters: neurosynth meta-analysis map of "Inhibition" (601 studies); Underlay brain: Mango Sample Image, Montreal Neurological Institute Coordinates. SN > SD = Same Numerator > Same Denominator, IC > CO = Incongruent > Congruent. IPS regions identified from the Juelich Histological Atlas, all other regions from the Talairach Daemon. *BG*, basal ganglia; *BS*, brainstem; *CG*, cingulated gyrus; *FLSG*, frontal lobe sub-gyral; *IOG*, inferior occipital gyrus; *IPS*, intraparietal sulcus; *LG*, lingual gyrus; *MFG*, medial frontal gyrus; *MOG*, medial occipital gyrus; *PFG*, precentral gyrus.

Considerable evidence points to the intraparietal sulcus as the key locus of whole number magnitude processing, both for symbolic and nonsymbolic stimuli (Arsalidou & Taylor, 2011; Sokolowski et al., 2017). The neural signature of magnitude process is sensitivity to the numerical distance between items, that is, greater activity for *near* (8 vs 9) than *far* comparisons (2 vs 9). Crucially, these effects are found in active comparison tasks, but also in passive viewing adaptation designs (Piazza, Pinel, Le Bihan, & Dehaene, 2007; Sokolowski et al., 2017), indicating that task difficulty alone does not drive neural distance effects. Developmental studies find that sensitivity to nonsymbolic distance is present early (Ansari & Dhital, 2006; Cantlon, Brannon, Carter, & Pelphrey, 2006) but develops slowly for symbolic stimuli (Ansari, Garcia, Lucas, Hamon, & Dhital, 2005; Bugden, Price, McLean, & Ansari, 2012; Rosenberg-Lee, Adair, & Menon, 2014). This pattern of results is consistent with the view that the ability of arbitrary number symbols to evoke magnitude representations in the IPS develops slowly with experience (Rosenberg-Lee, Tsang, & Menon, 2009) and suggests that accurate representations of rational number magnitudes should also engage the IPS in proficient adults.

This prediction is borne out in the handful of studies that have examined brain responses to rational numbers. Jacob and Nieder (2009b) used a passive viewing design where participants saw multiple instances of the same ratio (e.g., adapting to 1:5) represented either as pairs of unsegmented lines or sets of colored dots and then were presented with a deviant ratio (e.g., 4:5). They found that the recovery of BOLD signal scaled with the distance from the adapted ratio, in both the left and right IPS (Fig. 4). To assess the overlap with regions involved in nonsymbolic whole number processing, they also collected a dots adaptation task, finding surprisingly consistent activity between the two formats in their adult participants. A second study from this group used this same framework to examine symbolic processing, both of numerals and of fraction terms like "half" (Jacob & Nieder, 2009a). Again, they reported activity in the IPS which was sensitive to the numerical distance between the adapted and deviant stimuli, now in the symbolic domain (Fig. 5). Adaptation studies of this type are useful because they demonstrate that brain responses do not simply reflect behavioral difficulty effects but also abstract representation (Binder, Desai, Graves, & Conant, 2009). Further, in the current context, they demonstrate that the same regions employed for whole numbers are used for symbolic fraction and nonsymbolic symbolic ratios. However, neither study contrasted congruent and incongruent stimuli and so do not provide information on the brain loci needed to resolve whole number interference.

Only one study to date has attempted to identify brain regions sensitive to both rational and whole distance within a single task and sample ($n = 17$, Ischebeck et al., 2009). Using single-digit proper and improper fractions, they presented stimuli across four conditions. The common components' problems (i.e., same numerators, same denominators) followed typical practices in the field. However, for without common components problems, they assumed that individuals know that larger denominators indicate smaller numbers, so they considered problems such as 2/7 versus 3/6 as a congruent and 2/6 versus 3/7 as an incongruent, while the standard nomenclature would have designated them neutral and congruent, respectively (Fig. 1). Although their behavioral results were consistent with their designation (participants were slower on incongruent relative to congruent problems), it calls to question if these stimuli were tapping whole number interference, per se. Finally, rational distance varied continuously in their stimuli, but they explicitly manipulate whole distance into a two-level factor by making numerators or denominators 1 or 3 units apart, while keeping the other value the same (common components) or differing only by 1 (without common components).

Interestingly, whole distance drove accuracy, while rational distance explained response times, especially on their "incongruent" problems. Most crucially, with respect to brain activity, the IPS was sensitive to rational distance regardless of condition and no brain region was related to whole distance (Fig. 5). This is the only study to report neural congruency effects, finding greater activity for common numerator relative to common denominator comparisons, and incongruent relative to congruent comparisons, in regions which largerly overlapped those involved in inhibitory control, including medial and lateral prefrontal regions. Fig. 6 (Table 4) plots these locations with respect to "inhibition" studies identified by Neurosynth.org (Yarkoni et al., 2011).

More recently, DeWolf et al. (2016) applied representational similarity analysis (RSA) to examine relations between activity patterns for whole numbers and rational numbers represented as fractions or decimals. Stimuli were single-digit fractions, the equivalent values as decimals, rounded to 2 decimal points and the corresponding whole number quantities. For example, the fractions 1/9 versus 3/7 correspond to 0.11 versus 0.43 as decimals and 11 versus 43 as whole numbers. The authors employed a sequential presentation order and only analyzed the brain responses from the first stimuli. This approach is ideal for RSA because each number is presented in isolation and so its brain activity can be assessed without the potential confounding effects of processing the other number, or the brain responses to

**Table 4** Incongruency with whole number knowledge, wholebrain peaks for studies of rational numbers (plotted in Fig. 6).

| | | MNI coordinates | | |
|---|---|---|---|---|
| **Study** | **Contrast** | **x** | **y** | **z** |
| Ischebeck et al. (2009) | Same Numerator > Same Denominator[a] | −42 | 6 | 21 |
| | | −42 | 48 | 0 |
| | | 18 | 9 | 12 |
| | | −24 | −57 | 42 |
| | | 36 | −93 | −3 |
| | | −27 | −78 | 24 |
| | | −24 | −66 | 3 |
| | Incongruent > Congruent | −42 | 0 | 30 |
| | | 0 | 21 | 42 |
| | | −30 | 24 | −3 |
| | | 39 | 21 | −6 |
| | | −12 | 0 | 0 |
| | | 15 | 3 | −3 |
| | | −6 | −30 | −3 |
| | | 48 | −48 | 54 |

[a] In Table 3 of Ischebeck et al. (2009), this contrast is labeled as SD > SN (Same Denominator > Same Numerator). However, the text and Fig. 3 describe the contrast as Same Numerator > Same Denominator, thus we are using that label and interpretation here.

the comparison action (Dimsdale-Zucker & Ranganath, 2018). However, sequential analyses preclude assessment of numerical distance effects in brain responses.

RSA is a multivariate neuroimaging data analysis technique that involves extracting voxels for different experimental conditions and computing the similarity (e.g., Pearson correlation) between them (Kriegeskorte, Mur, & Bandettini, 2008). Using a searchlight approach over an IPS region of interest, Dewolf and colleagues found that activation patterns for fractions (e.g., 1/3) differed from decimal (e.g., 0.33) and whole (33) number patterns, which were indistinguishable from each other (DeWolf et al., 2016). Univariate analyses were convergent, finding greater activity for fractions in the IPS than for decimals or whole numbers. While these results confirmed that fractions are processed in the IPS, the experimental design of the decimals stimuli complicate interpretation of their results. Specifically, since the decimal stimuli were always presented with two significant digits, there is no need to process these stimuli with respect to their rational number magnitude, likely accounting for the high similarity between decimal and whole numbers.

To determine if magnitude processing regions are shared across rational number formats, Mock et al. (2018) examined two formats of symbolic (fractions and decimals) and nonsymbolic (unsegmented pie charts and dot sets) rational number representation. Base stimuli were created from single-digit fractions with numerators ranging from 1 to 8 and denominators from 2 to 9. While they manipulated congruency in the fraction stimuli using field typical labels (Fig. 1), they did not report behavioral or imaging contrasts based on this distinction. As in DeWolf et al. (2016), decimal stimuli were created by converting to decimals and truncating at two significant digits. Although the authors did not characterize their stimuli in this fashion, their nonsymbolic stimuli of pie charts and dot proportions end up creating a design which contrasts continuous versus discrete representations. Specifically, in the dot proportions stimuli, the incongruent stimuli would also be considered as counting incompatible as conceptualized in Jeong et al. (2007). Unfortunately, this factor was not contrasted in either their behavioral or brain imaging results. However, it's notable that the two factors which contained incongruent stimuli (fractions and dot proportions) also had the slowest speeds and lowest accuracies of the four formats.

In the imaging analyses, they sought brain regions sensitive to the rational distance between the numerical pairs. In a conjunction analysis, they found clusters in the right IPS sensitive to rational distance. While this result is consistent with the proposal that the IPS is expanded to process rational distance, because they do not orthogonalize whole distance, it's impossible to determine if these results simply reflect the whole distance, which would be driving all the activity, in, for example, the decimals conditions. Consistent with this contention are results from a follow up study from the same group using the same data (Mock et al., 2019). There, they found greater activity for the tasks which they say measure part-whole relationships (fractions, dot proportions, and pie charts) relative to decimals in bilateral IPS and frontal regions. They also compute region of interest-based RSA, finding greater similarity between dots and fractions in both the left and right IPS, while decimals and fractions were least similar. The authors interpreted this result to mean that part-whole relations have different activation patterns than decimals. Alternatively, both fractions and dots proportions draw on inhibitory control because of interference from whole number knowledge leading to similar activation patterns, whereas these specific decimals and continuous pie charts lacked these demands.

In summary, neuroimaging studies of rational numbers find activity and especially distance-based activity (Figs. 4 and 5) in the bilateral IPS and nearby parietal regions that have previously been associated with symbolic and nonsymbolic number processing (Sokolowski et al., 2017). However, insufficient attention has been paid to assessing rational distance effects independently from the confounding influence of whole distance. Moreover, while congruency with whole number knowledge is often manipulated in imaging studies, the contrast is rarely reported for symbolic studies, and never for nonsymbolic studies (Fig. 6). Therefore, the paltry evidence regarding the overlap of inhibitory control systems with regions involved in processing whole number interfering rational numbers means we cannot draw firm conclusions from imaging studies about the role of inhibitory control regions in rational number processing.

## Conclusions and directions for future research

In this chapter, I have considered less studied factors that make rational numbers difficult for many learners. First, I distinguished between properties of whole numbers that never apply to rationals (Category 1) from properties that sometimes, but not always, apply (Category 2). Implicit learning of the whole number properties may further exacerbate difficulties with Category 2 properties. Future research, probing the effects of explicitly highlighting the variable nature of these properties, could assess the effects of implicit understanding on rational number outcomes. A final category of properties (Category 3) does, in fact, apply to rationals, but learners struggle to extend them to this new number system. Thus, I contend that mastery of rational numbers requires refining and expanding one's numerical representation. In terms of refining, while many studies report whole number interference effects, only a handful explicitly test whether interference effects are related to individual differences in inhibitory control. Measuring inhibitory control is challenging, especially across development. I advocate for tasks which tap "semantic inhibition"; empirically these tasks show stronger relations with rational number outcomes compared to response inhibition tasks. Further, from a theoretical perspective, interference effects in rational numbers come from acquired understanding of number systems and thus should draw inhibition of semantic knowledge.

To assess expanding numerical representations, studies contrasting rational and whole distance were reviewed. For fractions, I highlight the inherent correlations between these dimensions (Fig. 1). and suggest that strategy variation further complicates matters, as factors like gap distance may also drive responses, but only among participants sophisticated enough to use this strategy. In contrast, I propose that decimals provide an ideal domain for contrasting rational and whole distance. Specifically, the property of *more digits = larger number* represents a Category 2 property with a well-defined behavioral effect, reducing the potential for strategy variation. From a practical perspective, decimals have only one whole number dimension (versus at least two dimensions for fractions) and there exists a stimuli set that equates whole and rational distance between congruent and incongruent problems, while keeping these factors orthogonal (Fig. 2). This set provides a platform for behavioral and imaging studies to establish the developmental trajectory of magnitude-based processing of rational numbers. In the domain of nonsymbolic representations of rational quantities, I highlight the distinction between continuous and discrete formats (Fig. 3). Further, nonsymbolic counting misleading discrete problems can be connected to symbolic incongruent fraction problems, suggesting that future work should examine individual differences in these effects and relate them to inhibitory control.

Finally, turning to neuroimaging studies, which we evaluated in the context of these theoretical and methodological issues. Notably, while many studies manipulate the congruence of rational number stimuli, only one explicitly reports the neural contrast (Ischebeck et al., 2009). Thus we know almost nothing about how brain regions involved in inhibitory control contribute to successful rational number processing. In terms of identifying distance-based processing of rational numbers, there is some evidence that these effects are found in the IPS, overlapping with, or near to, regions identified in a meta-analysis of whole number processing (Sokolowski et al., 2017). Notably, no studies have examined interference from whole number knowledge in decimals, and studies of symbolic fractions have not adequately addressed confounds across problem types. In terms of nonsymbolic formats, inadequate attention has been paid to the role of continuous versus discrete formats. Representational similarity analyses are an important tool in deciphering neural representations of rational numbers. The finding of greater similarity of fractions to discrete dots relative to continuous pie charts (Mock et al., 2019) supports the contention that whole number interference effects may be driving both impairments on discrete nonsymbolic comparisons and symbolic fractions. Another neuroimaging methodology, event-related potentials may be especially fruitful in this domain, as considerable work has already mapped the neural signatures of conflict processing in both the numerical (Szucs & Soltesz, 2007) and color-word (Atchley, Klee, & Oken, 2017) Stroop tasks, and a growing literature is now examining these signals in fraction comparison (Fu, Li, Xu, & Zeng, 2020; Zhang et al., 2012).

Missing from the imaging literature are examinations of the relations between individual variability in ability and brain responses: Do individuals with greater whole number interference show less magnitude processing for rational numbers? Finally, to date, there are no published imaging studies of rational number processing in children. It is vital to track the expansion and refining of number representations through the acquisition of this challenging material. If we can identify the end state of successful math instruction, then we can combine instruction and neuroimaging to track the efficacy of different instructional approaches in inducing the desired proficient performance profile in both brain and behavior (Rosenberg-Lee, 2018).

## Acknowledgments

The author wishes to thank Moriah Sokolowski, Daniel Ansari, and their coauthors for sharing the nonsymbolic and symbolic metaanalysis maps from Sokolowski et al. (2017). The author also thank Imran Khawaja, Portia Shaheed, and Roberto A. Abreu-Mendoza for figure preparation, and Shannon Cahalan for careful proof reading, and Arthur B. Powell for thoughtful comments on the manuscript. Finally, the author thanks Sheryl Adam for the suggestion of the connection with difficulties finding library books.

## References

Abreu-Mendoza, R. A., Coulanges, L., Ali, K., Powell, A., & Rosenberg-Lee, M. (2020). Children's discrete proportional reasoning is related to inhibitory control and enhanced by priming continuous representations. *Journal of Experimental Child Psychology*, *199*, 104931. https://doi.org/10.1016/j.jecp.2020.104931.

Ansari, D., & Dhital, B. (2006). Age-related changes in the activation of the intraparietal sulcus during nonsymbolic magnitude processing: An event-related functional magnetic resonance imaging study. *Journal of Cognitive Neuroscience*, *18*(11), 1820–1828. Retrieved from http://www.ncbi.nlm.nih.gov/entrez/query.fcgi?cmd=Retrieve&db=PubMed&dopt=Citation&list_uids=17069473.

Ansari, D., Garcia, N., Lucas, E., Hamon, K., & Dhital, B. (2005). Neural correlates of symbolic number processing in children and adults. *Neuroreport*, *16*(16), 1769–1773. Retrieved from http://www.ncbi.nlm.nih.gov/entrez/query.fcgi?cmd=Retrieve&db=PubMed&dopt=Citation&list_uids=16237324.

Arsalidou, M., & Taylor, M. J. (2011). Is 2+2=4? Meta-analyses of brain areas needed for numbers and calculations. *NeuroImage*, *54*(3), 2382–2393. https://doi.org/10.1016/j.neuroimage.2010.10.009.

Atchley, R., Klee, D., & Oken, B. (2017). EEG frequency changes prior to making errors in an easy Stroop task. *Frontiers in Human Neuroscience*, *11*, 521. https://doi.org/10.3389/fnhum.2017.00521.

Avgerinou, V. A., & Tolmie, A. (2019). Inhibition and cognitive load in fractions and decimals. *British Journal of Educational Psychology*. https://doi.org/10.1111/bjep.12321.

Baroody, A. J., & Ginsburg, H. P. (1983). The effects of instruction on children's understanding of the "equals" sign. *The Elementary School Journal*, *84*(2), 199–212. https://doi.org/10.1086/461356.

Best, J. R., & Miller, P. H. (2010). A developmental perspective on executive function. *Child Development, 81*(6), 1641–1660. Retrieved from http://www.jstor.org/stable/40925288.

Binder, J. R., Desai, R. H., Graves, W. W., & Conant, L. L. (2009). Where is the semantic system? A critical review and meta-analysis of 120 functional neuroimaging studies. *Cerebral Cortex, 19*(12), 2767–2796. https://doi.org/10.1093/cercor/bhp055.

Bonato, M., Fabbri, S., Umiltà, C., & Zorzi, M. (2007). The mental representation of numerical fractions: Real or integer? *Journal of Experimental Psychology: Human Perception and Performance, 33*(6), 1410–1419. https://doi.org/10.1037/0096-1523.33.6.1410.

Boyer, T. W., & Levine, S. C. (2015). Prompting children to reason proportionally: Processing discrete units as continuous amounts. *Developmental Psychology, 51*(5), 615–620. https://doi.org/10.1037/a0039010.

Boyer, T. W., Levine, S. C., & Huttenlocher, J. (2008). Development of proportional reasoning: Where young children go wrong. *Developmental Psychology, 44*(5), 1478–1490. https://doi.org/10.1037/a0013110.

Braithwaite, D. W., & Siegler, R. S. (2018). Developmental changes in the whole number bias. *Developmental Science, 21*(2), 13. https://doi.org/10.1111/desc.12541.

Bugden, S., Price, G. R., McLean, D. A., & Ansari, D. (2012). The role of the left intraparietal sulcus in the relationship between symbolic number processing and children's arithmetic competence. *Developmental Cognitive Neuroscience*. https://doi.org/10.1016/j.dcn.2012.04.001.

Cantlon, J. F., Brannon, E. M., Carter, E. J., & Pelphrey, K. A. (2006). Functional imaging of numerical processing in adults and 4-y-old children. *PLoS Biology, 4*(5). https://doi.org/10.1371/journal.pbio.0040125, e125.

Carey, S. (2011). Precis of 'The origin of concepts'. *The Behavioral and Brain Sciences, 34*(3), 113–124. discussion 124–162 https://doi.org/10.1017/S0140525X10000919.

Cavanaugh, K., Huizinga, M., Wallston, K.A., Gebretsadik, T., Shintani, A., Davis, D., … Rothman, R. L. (2008). Association of numeracy and diabetes control. *Annals of Internal Medicine, 148*(10), 737–746. https://doi.org/10.7326/0003-4819-148-10-200805200-00006.

Chakkalakal, R. J., Venkatraman, S., White, R. O., Kripalani, S., Rothman, R., & Wallston, K. (2017). Validating health literacy and numeracy measures in minority groups. *Health Literacy Research and Practice, 1*(2), e23–e30. https://doi.org/10.3928/24748307-20170329-01.

Chang, N., Koedinger, K. R., & Lovett, M. C. (2003). Learning spurious correlations instead of deeper relations. In *Paper presented at the proceedings of the 25th Cognitive Science Society* University of California.

Coulanges, L., Abreu-Mendoza, R. A., Varma, S., Uncapher, M. R., Gazzaley, A., Anguera, J., et al. (2021, January 19). Linking inhibitory control to math achievement via comparison of conflicting decimal numbers. *PsychArXiv Preprints*. https://doi.org/10.31234/osf.io/qsuz9.

DeWolf, M., Chiang, J. N., Bassok, M., Holyoak, K. J., & Monti, M. M. (2016). Neural representations of magnitude for natural and rational numbers. *NeuroImage, 141*, 304–312. https://doi.org/10.1016/j.neuroimage.2016.07.052.

DeWolf, M., & Vosniadou, S. (2015). The representation of fraction magnitudes and the whole number bias reconsidered. *Learning and Instruction, 37*, 39–49. https://doi.org/10.1016/j.learninstruc.2014.07.002.

Diamond, A. (2013). Executive functions. *Annual Review of Psychology, 64*, 135–168. https://doi.org/10.1146/annurev-psych-113011-143750.

Dimsdale-Zucker, H. R., & Ranganath, C. (2018). Chapter 27 – Representational similarity analyses: A practical guide for functional MRI applications. In D. Manahan-Vaughan (Ed.), *Vol. 28. Handbook of behavioral neuroscience* (pp. 509–525). Elsevier.

Eriksen, B. A., & Eriksen, C. W. (1974). Effects of noise letters upon the identification of a target letter in a nonsearch task. *Perception & Psychophysics, 16*(1), 143–149. https://doi.org/10.3758/BF03203267.

Friedman, N. P., & Miyake, A. (2004). The relations among inhibition and interference control functions: A latent-variable analysis. *Journal of Experimental Psychology: General, 133*(1), 101–135. https://doi.org/10.1037/0096-3445.133.1.101.

Friso-van den Bos, I., van der Ven, S. H. G., Kroesbergen, E. H., & van Luit, J. E. H. (2013). Working memory and mathematics in primary school children: A meta-analysis. *Educational Research Review, 10*, 29–44. https://doi.org/10.1016/j.edurev.2013.05.003.

Fu, X., Li, X., Xu, P., & Zeng, J. (2020). Inhibiting the whole number bias in a fraction comparison task: An event-related potential study. *Psychology Research and Behavior Management, 13*, 245–255. https://doi.org/10.2147/PRBM.S240263.

Fuchs, L. S., Malone, A. S., Schumacher, R. F., Namkung, J., & Wang, A. (2017). Fraction intervention for students with mathematics difficulties: Lessons learned from five randomized controlled trials. *Journal of Learning Disabilities, 50*(6), 631–639. https://doi.org/10.1177/0022219416677249.

Fuchs, L. S., Schumacher, R. F., Sterba, S. K., Long, J., Namkung, J., Malone, A., … Changas, P. (2014). Does working memory moderate the effects of fraction intervention? An aptitude-treatment interaction. *Journal of Educational Psychology, 106*(2), 499–514. https://doi.org/10.1037/a0034341.

Gelman, R., & Gallistel, C. R. (1978). *The child's understanding of number*. Cambridge, MA: Harvard University Press.

Gerardi, K., Goette, L., & Meier, S. (2013). Numerical ability predicts mortgage default. *Proceedings of the National Academy of Sciences of the United States of America, 110*(28), 11267–11271. https://doi.org/10.1073/pnas.1220568110.

Gómez, D. M., Jiménez, A., Bobadilla, R., Reyes, C., & Dartnell, P. (2015). The effect of inhibitory control on general mathematics achievement and fraction comparison in middle school children. *ZDM, 47*(5), 801–811. https://doi.org/10.1007/s11858-015-0685-4.

Handel, M. J. (2016). What do people do at work? *Journal for Labour Market Research, 49*(2), 177–197. https://doi.org/10.1007/s12651-016-0213-1.

Hansen, N., Jordan, N. C., Fernandez, E., Siegler, R. S., Fuchs, L., Gersten, R., & Micklos, D. (2015). General and math-specific predictors of sixth-graders' knowledge of fractions. *Cognitive Development, 35*, 34–49. https://doi.org/10.1016/j.cogdev.2015.02.001.

Harnishfeger, K. K. (1995). The development of cognitive inhibition: Theories, definitions, and research evidence. In F. N. Dempster, & C. J. Brainerd (Eds.), *Interference and inhibition in cognition*. San Diego: Academic Press.

Harnishfeger, K. K., & Pope, R. S. (1996). Intending to forget: The development of cognitive inhibition in directed forgetting. *Journal of Experimental Child Psychology, 62*(2), 292–315. https://doi.org/10.1006/jecp.1996.0032.

Henik, A., & Tzelgov, J. (1982). Is 3 greater than 5 – The relation between physical and semantic size in comparison tasks. *Memory & Cognition, 10*(4), 389–395. Retrieved from <Go to ISI>://A1982PE73100011.

Huber, S., Klein, E., Willmes, K., Nuerk, H. C., & Moeller, K. (2014). Decimal fraction representations are not distinct from natural number representations – evidence from a combined eye-tracking and computational modeling approach. *Frontiers in Human Neuroscience, 8*, 172. https://doi.org/10.3389/fnhum.2014.00172.

Hurst, M. A., & Cordes, S. (2018). Attending to relations: Proportional reasoning in 3-to 6-year-old children. *Developmental Psychology, 54*(3), 428–439. https://doi.org/10.1037/dev0000440.

Ischebeck, A., Schocke, M., & Delazer, M. (2009). The processing and representation of fractions within the brain: An fMRI investigation. *NeuroImage, 47*(1), 403–413. https://doi.org/10.1016/j.neuroimage.2009.03.041.

Jacob, S. N., & Nieder, A. (2009a). Notation-independent representation of fractions in the human parietal cortex. *Journal of Neuroscience, 29*(14), 4652–4657. https://doi.org/10.1523/jneurosci.0651-09.2009.
Jacob, S. N., & Nieder, A. (2009b). Tuning to non-symbolic proportions in the human frontoparietal cortex. *European Journal of Neuroscience, 30*(7), 1432–1442. https://doi.org/10.1111/j.1460-9568.2009.06932.x.
Jeong, Y., Levine, S. C., & Huttenlocher, J. (2007). The development of proportional reasoning: Effect of continuous versus discrete quantities. *Journal of Cognition and Development, 8*(2), 237–256. https://doi.org/10.1080/15248370701202471.
Kainulainen, M., McMullen, J., & Lehtinen, E. (2017). Early developmental trajectories toward concepts of rational numbers. *Cognition and Instruction, 35*(1), 4–19. https://doi.org/10.1080/07370008.2016.1251287.
Kallai, A. Y., & Tzelgov, J. (2014). Decimals are not processed automatically, not even as being smaller than one. *Journal of Experimental Psychology. Learning, Memory, and Cognition, 40*(4), 962–975. https://doi.org/10.1037/a0035782.
Kaplan, A., & Great Source Education Group. (2003). *Math on call: A mathematics handbook*. Perfection Learning Corporation.
Kipp, K. (2005). A developmental perspective on the measurement of cognitive deficits in attention-deficit/hyperactivity disorder. *Biological Psychiatry, 57*(11), 1256–1260. https://doi.org/10.1016/j.biopsych.2005.03.012.
Kriegeskorte, N., Mur, M., & Bandettini, P. (2008). Representational similarity analysis - connecting the branches of systems neuroscience. *Frontiers in Systems Neuroscience, 2*, 4. https://doi.org/10.3389/neuro.06.004.2008.
Lee, M. (2010). The advent of decimalisation in Britain: 1971. *The Historian, 106*, 20–23.
Lee, K., & Lee, H. W. (2019). Inhibition and mathematical performance: Poorly correlated, poorly measured, or poorly matched? *Child Development Perspectives, 13*(1), 28–33. https://doi.org/10.1111/cdep.12304.
Lerner, M. D., & Lonigan, C. J. (2014). Executive function among preschool children: Unitary versus distinct abilities. *Journal of Psychopathology and Behavioral Assessment, 36*(4), 626–639. https://doi.org/10.1007/s10862-014-9424-3.
Matthes, B., Christoph, B., Janik, F., & Ruland, M. (2014). Collecting information on job tasks—An instrument to measure tasks required at the workplace in a multi-topic survey. *Journal for Labour Market Research, 47*(4), 273–297. https://doi.org/10.1007/s12651-014-0155-4.
Matthews, P. G., Lewis, M. R., & Hubbard, E. M. (2016). Individual differences in nonsymbolic ratio processing predict symbolic math performance. *Psychological Science, 27*(2), 191–202. https://doi.org/10.1177/0956797615617799.
McNeil, N. M., & Alibali, M. W. (2005). Knowledge change as a function of mathematics experience: All contexts are not created equal. *Journal of Cognition and Development, 6*(2), 285–306. https://doi.org/10.1207/s15327647jcd0602_6.
Meert, G., Gregoire, J., & Noel, M. P. (2009). Rational numbers: Componential versus holistic representation of fractions in a magnitude comparison task. *Querterly Journal of Experimental Psychology (Colchester), 62*(8), 1598–1616. https://doi.org/10.1080/17470210802511162.
Meert, G., Gregoire, J., & Noel, M. P. (2010). Comparing the magnitude of two fractions with common components: Which representations are used by 10- and 12-year-olds? *Journal of Experimental Child Psychology, 107*(3), 244–259. https://doi.org/10.1016/j.jecp.2010.04.008.
Merkley, R., Thompson, J., & Scerif, G. (2015). Of huge mice and tiny elephants: Exploring the relationship between inhibitory processes and preschool math skills. *Frontiers in Psychology, 6*, 1903. https://doi.org/10.3389/fpsyg.2015.01903.
Miyake, A., Friedman, N. P., Emerson, M. J., Witzki, A. H., Howerter, A., & Wager, T. D. (2000). The unity and diversity of executive functions and their contributions to complex "frontal lobe" tasks: A latent variable analysis. *Cognitive Psychology, 41*(1), 49–100. https://doi.org/10.1006/cogp.1999.0734.

Mock, J., Huber, S., Bloechle, J., Bahnmueller, J., Moeller, K., & Klein, E. (2019). Processing symbolic and non-symbolic proportions: Domain-specific numerical and domain-general processes in intraparietal cortex. *Brain Research, 1714*, 133–146. https://doi.org/10.1016/j.brainres.2019.02.029.

Mock, J., Huber, S., Bloechle, J., Dietrich, J. F., Bahnmueller, J., Rennig, J., … Moeller, K. (2018). Magnitude processing of symbolic and non-symbolic proportions: An fMRI study. *Behavioral and Brain Functions, 14*(1), 9. https://doi.org/10.1186/s12993-018-0141-z.

Moyer, R. S., & Landauer, T. K. (1967). Time required for judgements of numerical inequality. *Nature, 215*(5109), 1519–1520. Retrieved from http://www.ncbi.nlm.nih.gov/pubmed/6052760.

Murphy, J., Long, D., & MacDonald, J. B. (2013). Students' understanding of the Library of Congress call number system. *The Reference Librarian, 54*(2), 103–117. https://doi.org/10.1080/02763877.2013.755418.

National Governors Association Center for Best Practices, & Council of Chief State School Officers. (2010). *Common core state standards for mathematics*. Washington, DC: Authors.

National Mathematics Advisory Panel. (2008). *Foundations for success: The final report of the National Mathematics Advisory Panel*. Retrieved from Washington, DC.

Ni, Y., & Zhou, Y.-D. (2005). Teaching and learning fraction and rational numbers: The origins and implications of whole number bias. *Educational Psychologist, 40*(1), 27–52. https://doi.org/10.1207/s15326985ep4001_3.

Nuerk, H.-C., Moeller, K., Klein, E., Willmes, K., & Fischer, M. H. (2011). Extending the mental number line. *Zeitschrift für Psychologie, 219*(1), 3–22. https://doi.org/10.1027/2151-2604/a000041.

Obersteiner, A., Dresler, T., Bieck, S. M., & Moeller, K. (2019). Understanding fractions: Integrating results from mathematics education, cognitive psychology, and neuroscience. In A. Norton, & M. W. Alibali (Eds.), *Constructing number: Merging perspectives from psychology and mathematics education* (pp. 135–162). Cham: Springer International Publishing.

Obersteiner, A., & Tumpek, C. (2016). Measuring fraction comparison strategies with eye-tracking. *ZDM—The International Journal on Mathematics Education, 48*(3), 255–266. https://doi.org/10.1007/s11858-015-0742-z.

Obersteiner, A., Van Dooren, W., Van Hoof, J., & Verschaffel, L. (2013). The natural number bias and magnitude representation in fraction comparison by expert mathematicians. *Learning and Instruction, 28*, 64–72. https://doi.org/10.1016/j.learninstruc.2013.05.003.

Osborn, C. Y., Wallston, K. A., Shpigel, A., Cavanaugh, K., Kripalani, S., & Rothman, R. L. (2013). Development and validation of the general health numeracy test (GHNT). *Patient Education and Counseling, 91*(3), 350–356. https://doi.org/10.1016/j.pec.2013.01.001.

Peng, P., Namkung, J., Barnes, M., & Sun, C. (2016). A meta-analysis of mathematics and working memory: Moderating effects of working memory domain, type of mathematics skill, and sample characteristics. *Journal of Educational Psychology, 108*(4), 455–473. https://doi.org/10.1037/edu0000079.

Piazza, M., Pinel, P., Le Bihan, D., & Dehaene, S. (2007). A magnitude code common to numerosities and number symbols in human intraparietal cortex. *Neuron, 53*(2), 293–305. Retrieved from http://www.ncbi.nlm.nih.gov/entrez/query.fcgi?cmd=Retrieve&db=PubMed&dopt=Citation&list_uids=17224409.

Pincus, M., Coulanges, L., Abreu-Mendoza, R. A., Mill, R., Cole, M. W., & Rosenberg-Lee, M. (2020). Whole vs. rational: A preliminary study of neural distance effects in decimal number comparison. In *Paper presented at the Organization for Human Brain Mapping, Montreal, QC, Canada, June 2020.*

Roell, M., Viarouge, A., Houde, O., & Borst, G. (2017). Inhibitory control and decimal number comparison in school-aged children. *PLoS One, 12*(11). https://doi.org/10.1371/journal.pone.0188276, e0188276.

Roell, M., Viarouge, A., Houde, O., & Borst, G. (2019). Inhibition of the whole number bias in decimal number comparison: A developmental negative priming study. *Journal of Experimental Child Psychology*, *177*, 240–247. https://doi.org/10.1016/j.jecp.2018.08.010.
Rosenberg-Lee, M. (2018). Training studies: An experimental design to advance educational neuroscience. *Mind, Brain and Education*, *12*(1), 12–22. https://doi.org/10.1111/mbe.12166.
Rosenberg-Lee, M., Adair, E., & Menon, V. (2014). Neural and behavioral dissociations between symbolic and non-symbolic magnitude processing and mathematical ability in children. In *Paper presented at the EARLI SIG 22 educational neuroscience, Gottigen, Germany*.
Rosenberg-Lee, M., Tsang, J. M., & Menon, V. (2009). Symbolic, numeric, and magnitude representations in the parietal cortex. *Behavioral and Brain Sciences*, *32*(3–4), 350–351. https://doi.org/10.1017/S0140525x09990860.
Schneider, M., Beeres, K., Coban, L., Merz, S., Susan Schmidt, S., Stricker, J., & De Smedt, B. (2017). Associations of non-symbolic and symbolic numerical magnitude processing with mathematical competence: A meta-analysis. *Developmental Science*, *20*(3), 1–16. https://doi.org/10.1111/desc.12372.
Siegler, R. S. (1999). Strategic development. *Trends in Cognitive Sciences*, *3*(11), 430–435.
Siegler, R. S., Duncan, G. J., Davis-Kean, P. E., Duckworth, K., Claessens, A., Engel, M., … Chen, M. C. (2012). Early predictors of high school mathematics achievement. *Psychological Science*, *23*(7), 691–697. https://doi.org/10.1177/0956797612440101.
Siegler, R. S., Thompson, C. A., & Schneider, M. (2011). An integrated theory of whole number and fractions development. *Cognitive Psychology*, *62*(4), 273–296. https://doi.org/10.1016/j.cogpsych.2011.03.001.
Sokolowski, H. M., Fias, W., Mousa, A., & Ansari, D. (2017). Common and distinct brain regions in both parietal and frontal cortex support symbolic and nonsymbolic number processing in humans: A functional neuroimaging meta-analysis. *NeuroImage*, *146*, 376–394. https://doi.org/10.1016/j.neuroimage.2016.10.028.
Stroop, J. (1992). "Reprint of" studies of interference in serial verbal reactions: Original work published 1935 (Vol. 18, pp. 643–662). *Journal of Experimental Psychology; General*, *121*(1), 15.
Szucs, D., & Soltesz, F. (2007). Event-related potentials dissociate facilitation and interference effects in the numerical Stroop paradigm. *Neuropsychologia*, *45*(14), 3190–3202. https://doi.org/10.1016/j.neuropsychologia.2007.06.013.
Vamvakoussi, X., Van Dooren, W., & Verschaffel, L. (2012). Naturally biased? In search for reaction time evidence for a natural number bias in adults. *The Journal of Mathematical Behavior*, *31*(3), 344–355. https://doi.org/10.1016/j.jmathb.2012.02.001.
Van Hoof, J., Degrande, T., Ceulemans, E., Verschaffel, L., & Van Dooren, W. (2018). Towards a mathematically more correct understanding of rational numbers: A longitudinal study with upper elementary school learners. *Learning and Individual Differences*, *61*, 99–108. https://doi.org/10.1016/j.lindif.2017.11.010.
Van Hoof, J., Verschaffel, L., & Van Dooren, W. (2017). Number sense in the transition from natural to rational numbers. *British Journal of Educational Psychology*, *87*(1), 43–56. https://doi.org/10.1111/bjep.12134.
Varma, S., & Karl, S. R. (2013). Understanding decimal proportions: Discrete representations, parallel access, and privileged processing of zero. *Cognitive Psychology*, *66*(3), 283–301. https://doi.org/10.1016/j.cogpsych.2013.01.002.
Vera, H. (2012). In E. Ikegami, M. Donnelly, C. Forment, & O. Frankel (Eds.), *The social life of measures metrication in the United States and Mexico, 1789–2004* ProQuest Dissertations Publishing.
Weber, K. (2001). In J. R. Hayes (Ed.), *Investigating and teaching the strategic knowledge needed to construct proofs* ProQuest Dissertations Publishing.

Werner, H. (1957). *Comparative psychology of mental development* (Rev. ed.). New York: International Universities Press.

Woodcock, R. W., McGrew, K. S., & Mather, N. (2001). *Woodcock–Johnson III tests of achievement*. Istasca, IL: Riverside.

Wright, A., & Diamond, A. (2014). An effect of inhibitory load in children while keeping working memory load constant. *Frontiers in Psychology*, *5*, 213. https://doi.org/10.3389/fpsyg.2014.00213.

Yarkoni, T., Poldrack, R. A., Nichols, T. E., Van Essen, D. C., & Wager, T. D. (2011). Large-scale automated synthesis of human functional neuroimaging data. *Nature Methods*, *8*(8), 665–670. https://doi.org/10.1038/nmeth.1635.

Ye, A., Resnick, I., Hansen, N., Rodrigues, J., Rinne, L., & Jordan, N. C. (2016). Pathways to fraction learning: Numerical abilities mediate the relation between early cognitive competencies and later fraction knowledge. *Journal of Experimental Child Psychology*, *152*, 242–263. https://doi.org/10.1016/j.jecp.2016.08.001.

Zhang, L., Chen, M., Lin, C., & Szucs, D. (2014). Flexible and unique representations of two-digit decimals. *Acta Psychologica*, *151*, 89–97. https://doi.org/10.1016/j.actpsy.2014.06.002.

Zhang, L., Xin, Z., Li, F., Wang, Q., Ding, C., & Li, H. (2012). An ERP study on the processing of common fractions. *Experimental Brain Research*, *217*(1), 25–34. https://doi.org/10.1007/s00221-011-2969-4.

CHAPTER 8

# Learning and education in numerical cognition: We do need education

**Bert De Smedt**
Parenting and Special Education Research Unit, Faculty of Psychology and Educational Sciences, KU Leuven, University of Leuven, Leuven, Belgium

## Introduction

In this contribution, I provide some reflections as to why it is important to consider the educational context in research on numerical cognition. Mathematics is in essence a symbolic activity that can only be learned via education. This sounds obvious for the learning of calculation, such as learning to multiply, but the same applies to the seemingly basic processing of symbolic number, which is also acquired via education; and although some of this learning process starts before formal schooling, it still requires instruction, i.e., from preschool teachers, caregivers, or parents. There are huge differences in the educational contexts around the globe in which these number processing and mathematical skills are acquired. For example, learning multidigit numbers will be easier in languages where the decade-unit structure of the Arabic digits is consistent with the number words (e.g., in English 86 reads as eighty-six) compared to languages where this is not the case (e.g., in Dutch 86 reads as six-and-eighty; in French 86 reads as four-times-twenty-and-six) (e.g., Dowker & Nuerk, 2016). The use of finger counting to answer a calculation problem will depend on whether children are allowed to do so during primary school (as is the case in North-America) or not (as is the case in Belgium) (e.g., De Smedt, 2016). The methods taught in school for calculations such as 612 − 591 (e.g., horizontal vs. vertical) and the primary strategies for these calculations (e.g., separately *subtracting* hundreds, decades, and units vs. determining how much needs to be *added* to the second number to arrive at the first) (e.g., Verschaffel et al., 2021) vary across countries and curricula. These curricula are subject to heated controversies (e.g., reform-oriented vs. traditional; see Schoenfeld, 2004), which impacts how mathematics, even something as simple as basic

*Heterogeneous Contributions to Numerical Cognition*
https://doi.org/10.1016/B978-0-12-817414-2.00008-7

number processing, is taught in the classroom. These educational contexts are relatively understudied or acknowledged in neurocognitive studies on numerical cognition but they might have a massive impact on the findings in this research field, and they might explain differences between seemingly similarly designed studies executed in different countries.

In this chapter, I will first illustrate that through (mathematics) education children learn different meanings of Arabic number symbols and that this learning process might explain why Arabic numerals are not so tightly coupled to their quantity referents. I further describe studies that have shown how the association between number processing and arithmetic is moderated by the calculation methods that are taught in the classroom. I next show that learning mathematics in the classroom by itself changes children's processing of number and argue that rather than assuming a unidirectional association between number processing and mathematics learning, this association should be considered as reciprocal. I emphasize that there is a need to conduct experimental studies that manipulate learning to test causal associations between numerical cognition and mathematics learning. I next present the possibility that education can be used as a natural experiment, via the school cutoff design and cross-national comparisons, to understand the broad impact of education on numerical cognition. This chapter ends with some brief concluding thoughts and the suggestion to take education more seriously in future research on numerical cognition.

## The complexity of *basic* symbolic number processing

Symbolic numbers, particularly Arabic numerals, are the crux of mathematics and they have to be learned via education. This learning of Arabic numerals already begins early in development before the start of formal schooling (e.g., Bakker, Torbeyns, Wijns, Verschaffel, & De Smedt, 2019; Merkley & Ansari, 2016; Yuan, Prather, Mix, & Smith, 2019) albeit in nonformal contexts, such as in preschool or the home environment. The acquisition of Arabic numerals has been extensively investigated in the field of numerical cognition, yet it also entails one of its most debated areas (e.g., Merkley & Ansari, 2016). One outstanding question is how these Arabic numerals acquire their semantic meaning, that is, the symbol-grounding problem (e.g., Leibovich & Ansari, 2016): Are these Arabic numerals grounded in children's understanding of (nonsymbolic) quantities, as is assumed by the majority of studies in the field? Or, alternatively, do they acquire their meaning via their relations to other numerical symbols, without (initially) being grounded in the quantities they represent (e.g., Yuan et al., 2019)? Whatever the answer to this conundrum might be, once children are in school, they are quickly introduced to more advanced symbolic numbers that go beyond the small natural numbers that have received the most of the research attention in studies of (early) numerical cognition. It is plausible, as yet untested, that these educational experiences with more complex numbers change the original symbolic representations of these small numbers.

A key example of such an advanced symbolic representation that children quickly learn are fractions. Preschoolers and early elementary school children have a rudimentary understanding of simple fractions, which can be observed in the context of sharing or measurement (e.g., National Mathematics Advisory Panel, 2008; Singer-Freeman & Goswami, 2001). Symbolic fraction instruction typically begins in most countries around third grade at the age 8–9 (Katholiek Onderwijs Vlaanderen, 2021; NCTM, 2006; Vlaams Ministerie van Onderwijs en Vorming 2021). This learning of symbolic fractions requires a revision of prior knowledge of natural symbolic numbers (e.g., Rosenberg-Lee, 2021; Siegler, Fazio, Bailey, & Zhou, 2013, for a discussion). Consider the magnitude and order of the numbers 2 and 4. When children learn the fractions 1/2 and 1/4, they have to understand that 1/2 is larger than 1/4 and that 1/2 comes after 1/4, even though 2 is smaller than 4 and 2 comes before 4. As a result, the symbols 2 and 4 have a different meaning in 1/2 and 1/4 compared to the natural numbers 2 and 4: The previously learned knowledge of natural numbers is not readily applicable anymore to fractions. These educational experiences undoubtedly impact on how we process symbolic numbers, even on numbers as simple as single digits.

This learning of multiple symbolic representations might explain why, in educated adults, symbolic numbers may not be so tightly coupled to their corresponding representations of quantity. For example, Lyons, Ansari, and Beilock (2012) showed in a series of experiments in adults using nonsymbolic, symbolic, and mixed format comparison tasks that Arabic numerals (of natural numbers) do not provide direct access to the quantity they represent. They put forward the hypothesis of *symbolic estrangement*, which posits that over development and with increasing expertise, Arabic numerals become *estranged* from their perceptual nonsymbolic representation of quantity. This hypothesis nicely aligns with a recent analysis of multivariate brain-imaging data (Bulthe, De Smedt, & Op de Beeck, 2018), which combined data from different samples of adults who varied in their arithmetical expertise (Bulthe, De Smedt, & Op de Beeck, 2014; Bulthe et al., 2019). Specifically, Bulthe et al. compared adults with high expertise in arithmetic (e.g., engineers and PhD students in Psychology) (Bulthe et al., 2014), adults with average expertise in arithmetic (e.g., undergraduates in social sciences), and adults with dyscalculia with low expertise in arithmetic, who matched with the latter group in terms of their general education (Bulthe et al., 2019). In all studies, participants had to perform a magnitude comparison task with the numerosities 2, 4, 6, and 8 in both symbolic and nonsymbolic formats, while their brain activity was recorded via fMRI. The authors used MultiVariate Pattern Analysis (MVPA) to test whether brain activity during a given numerosity in the symbolic format (e.g., 4) could be predicted from brain activity in the corresponding nonsymbolic format (e.g., 4 dots) and vice versa, with high generalization from one to the other format indicating a tight coupling between the two representations and a low generalization indicating little or no coupling between the two (see Bulthe et al., 2014, for a more detailed description). The symbolic estrangement hypothesis predicts that with increasing expertise Arabic numerals become estranged from their underlying quantity. If this is the case, then one should observe a negative correlation between the degree of MVPA generalization between symbolic and nonsymbolic number and arithmetic expertise. This is exactly what Bulthe et al. (2018) observed: Individuals with higher arithmetical expertise had weaker associations between symbolic and nonsymbolic representations in the parietal cortex compared with individuals with lower arithmetical expertise. These data all indicate that with increasing experience Arabic numerals may become dissociated from their underlying quantity referent.

The learning of different types of symbolic numbers through education might gradually lead to an associative system of symbolic numbers in which their meaning is defined in terms of relations with other symbols. For example, the following numbers all include the symbol 3, but in each number the symbol 3 denotes different information: 34, 0.31, $-3$, 1/3, $6^3$, $\sqrt[3]{27}$, $\log_3 56$. The more different meanings of a symbolic number people learn through mathematics education, the more such a symbolic number might become disengaged from its quantity referent. This possibility that educational experiences change the symbolic representation of number raises a series of questions. How are these different symbolic representations interconnected? How do these representations develop and (how) do newly learned representations change previously acquired ones? How should we interpret the (parietal) brain activity that is consistently observed during the presentation of Arabic numerals, which might not necessarily reflect the processing of quantity? And, how should we interpret associations between these symbolic representations and mathematics learning in the classroom, when they do not only reflect an understanding of quantity?

## How is numerical cognition related to classroom learning?

Various studies have examined the association between number processing and mathematics learning in the classroom, showing robust and consistent associations between these two (e.g., De Smedt, Noel, Gilmore, & Ansari, 2013; Schneider et al., 2017). In many of these studies, mathematics learning is operationalized by means of general standardized achievement tests. The use of such general tests is only a first step in understanding the complex interrelations between number processing and mathematics learning in the classroom. Such general achievement tests are important to understand *what* predicts mathematics outcomes in the shorter or longer run, and consequently, to inform the development of screening measures that can identify children who are at risk for developing difficulties in mathematics learning (e.g., Brankaer, Ghesquière, & De Smedt, 2017; Nosworthy, Bugden, Archibald, Evans, & Ansari, 2013). These general achievement tests, however, are not informative to understand *why* number processing is related to mathematics learning in the classroom. This is because such general tests typically average performance across a wide range of mathematical abilities, all of which might show differential associations with number processing. The content of these general achievement tests further varies across ages, which all makes it hard to interpret differences in correlations between different ages and different studies.

One step further in understanding this association between number processing and mathematics learning is to narrow down mathematics learning to specific mathematical abilities (e.g., Vanbinst & De Smedt, 2016, for a discussion). This allows researchers to understand on a more functional level how number processing is related to different types of mathematics learning. Illustrative for this approach, an increasing number of studies has therefore investigated the associations between number processing and development of (single-digit) arithmetic (e.g., Bellon, Fias, & De Smedt, 2019; Chu, VanMarle, Rouder, & Geary, 2018; Traff, 2013; Vanbinst, Ansari, Ghesquière, & De Smedt, 2016; Vanbinst, Ghesquière, & De Smedt, 2012, 2015).

In one of the first studies taking this approach, Vanbinst et al. (2012) investigated how number processing was correlated with the use of different strategies during single-digit arithmetic, i.e., whether children directly retrieved the answer to an arithmetic problem from their long-term memory or whether they used a procedure, such as counting or decomposing a problem into smaller problems. Vanbinst et al. observed that children with better symbolic magnitude processing skills retrieved more facts from their long-term memory and were faster in executing retrieval as well as procedural strategies. The authors suggested that a better understanding of symbolic number provides a scaffold for learning arithmetic facts, as facts with meaningful numbers are easier to store and to recall. A good understanding of symbolic number might also help to develop more advanced procedural strategies (i.e., decompositions of numbers to cross 10), which in turn might lead to higher arithmetical fluency and faster transitions from procedural strategies to arithmetic fact retrieval. This prediction was further tested and confirmed in a longitudinal study that followed children from the start of schooling up to second grade (Vanbinst et al., 2015). In this study, symbolic magnitude processing at the start of schooling predicted arithmetic in the middle of first grade and arithmetic at the start of second grade. Symbolic magnitude processing at the start of schooling also predicted the development in arithmetic from first to second grade: Children with better symbolic magnitude processing skills showed a larger development in arithmetic fluency and a larger increase in the reliance on fact retrieval. More recently, Chu et al. (2018) further showed that children's sophistication of arithmetic strategies at the start of elementary school was predicted by their number processing skills in kindergarten, which in turn were predicted by their knowledge of cardinality in preschool. These data all suggest a link between individual differences in children's understanding of number and an increasing sophistication of the strategies they use during arithmetic.

When children learn arithmetic in the classroom, they are also instructed in different methods for doing these calculations (e.g., Kilpatrick, Swafford, & Findell, 2001; Thompson, 1999; Verschaffel, Greer, & De Corte, 2007). The introduction of these methods varies across curricula and countries and, critically, these methods involve different cognitive processes that might differentially rely on children's number processing skills. This hypothesis has been posited by many mathematics educators (e.g., Thompson, 1999; Van den Heuvel-Panhuizen, 2001; Verschaffel et al., 2007), but only very few studies have empirically tested this.

One exception to this trend is a study by Linsen, Verschaffel, Reynvoet, and De Smedt (2015), who investigated this prediction by focusing on children's multidigit subtraction in Flemish primary school children. In Flanders, children are instructed in two major methods for performing multidigit subtraction, namely, mental arithmetic and algorithmic computation. Mental arithmetic hinges on (horizontally) performing operations on numbers: 98–33 can be solved as 98–30 = 68, 68–3 = 65, or by noticing that 98 is close to 100 and 100–33 = 67, 67–2 = 65. Algorithmic computation uses the written algorithm, which operates on digits: 98–33 is then (vertically) solved by first subtracting the units (8–3 = 5) and then the decades (9–3 = 6). The algorithm depends on a fixed sequence of well-defined elementary calculation steps that have to be performed on the digits in the problem and when one strictly follows this sequence, a correct response is guaranteed. In mental arithmetic, however, one does not always routinely follow a fixed solution path, yet one adapts the solution strategy to the specific numerical features of the problem (e.g., 83–49 can be solved by first subtracting the number 50 from 83—which is computationally much easier – and then adding the remainder 1; 83–50 + 1). Against this background, several mathematics educators have posited that there will be a stronger association between number processing and mental arithmetic, as children use their understanding of numbers to select various strategies, whereas such association with algorithmic computation will be much smaller or even absent (Thompson, 1999; Van den Heuvel-Panhuizen, 2001; Verschaffel et al., 2007). Linsen et al. (2015) investigated this hypothesis by correlating performance during mental arithmetic and algorithmic computation with measures of number processing (symbolic and nonsymbolic magnitude comparison). They observed that individual differences in number processing correlated with individual differences in multidigit subtraction. Critically, and in line with the abovementioned prediction from various mathematics educators, Linsen et al. found that these associations were stronger for mental arithmetic than for algorithmic computation.

Taken together, the data by Linsen et al. indicate that the method that is used during multidigit subtraction determines the strength of the association between number processing and arithmetic. It needs to be acknowledged that curricula vary highly in their instructional focus on mental arithmetic (i.e., more reform-oriented approaches) vs algorithmic computation (i.e., more traditional approaches). It is therefore critical for studies that investigate the association between number processing and arithmetic—and by extension those who investigate associations between number processing and other mathematical abilities—to know, report, and consider how arithmetic is learned in the classroom. This will help to contextualize and understand similarities and differences between studies that report on these associations between number processing and mathematics learning at different ages and in different countries (e.g., see "Cross-national comparisons" section).

It is important to note that most of the studies on number processing and mathematics achievement, including the studies that have focused on specific mathematical abilities, are cross-sectional. Such studies cannot determine the direction of the association between number processing and mathematics. It is often (albeit implicitly) assumed that so-called (neuro) cognitive variables, such as number processing or working memory, are, by their (neuro)cognitive nature, more predictive or foundational of academic achievement. However, the predictive value of a particular (neuro)cognitive variable such as number processing is not an inherent characteristic of the variable itself. Instead, this predictive value can only be inferred by the research design through which that (neuro)cognitive variable is investigated. This design needs at least to be longitudinal to verify the extent to which a (neuro)cognitive variable remains to be a predictor of later mathematics achievement once critical confounding variables are controlled for. Panel longitudinal designs, which measure all variables under study at all time points, are the preferred way to do this. Such panel designs have the potential to also investigate the equally plausible, as yet uninvestigated hypothesis, that mathematics achievement itself *predicts* its so-called (neuro)cognitive predictors, such as number processing, as I will discuss in the next section.

## Classroom learning changes its (neuro)cognitive predictors

Most studies in numerical cognition have started from the conventional unidirectional hypothesis that basic (neuro)cognitive numerical abilities, such as numerical magnitude processing or knowledge of the Arabic number system, are foundational and predictive of later mathematics achievement. Although there is ample evidence that supports this idea (e.g., De Smedt et al., 2013; Gobel, Watson, Lervag, & Hulme, 2014; Schneider et al., 2017), it is also plausible that learning mathematics via education by itself changes its so-called foundational predictors. For example, when children receive arithmetic instruction in first grade and understand that $3+4=7$, they learn, perhaps implicitly, that as a result of the addition process, 7 must be larger than 3 or 4 (magnitude) and 7 must come after 3 or 4 (order). This suggests that there is also an impact of mathematics education on so-called foundational (neuro)cognitive abilities and that there might be a bidirectional relation between (neuro)cognitive predictors and academic skill learning in the classroom.

This hypothesis of bidirectional associations between neurocognitive predictors and academic achievement has been well documented in the field of reading. Phonological awareness, a critical cognitive precursor of reading ability, and learning to read are reciprocal, such that learning to read also improves children's phonological abilities (e.g., Perfetti, Beck, Bell, & Hughes, 1987). There is even an increasing amount of evidence that such bidirectional relations also exist for domain-general predictors of academic performance (e.g., Peng & Kievit, 2020, for a review), such as working memory (Miller-Cotto & Byrnes, 2020). Is this hypothesis of bidirectionality also true for the learning of arithmetic?

Vanbinst, Ghesquière, and De Smedt (2019) were the first to test this possibility of reciprocal associations and investigated whether learning to calculate had an impact on children's symbolic magnitude processing skills. They followed 56 children from the beginning (Grade 1, $M_{age} = 6.3$ years) up to the end (Grade 6, $M_{age} = 11.1$ years) of primary education. Vanbinst et al. examined the cross-lagged associations between children's arithmetic abilities and their symbolic magnitude processing skills, which were assessed at both time points. These cross-lagged associations were moderate and they were similar in size. After controlling for processing speed and IQ, arithmetic in grade 1 predicted symbolic magnitude processing in Grade 6 ($\beta = 0.425$, $p < 0.01$), and symbolic magnitude processing in Grade 1 predicted arithmetic in Grade 6 ($\beta = 0.430$, $p < 0.01$), which replicated a large number of studies in the numerical cognition field. These cross-lagged associations already provided some first indication of a reciprocal association. Vanbinst et al. further performed a very stringent, yet critical, regression analysis that additionally controlled for the autoregressive effects of prior ability. Arithmetic in grade 1 predicted symbolic magnitude processing skills in grade 6, controlling for symbolic magnitude processing skills in grade 1, but the effect was small and marginally significant ($\beta = 0.302$, $p = 0.065$). Turning to the reverse direction, symbolic magnitude processing in Grade 1 did not predict arithmetic in Grade 6 anymore, once the autoregressive effect of arithmetic in Grade 1 was controlled for ($\beta = 0.110$, $p = 0.528$). Taken together, these data provide a first indication that learning to calculate predicts subsequent symbolic magnitude processing.

The observation that learning to calculate might impact on its precursors has important implications for our understanding of atypical mathematical development or dyscalculia (American Psychiatric Association, 2013). It suggests that children with dyscalculia may be twice at risk for difficulties in learning mathematics. First, they are at risk due to difficulties in precursors of mathematics learning, such as numerical magnitude processing (e.g., Schwenk et al., 2017, for a meta-analysis). Second, they are at risk due to their difficulties in learning to calculate, which is known to be the hallmark of children with dyscalculia (e.g., American Psychiatric Association, 2013; Geary, 2011), and which may lead to poor symbolic magnitude processing skills. In other words, the symbolic magnitude processing skills of children with dyscalculia may not benefit as much from their calculation development as compared to their typically developing peers. Longitudinal studies are needed to verify this possibility. This hypothesis may not only be applicable to symbolic magnitude processing but also to other so-called domain-specific predictors of mathematics learning, such as knowledge of Arabic numerals or ordinality (e.g., Gobel et al., 2014; Lyons, Price, Vaessen, Blomert, & Ansari, 2014).

The data by Vanbinst et al. (2019) are only but a first step in showing that learning to calculate might change children's subsequent numerical cognition and they need replication in a much larger sample. It is also critical to note that the developmental period investigated in Vanbinst et al. constituted a very large time window of 6 years. Studies that investigate this issue over a shorter amount of time, e.g., within 1 year, are needed to precisely know how and when these effects are operating. In addition to this correlational approach, it is also important to conduct experimental studies in which arithmetic learning is trained via explicit direct instruction and the effects of this training on children's symbolic magnitude processing as well as their other basic numerical abilities are investigated. Such experimental data further help us to unpack the codevelopment of basic numerical processing and arithmetic learning, as I will discuss in the next section.

## The possibilities of experimental studies

Studies with an experimental design, in which a given process is manipulated and its effect on behavior is investigated, are the most valid and straightforward approach to understand causal associations between different variables. They allow asking basic questions about the causal role of, for example, numerical processing in arithmetic skill acquisition and vice versa. What is being manipulated in these studies can take various forms, ranging from something very specific to something very broad. Consequently, the questions that such experimental studies are able to address vary from questions on *why* an approach works (i.e., questions about mechanisms), as is the case in the first category of studies discussed below, to questions on *if* an approach works (i.e., questions about effectiveness), as is the case in the second category of studies discussed below. It is critical to acknowledge that the underlying aim of these two types of questions and the findings they can deliver differs: The first type are cognitive experiments that aim to understand causal mechanisms, the second type are educational interventions that aim to improve learning. Understanding mechanisms does not necessarily generalize to the best ways of improving learning and vice versa. Therefore data from cognitive experiments are not typically generalizable to educational interventions and vice versa.

A first category of experimental studies involves cognitive and laboratory-based within-subject designs in which a particular cognitive process is manipulated. These studies can include a manipulation within a task or a single session or they can consist of a short amount of training (usually over a few days or weeks) of a very specific cognitive process. For example, studies that investigate how people process fractions (Rosenberg-Lee, 2021) have compared performance on problems that are either congruent and incongruent with natural number knowledge to investigate the effect of natural number knowledge on fraction performance (e.g., Obersteiner, Van Dooren, Van Hoof, & Verschaffel, 2013). Studies that investigate how people select certain types of calculation strategies have used the choice/no-choice method (Lemaire & Siegler, 1995) in which participants are forced to use certain types of strategies, to examine their efficiency and selection processes (see Verschaffel et al., 2021). Other studies in this category are cognitive training studies that consist of short amounts of training of a particular mathematical process (e.g., Bugden, DeWind, & Brannon, 2016; Rosenberg-Lee, 2018 for a discussion). One example are arithmetic training studies that have manipulated the transition from procedural to fact retrieval strategies, by repeatedly presenting participants sets of arithmetic problems (e.g., Rosenberg-Lee, 2018; Zamarian & Delazer, 2015, for a discussion). Another example are studies that have manipulated individuals processing of nonsymbolic number to investigate the potential causal association between nonsymbolic number processing and symbolic mathematics (e.g., Bugden et al., 2016; Szucs & Myers, 2017; for a review). These specific manipulations do not necessarily lead to improved educational outcomes. In fact, most of these training studies showed very specific effects on isolated numerical processes, yet they generally did not result in improved mathematics achievement outcomes (Szucs & Myers, 2017). The latter was, however, not their primary aim, even though, many of these studies inappropriately claim that their findings will directly lead to improved mathematics education.

A second category of studies consists of educational interventions, which aim to improve children's mathematics learning. These studies are based on cognitive and developmental research and they are further grounded in instructional design principles (e.g., Clements & Sarama, 2008; Jordan, Glutting, Dyson, Hassinger-Das, & Irwin, 2012). These studies typically have longer durations, they include a wide variety of tasks and activities, and they are implemented outside the laboratory in the classroom by trained researchers or teachers. For example, Jordan et al. (2012) investigated the effect of a number sense intervention, which was developed against the background of cognitive theories on children's number development. The intervention consisted of 24 lessons of 30 min and included counting, numeral identification, number comparison, and small addition and subtraction. Jordan et al. compared the effect of the intervention to a language intervention and a business-as-usual control group, revealing that the number sense intervention leads to higher numerical and mathematical performance compared to the other groups. This educational intervention clearly resulted in better educational outcomes, as was its aim. However, this intervention study does not allow one to draw strong conclusions on the very precise mechanisms of changes in numerical processes that occurred as a result of the intervention, which would be of interest for informing cognitive theory on children's number development.

These methodological and practical considerations create a tension in decisions on the design of experimental studies. The more a study is concentrated on understanding cognitive mechanisms, the more likely it will need to include tasks that focus on very isolated cognitive processes (Szucs & Myers, 2017, for an elaboration). Such design results in high chance of finding the precise cognitive mechanism that causes an effect, but it will lead to a lower chance that the training of that cognitive process results in improved educational outcomes. Conversely, the more a study is concentrated on improving educational outcomes, the more likely it will need to include a broad variety of activities, and consequently, the more likely it will tap into a variety of cognitive processes. Such research design results in a higher chance of improving educational outcomes, yet it provides much less precise knowledge on the specific cognitive processes that improve academic performance. These methodological and practical considerations do not mean that one type of experimental design is preferred over the other. They mean that experimental studies need to be clear and transparent about their aims and scope: understanding precise cognitive mechanisms vs improving education. One single study cannot fully deliver both aims and the extent of emphasizing one of these aims will impact the capacity to address the other. The field of numerical cognition needs these two types of studies to build a rich knowledge base on how numerical abilities develop and how they can be optimally fostered in the classroom, an endeavor that will require different forms of expertise and collaboration between cognitive scientists and mathematics educators.

## Using the educational context as a natural experiment

The experimental studies that were discussed in the previous section provide rather limited manipulations of learning and they do not fully grasp how the more entrenched experience of education hones numerical cognition. This raises the question of how very broad interventions, such as education itself, impact on numerical cognition. In this section, I discuss two forms of quasi-experimental studies in which education is used as a natural experiment to investigate its impact on numerical cognition: the school cutoff design and cross-national studies.

### The school cutoff design

A clever way of studying the effects of education on numerical cognition is by taking advantage of the fact that children of similar ages can be in different grades. This quasi-experimental approach allows one to disentangle schooling-related from age-related maturational changes in cognitive performance (Morrison, Kim, Connor, & Grammer, 2019). This approach, also known as the *school cutoff design*, starts from the nearly universal practice of using an arbitrary fixed cutoff date for school entry (e.g., in Belgium January 1st). It compares the abilities of children who fall closely on either side of this official cutoff date for school entry, typically in a two- or three-month window before and after the cutoff. This creates two groups of children who are close in age (e.g., 6-year olds) but are enrolled in different school years (e.g., grade 1 vs. grade 2; preschool year 1 vs. preschool year 2) and/or systems (e.g., kindergarten vs. first grade). The comparison of these groups allows researchers to draw valid inferences on the general effect of education on children's cognitive performance. In a typical implementation of the school cutoff design, children are tested at two time points, i.e., at baseline (T1) and 1 year later (T2), when one group of children will have received one more year of schooling and the other will not. This implementation makes the design similar to a classic pretest-posttest design. When there are no systematic practices of pushing forward or holding back children, as is the case in many countries, this results in a (quasi-)random assignment of children to the conditions or educational environments. Comparing the growth in cognitive performance from T1 to T2 in children who just missed or made the cutoff for school entry allows one to assess the impact of 1 year of schooling. Using this school cutoff approach, schooling effects have been observed in many cognitive domains (Morrison et al., 2019, for a review), including reading (e.g., Morrison, Smith, & Dow-Ehrensberger, 1995), reasoning (e.g., Cahan & Cohen, 1989), and executive function (Zhang et al., 2019). These schooling effects have also been observed in numerical cognition, as I elaborate below.

Bisanz, Dunn, and Morrison (1995) reported the first study in the field of numerical cognition, which was later replicated by Naito and Miura (2001), in which these schooling effects were investigated. Using the school cutoff design, both studies compared three groups of children: *old* kindergartners (Bisanz et al.: $Mdn_{age} = 5.6$ years; $n = 19$; Naito and Miura: $Mdn_{age} = 5.9$ years; $n = 22$), *young* first graders (Bisanz et al.: $Mdn_{age} = 5.8$ years; $n = 19$; Naito and Miura: $Mdn_{age} = 6.1$ years; $n = 22$), and *old* first graders (Bisanz et al. $Mdn_{age} = 6.6$ years; $n = 18$; Naito and Miura: $Mdn_{age} = 6.9$ years; $n = 26$). The *old* kindergarteners and *young* first graders differed in the amount of schooling but they were similar in age. The two groups of first graders differed in age but they were similar in the amount of schooling. All children completed a number conservation task and a mental addition task. Both studies revealed a clear effect of schooling on mental addition accuracy. Schooling affected children's addition strategy use in Naito and Miura (2001), but not in Bisanz et al. (1995). Both studies showed that schooling did not influence children's conservation of number, which was only affected by age. However, Naito and Miura (2001) additionally included a place value task in their design, which measured children's understanding of their base 10 number concepts, and found that this understanding of place value was determined by both schooling and age: First graders showed higher performance than kindergartners, but *old* first graders were better than the *young* ones. In all, both studies showed clear effects of schooling, and they both revealed that such effects of schooling on numerical cognition can be very specific.

This first evidence for schooling effects in numerical cognition, albeit obtained in very small samples, did not thoroughly investigate children's number processing abilities that emerged over the last decade as critical predictors of their mathematical development, such as numerical magnitude processing (e.g., Schneider et al., 2017), numeral identification (e.g., Gobel et al., 2014), or number ordering (e.g., Lyons et al., 2014). These number processing abilities show massive developments in the ages 5–7 years (e.g., Geary et al., 2018; Hutchison, Ansari, Zheng, De Jesus, & Lyons, 2020; Purpura, Baroody, & Lonigan, 2013), but developmental studies in numerical cognition have not attempted to separate effects of schooling from effects of age in this developmental window. Disentangling these effects is important for constructing and evaluating theories about the mechanisms that contribute to the normal course of numerical development and its enhancement via educational interventions. In view of the potential specificity of schooling effects on children's mathematical cognition (Bisanz et al., 1995; Naito & Miura, 2001), it is likely that similar differential and specific effects can be predicted and observed in studies that include fine-grained measures of children's number processing abilities. It further needs to be emphasized that these number processing abilities already show large developments before formal schooling starts in grade 1. It is therefore important that such effects of schooling should also be considered at the preschool level.

In a first attempt to address this issue, Aravindran Rajagopal, Vandecruys, and De Smedt (2021) compared different groups of preschoolers in a secondary data analysis of a longitudinal project that followed more than 400 children through different years of preschool in Belgium (e.g., Bakker et al., 2019). In the Flanders region of Belgium, nearly all children (98%) attend the fully government-subsidized preschool, which starts at the age of 2.5 years up to the year children turn 6 years. Preschool consists of 3 years (Y1, Y2, and Y3) during which children receive playful nonformal introductions to various domains, including mathematics. This education is delivered by trained preschool teachers, who have to focus in their teaching on specific core learning goals that are specified by the government (Vlaams Ministerie van Onderwijs en Vorming, 2021), including goals for mathematics learning. In the longitudinal project, all children completed the same set of numerical measures when they were in the second year of preschool (Y2, $Mdn_{age} = 4.8$ years) and third year of preschool (Y3, $Mdn_{age} = 5.8$ years). These measures were nonsymbolic comparison, symbolic comparison, numeral identification, number order, dot enumeration, object counting, and symbolic calculation (Bakker et al., 2019). To examine the effects of schooling, Aravindran Rajagopal et al. (2021) compared the performance of *old* children in Y2 ($\text{Mdn}_{age} = 5.2$ years, $SD = 0.95$ months; $n = 111$, all born in January–March) with the *young* children in Y3 ($\text{Mdn}_{age} = 5.4$, $SD = 0.99$ months; $n = 71$, all born in October–December). This comparison reflects the effect of 1 year of preschool education, as these two groups differed in the amount of preschool but were similar in age. Young children in Y3 significantly outperformed the old children in Y2 in numeral identification, object counting, number ordering, and calculation, which indicates strong influences of schooling on these number processing abilities at this age (i.e., in Y3). The two groups did not differ in nonsymbolic comparison, symbolic comparison, and dot enumeration, which suggests that these numerical abilities were less affected by schooling at this age (i.e., in Y3).

The abovementioned studies clearly show that education itself has effects on children's development of numerical cognition independent of children's maturational age. These schooling effects not only occur for mathematical abilities that are formally taught at school, such as arithmetic. They also appear for numerical abilities that show massive development before formal schooling starts, and which have been put forward as predictors for later mathematical achievement. It is therefore critical for studies in numerical cognition, specifically at the preschool age, to also consider these effects of education when studying development at this age.

## Cross-national comparisons

Another way to grasp the more general effect of education on numerical cognition is to compare groups of children in different countries, who experience different mathematical curricula. Being included in one or the other curriculum can be considered as a random process, and if such groups are well matched (e.g., Geary, Salthouse, Chen, & Fan, 1996), if the compared countries are similar in a series of background variables (e.g., overall level of education), and if the differences between the two curricula can be precisely defined, such comparison might resemble with a (quasi-)experimental design.

Many studies have focused on differences between Chinese and North American children, finding that Chinese children have superior arithmetic skills compared with North American ones (e.g., Campbell & Xue, 2001; Geary et al., 1996). Fine-grained analyses of the cognitive processes in arithmetic across these two different sociocultural settings suggest that Chinese children transition more quickly from counting to fact retrieval (Geary et al., 1996) and that Chinese students use fact retrieval more frequently across all arithmetic operations compared to their North American peers (Campbell & Xue, 2001). These differences are, at least partially, explained by different instructional practices in these two countries (Geary et al., 1996). Evidently, if children are taught to use different strategies, they will differ in the way they recruit these strategies during calculation. A more intriguing question, however, is how such differences in learning approaches impact on children's numerical cognition and particularly on its association with arithmetic performance.

This question was investigated by Vanbinst, Peters, Ansari, and De Smedt (2021) by comparing Belgian and Canadian first- and second-grade children. Vanbinst et al. focused on these countries, not only because they have strong traditions in numerical cognition research, but also because they are comparable in their general level of mathematics education as is expressed by the PISA 2018 scores for mathematics (Belgium: $M=508$, $SD=95$; Canada; $M=512$, $SD=92$; Schleicher, 2019) and their start of formal schooling (September of the year children turn 6) and they both have state-funded kindergarten (2 years in Canada, 3.5 years in Belgium). Critically, arithmetic instruction in these two countries in the early grades of primary school differs in its relative focus on counting vs arithmetic fluency. Specifically, in Belgium, teachers quickly move away from counting strategies, as early as in first grade, and, from an early point on, they do not allow children to count on their fingers. Teachers put a strong emphasis on retrieving very basic arithmetic facts, which are frequently practiced. For larger problems (e.g., problems $> 10$), children systematically learn to decompose them into smaller ones, for example by decomposing the second number such that one part can be used to make 10 and the other part constitutes the remainder that needs to be added/subtracted (e.g., $8+5$ is solved by decomposing 5 into 2 and 3; this results in $8+2=10$ and $10+3=13$). To learn this decomposition strategy, children in first grade receive a lot of practice in decomposing single-digit numbers, and these decompositions need to be memorized. In Canada, teachers put a lot more emphasis on counting, and children can use their fingers for a much longer time to solve single-digit problems. There is much less emphasis on the use of fact retrieval and on arithmetic fluency, as compared with Belgium. Do these differences in arithmetic instruction moderate the association between arithmetic and its neurocognitive predictors, such as symbolic magnitude processing?

Vanbinst et al. compared 100 Belgian and 100 Canadian first and second graders (50 children per country per grade). As predicted, the groups differed in their level of arithmetic fluency, as measured by the Tempotest Arithmetic (De Vos, 1992) and Woodcock-Johnson Math Fluency (Schrank, Mather, & McGrew, 2014) standardized tests: Belgian children outperformed their Canadian peers, confirming the idea of a stronger emphasis on arithmetic fluency in Belgium than in Canada. As also expected, the groups differed in their strategy use: Belgian children used more fact retrieval than their Canadian peers, while Canadian children had higher frequencies of counting. Belgian children also had higher frequencies of decomposition strategies, but this was only observed in grade 2. These differences in strategies mirrored the differences in instructional focus. Interestingly, the countries did not differ in their accuracy of arithmetic, but Belgian children were faster than their Canadian peers, which was explained by their higher frequencies of retrieval use. The critical question was whether these country differences in arithmetic moderated the association between arithmetic and children's symbolic magnitude processing. The authors found that this association was generally stronger for Canadian children than for their Belgian peers. This might be explained by the larger amount of counting strategy use in Canadians, which may relate more to the processing of quantity. Most interestingly, strong associations occurred between fact retrieval use and symbolic magnitude processing in Canada, such that children with higher frequencies of retrieval were better at symbolic magnitude processing, but these associations were absent in Belgium. Importantly, this absence of an association was not due to a ceiling effect or to less variation in the frequency of retrieval use in the Belgian children. These findings suggest different cognitive numerical processes underlying arithmetic fact retrieval in Canadian vs Belgian children. One possibility could be that this difference reflects meaningful semantic retrieval (based on quantity) vs rote fact retrieval, although such possibility needs to be verified in future research.

In all, these data show that the educational context impacts on the well-established association between numerical magnitude processing and arithmetic (e.g., Schneider et al., 2017). This should be acknowledged when comparing data from studies across different countries. It urges researchers in the field of numerical cognition to carefully document the educational context in which their data were collected as this context has implications for data interpretation and generalizability.

## Conclusion

The current chapter provided some reflections on the potential impact of education on numerical cognition. Because mathematics is in essence a symbolic activity that is learned via education, this learning process will have an impact on how we process number. The impact of this learning process was illustrated by the fact that the processing of natural Arabic numerals might be changed by learning other symbolic representations, such as fractions, which already starts in the early grades of primary school education. This impact of learning was also bolstered by the observation that learning to calculate, by itself, affects children's numerical processing. The studies that have used the educational context as a natural experiment, i.e., studies with the school cutoff design or cross-national comparisons, additionally provided more direct evidence on the broad impact of the educational context on numerical cognition.

Against the background of this data, it is critical for future studies in numerical cognition to report on the educational context in which their data were collected. On the one hand, this adds to the external validity of the findings of these studies. It allows us to better understand to which types of contexts the findings of a particular study are generalizable. This is especially important for developmental research in young children, for example during the transition from nonformal to formal educational systems, as children are educated in systems that at this point vary considerably across countries. It will also help us to better understand differences between various studies with similar paradigms that have been executed in different contexts. On the other hand, considering the educational context will also add to the internal validity of our findings. The consideration of how mathematics is learned in the classroom, such as the strategies children learn to calculate, might allow us to make more fine-grained predictions on how the processing of number is related to this classroom learning, and vice versa. As a result, it might modify our theories on how the processing of number develops.

In addition to considering the context in which mathematics is learned, there is also a need for studies that experimentally manipulate the context in which numbers and mathematics are learned. This research is needed to go beyond the insights that are gleaned from correlational studies to understand the causal associations between numerical cognition and classroom learning. Such experimental studies should be clear in their scope and aims, i.e., whether they aim to understand a particular cognitive processing mechanism or whether they aim to improve children's learning and achievement. Both types of research are needed to build a rich knowledge base on how numerical abilities develop and how they can be optimally fostered in the classroom.

## Acknowledgments

This work was supported by a project of the Fund for Scientific Research Flanders (G.0707.20).

## References


American Psychiatric Association. (2013). *Diagnostic and statistical manual of mental disorders* (5th ed.). American Psychiatric Association.

Aravindran Rajagopal, A., Vandecruys, F., & De Smedt, B. (2021). Numerical development in kindergarten: Effect of age or school? Manuscript in preparation.

Bakker, M., Torbeyns, J., Wijns, N., Verschaffel, L., & De Smedt, B. (2019). Gender equality in 4-to 5-year-old preschoolers' early numerical competencies. *Developmental Science, 22*(1), e12718. https://doi.org/10.1111/desc.12718.

Bellon, E., Fias, W., & De Smedt, B. (2019). More than number sense: The additional role of executive functions and metacognition in arithmetic. *Journal of Experimental Child Psychology, 182*, 38–60. https://doi.org/10.1016/j.jecp.2019.01.012.

Bisanz, J., Dunn, M., & Morrison, F. J. (1995). Effects of age and schooling on the acquisition of elementary quantitative skills. *Developmental Psychology, 31*, 221–236. https://doi.org/10.1037/0012-1649.31.2.221.

Brankaer, C., Ghesquière, P., & De Smedt, B. (2017). Symbolic magnitude processing in elementary school children: A group administered paper-and-pencil measure (SYMP test). *Behavior Research Methods, 49*(4), 1361–1373. https://doi.org/10.3758/s13428-016-0792-3.

Bugden, S., DeWind, N. K., & Brannon, E. M. (2016). Using cognitive training studies to unravel the mechanisms by which the approximate number system supports symbolic math ability. *Current Opinion in Behavioral Sciences, 10*, 73–80. https://doi.org/10.1016/j.cobeha.2016.05.002.

Bulthe, J., De Smedt, B., & Op de Beeck, H. P. (2014). Format-dependent representations of symbolic and non-symbolic numbers in the human cortex as revealed by multi-voxel pattern analyses. *NeuroImage, 87*, 311–322. https://doi.org/10.1016/j.neuroimage.2013.10.049.

Bulthe, J., De Smedt, B., & Op de Beeck, H. P. (2018). Arithmetic skills correlate negatively with the overlap of symbolic and non-symbolic number representations in the brain. *Cortex, 101*, 306–308. https://doi.org/10.1016/j.cortex.2018.01.008.

Bulthe, J., Prinsen, J., Vanderauwera, J., Duyck, S., Daniels, N., Gillebert, C. R., et al. (2019). Multi-method brain imaging reveals impaired representations of number as well as altered connectivity in adults with dyscalculia. *NeuroImage, 190*, 289–302. https://doi.org/10.1016/j.neuroimage.2018.06.012.

Cahan, S., & Cohen, N. (1989). Age versus schooling effects on intelligence development. *Child Development, 60*, 1239–1249. https://doi.org/10.2307/1130797.

Campbell, J. I. D., & Xue, Q. L. (2001). Cognitive arithmetic across cultures. *Journal of Experimental Psychology-General, 130*(2), 299–315. https://doi.org/10.1037//0096-3445.130.2.299.

Chu, F. W., VanMarle, K., Rouder, J., & Geary, D. C. (2018). Children's early understanding of number predicts their later problem-solving sophistication in addition. *Journal of Experimental Child Psychology, 169*, 73–92. https://doi.org/10.1016/j.jecp.2017.12.010.

Clements, D. H., & Sarama, J. (2008). Experimental evaluation of the effects of a research-based preschool mathematics curriculum. *American Educational Research Journal, 45*(2), 443–494. https://doi.org/10.3102/0002831207312908.

De Smedt, B. (2016). Individual differences in arithmetic fact retrieval. In D. Berch, D. Geary, & K. Mann-Koepke (Eds.), *Mathematical Cognition and Learning: Vol. 2* (pp. 219–243). Elsevier Academic Press.

De Vos, T. (1992). *Tempo-test rekenen*. Nijmegen: Berkhout.

De Smedt, B., Noel, M. P., Gilmore, C., & Ansari, D. (2013). The relationship between symbolic and non-symbolic numerical magnitude processing and the typical and atypical development of mathematics: A review of evidence from brain and behavior. *Trends in Neuroscience and Education, 2*, 48–55. https://doi.org/10.1016/j.tine.2013.06.001.

Dowker, A., & Nuerk, H. C. (2016). Linguistic influences on mathematics. *Frontiers in Psychology*, 7, 1035. https://doi.org/10.3389/fpsyg.2016.01035.

Geary, D. C. (2011). Consequences, characteristics, and causes of mathematical learning disabilities and persistent low achievement in mathematics. *Journal of Developmental and Behavioral Pediatrics*, *32*(3), 250–263. https://doi.org/10.1097/DBP.0b013e318209edef.

Geary, D. C., Salthouse, T. A., Chen, G. P., & Fan, L. (1996). Are east Asian versus American differences in arithmetical ability a recent phenomenon? *Developmental Psychology*, *32*(2), 254–262. https://doi.org/10.1037/0012-1649.32.2.254.

Geary, D. C., vanMarle, K., Chu, F. W., Rouder, J., Hoard, M. K., & Nugent, L. (2018). Early conceptual understanding of cardinality predicts superior school-entry number-system knowledge. *Psychological Science*, *29*(2), 191–205. https://doi.org/10.1177/0956797617729817.

Gobel, S. M., Watson, S. E., Lervag, A., & Hulme, C. (2014). Children's arithmetic development it is number knowledge, not the approximate number sense, that counts. *Psychological Science*, *25*(3), 789–798. https://doi.org/10.1177/0956797613516471.

Hutchison, J. E., Ansari, D., Zheng, S., De Jesus, S., & Lyons, I. M. (2020). The relation between subitizable symbolic and non-symbolic number processing over the kindergarten school year. *Developmental Science*, *23*(2), e12884. https://doi.org/10.1111/desc.12884.

Jordan, N. C., Glutting, J., Dyson, N., Hassinger-Das, B., & Irwin, C. (2012). Building kindergartners' number sense: A randomized controlled study. *Journal of Educational Psychology*, *104*(3), 647–660. https://doi.org/10.1037/a0029018.

Katholiek Onderwijs Vlaanderen. (2021). *Didactiek en leerplannen* [Curricula]. Retrieved from https://pro.katholiekonderwijs.vlaanderen/didactiek-en-leerplannen?tab=Bao.

Kilpatrick, J., Swafford, J., & Findell, B. (2001). *Adding it up: Helping children learn mathematics*. National Academies Press.

Leibovich, T., & Ansari, D. (2016). The symbol-grounding problem in numerical cognition: A review of theory, evidence, and outstanding questions. *Canadian Journal of Experimental Psychology-Revue Canadienne De Psychologie Experimentale*, *70*(1), 12–23. https://doi.org/10.1037/cep0000070.

Lemaire, P., & Siegler, R. S. (1995). Four aspects of strategic change - contributions to children's learning of multiplication. *Journal of Experimental Psychology-General*, *124*(1), 83–97. https://doi.org/10.1037/0096-3445.124.1.83.

Linsen, S., Verschaffel, L., Reynvoet, B., & De Smedt, B. (2015). The association between numerical magnitude processing and mental versus algorithmic multi-digit subtraction in children. *Learning and Instruction*, *35*, 42–50. https://doi.org/10.1016/j.learninstruc,2014.09.003.

Lyons, I. M., Ansari, D., & Beilock, S. L. (2012). Symbolic estrangement: Evidence against a strong association between numerical symbols and the quantities they represent. *Journal of Experimental Psychology-General*, *141*(4), 635–641. https://doi.org/10.1037/a0027248.

Lyons, I. M., Price, G. R., Vaessen, A., Blomert, L., & Ansari, D. (2014). Numerical predictors of arithmetic success in grades 1-6. *Developmental Science*, *17*(5), 714–726. https://doi.org/10.1111/desc.12152.

Merkley, R., & Ansari, D. (2016). Why numerical symbols count in the development of mathematical skills: Evidence from brain and behavior. *Current Opinion in Behavioral Sciences*, *10*, 14–20. https://doi.org/10.1016/j.cobeha.2016.04.006.

Miller-Cotto, D., & Byrnes, J. P. (2020). What's the best way to characterize the relationship between working memory and achievement?: An initial examination of competing theories. *Journal of Educational Psychology*, *112*, 1074–1084. https://doi.org/10.1037/edu0000395.

Morrison, F. J., Smith, L., & Dow-Ehrensberger, M. (1995). Education and cognitive development—A natural experiment. *Developmental Psychology*, *31*(5), 789–799. https://doi.org/10.1037/0012-1649.31.5.789.

Morrison, F. J., Kim, M. H., Connor, C. M., & Grammer, J. K. (2019). The causal impact of schooling on children's development: Lessons for developmental science. Current Directions in Psychological Science, 28(5), 441-449, Article Unsp 0963721419855661. https://doi.org/10.1177/0963721419855661.

Naito, M., & Miura, H. (2001). Japanese children's numerical competencies: Age- and schooling-related influences on the development of number concepts and addition skills. *Developmental Psychology*, *37*(2), 217–230. https://doi.org/10.1037//0012-1649.37.2.217.

National Council for Teaching Mathematics (NCTM). (2006). *Curriculum focal points for pre-kindergarten through grade 8 mathematics*. NCTM.

National Mathematics Advisory Panel. (2008). *Foundations for success: The final report of the National Mathematics Advisory Panel*. US Department of Education.

Nosworthy, N., Bugden, S., Archibald, L., Evans, B., & Ansari, D. (2013). A two-minute paper-and-pencil test of symbolic and nonsymbolic numerical magnitude processing explains variability in primary school children's arithmetic competence. *PLos One*, *8*(7), e67918. https://doi.org/10.1371/journal.pone.0067918.

Obersteiner, A., Van Dooren, W., Van Hoof, J., & Verschaffel, L. (2013). The natural number bias and magnitude representation in fraction comparison by expert mathematicians. *Learning and Instruction*, *28*, 64–72. https://doi.org/10.1016/j.learninstruc.2013.05.003.

Peng, P., & Kievit, R. A. (2020). The development of academic achievement and cognitive abilities: A bidirectional perspective. *Child Development Perspectives*, *14*(1), 15–20. https://doi.org/10.1111/cdep.12352.

Perfetti, C. A., Beck, I., Bell, L. C., & Hughes, C. (1987). Phonemic knowledge and learning to read are reciprocal—A longitudinal study of 1st grade children. *Merrill-Palmer Quarterly-Journal of Developmental Psychology*, *33*(3), 283–319.

Purpura, D. J., Baroody, A. J., & Lonigan, C. J. (2013). The transition from informal to formal mathematical knowledge: Mediation by numeral knowledge. *Journal of Educational Psychology*, *105*(2), 453–464. https://doi.org/10.1037/a0031753.

Rosenberg-Lee, M. (2018). Training studies: An experimental design to advance educational neuroscience. *Mind Brain and Education*, *12*(1), 12–22. https://doi.org/10.1111/mbe.12166.

Rosenberg-Lee, M. (2021). Probing the neural basis rational numbers: The role of inhibitory control and magnitude representations. In W. Fias, & A. Henik (Eds.), *Heterogeneous Contributions to Numerical Cognition: Learning and Education in Mathematical Cognition* (pp. 143-180). Elsevier.

Schleicher, A. (2019). *PISA 2018. Insights and interpretations*. OECD.

Schneider, M., Beeres, K., Coban, L., Merz, S., Schmidt, S., Stricker, J., et al. (2017). Associations of non-symbolic and symbolic numerical magnitude processing with mathematical competence: A meta-analysis. *Developmental Science*, *20*, e12372. https://doi.org/10.1111/desc.12372.

Schoenfeld, A. L. (2004). The math wars. *Educational Policy*, *18*(1), 253–286. https://doi.org/10.1177/0895904803260042.

Schrank, F. A., Mather, N., & McGrew, K. S. (2014). *Woodcock-Johnson IV tests of achievement. Riverside.*

Schwenk, C., Sasanguie, D., Kuhn, J. T., Kempe, S., Doebler, P., & Holling, H. (2017). (Non-)symbolic magnitude processing in children with mathematical difficulties: A meta-analysis. *Research in Developmental Disabilities*, *64*, 152–167. https://doi.org/10.1016/j.ridd.2017.03.003.

Siegler, R. S., Fazio, L. K., Bailey, D. H., & Zhou, X. L. (2013). Fractions: The new frontier for theories of numerical development. *Trends in Cognitive Sciences*, *17*(1), 13–19. https://doi.org/10.1016/j.tics.2012.11.004.

Singer-Freeman, K. E., & Goswami, U. (2001). Does half a pizza equal half a box of chocolates? Proportional matching in an analogy task. *Cognitive Development*, *16*(3), 811–829. https://doi.org/10.1016/s0885-2014(01)00066-1.

Szucs, D., & Myers, T. (2017). A critical analysis of design, facts, bias and inference in the approximate number system training literature: A systematic review. *Trends in Neuroscience and Education, 6*, 187–203. https://doi.org/10.1016/j.tine.2016.11.002.

Thompson, I. (1999). *Issues in teaching numeracy in primary schools*. Open University Press.

Traff, U. (2013). The contribution of general cognitive abilities and number abilities to different aspects of mathematics in children. *Journal of Experimental Child Psychology, 116*(2), 139–156. https://doi.org/10.1016/j.jecp.2013.04.007.

Van den Heuvel-Panhuizen, M. (2001). *Children learn mathematics*. Freudenthal Institute, University of Utrecht.

Vanbinst, K., & De Smedt, B. (2016). Individual differences in children's mathematics achievement: The roles of symbolic numerical magnitude processing and domain-general cognitive functions. In M. Cappelletti, & W. Fias (Eds.), *Mathematical brain across the lifespan: Vol. 227* (pp. 105–130). https://doi.org/10.1016/bs.pbr.2016.04.001.

Vanbinst, K., Ghesquière, P., & De Smedt, B. (2012). Numerical magnitude representations and individual differences in children's arithmetic strategy use. *Mind Brain and Education, 6*(3), 129–136. https://doi.org/10.1111/j.1751-228X.2012.01148.x.

Vanbinst, K., Ghesquière, P., & De Smedt, B. (2015). Does numerical processing uniquely predict first graders' future development of single-digit arithmetic? *Learning and Individual Differences, 37*, 153–160. https://doi.org/10.1016/j.lindif.2014.12.004.

Vanbinst, K., Ansari, D., Ghesquière, P., & De Smedt, B. (2016). Symbolic numerical magnitude processing is as important to arithmetic as phonological awareness is to reading. *PLoS One, 11*(3), e0151045. https://doi.org/10.1371/journal.pone.0151045.

Vanbinst, K., Ghesquière, P., & De Smedt, B. (2019). Is the a long-term association between symbolic numerical magnitude processing and arithmetic bi-directional? *Journal of Numerical Cognition, 5*(3), 358–370. https://doi.org/10.5964/jnc.v5i3.202.

Vanbinst, K., Peters, L., Ansari, D., & De Smedt, B. 2021 The effect of formal math education on learning arithmetic: A cross-national study. Manuscript in preparation.

Verschaffel, L., Greer, B., & De Corte, E. (2007). Whole number concepts and operations. In F. K. Lester (Ed.), *Second handbook of research on mathematics teaching and learning* (pp. 557–628). Information Age Publishing.

Verschaffel, L., De Smedt, B., Van Der Auwera, S., & Torbeynsa, J. (2021). Subtraction by addition: A remarkably natural and clever way to subtract? In W. Fias, & A. Henik (Eds.), *Heterogeneous Contributions to Numerical Cognition: Learning and Education in Mathematical Cognition* (pp. 117-141). Elsevier.

Vlaams Ministerie van Onderwijs en Vorming. (2021). Onderwijsdoelen [Curricular goals]. Retrieved from https://onderwijsdoelen.be/.

Yuan, L., Prather, R. W., Mix, K. S., & Smith, L. B. (2019). Preschoolers and multi-digit numbers: A path to mathematics through the symbols themselves. *Cognition, 189*, 89–104. https://doi.org/10.1016/j.cognition.2019.03.013.

Zamarian, L., & Delazer, M. (2015). Arithmetic learning in adults—Evidence from brain imaging. In R. Cohen-Kadosh, & A. Dowker (Eds.), *The Oxford Handbook of Numerical Cognition* Oxford University Press.

Zhang, Q., Wang, C. P., Zhao, Q. W., Yang, L., Buschkuehl, M., & Jaeggi, S. M. (2019). The malleability of executive function in early childhood: Effects of schooling and targeted training. *Developmental Science, 22*(2), e12748. https://doi.org/10.1111/desc.12748.

SECTION III

# Intervention studies

CHAPTER 9

# Supporting early numeracy: The role of spontaneous mathematical focusing tendencies in learning and instruction

Minna Hannula-Sormunen[a], Jake McMullen[a], and Erno Lehtinen[a,b]
[a]Department of Teacher Education, University of Turku, Turku, Finland
[b]Vytautas Magnus University, Kaunas, Lithuania

## Introduction

In 2009 in the United States, the National Research Council's Committee on Early Childhood Mathematics stated "most early childhood programs spend little focused time on mathematics, and most of it is low in instructional quality" (p. 339) (NRC, 2009). According to Hachey (2013), many early educators do not know how early mathematical thinking can be systematically supported in any other way than by decontextualized formal tasks and thus often have negative attitudes about early mathematics education. This is in sharp contrast with great emphasis in many countries, including Finland, on supporting of preschool reading readiness skills and development. For example, the latest national evaluation of 5-year-old children's early education in Finland excluded early mathematical skills and focused only on linguistic and social skills (Repo et al., 2019). Not only in Finland, but also internationally, early math activities have been considered less important than language, literacy, and social skills, both at home and preschool (Blevins-Knabe, Austin, Musun, Eddy, & Jones, 2000). Even though preschool teachers regularly engage children in communication about mathematical phenomena to some extent, they do not systematically use the physical environment or daily activities as a point of departure for directing children's attention to specific mathematical concepts or principles (Björklund & Barendregt, 2016). It also seems that many existing teacher support materials for early mathematics education focus on the separate early numeracy skills of children, which are often taught through traditional formal exercises with very little guidance for using the skills in

*Heterogeneous Contributions to Numerical Cognition*
https://doi.org/10.1016/B978-0-12-817414-2.00013-0

everyday life of children. Typical mathematics textbooks in the early grades also seem to have the same limitation. There are extremely few tasks for mathematizing the real world outside of the textbook. Even word problems, which are aimed at linking the real world and formal mathematical skills and knowledge in textbooks, have been described as too simple or straightforward and solved easily by using superficial strategies (Wyndhamn & Säljö, 1997). They mostly ask for a precise numerical answer, which leaves little room for realistic considerations to be integrated into the solution process (Freudenthal, 1991; Pongsakdi, Brezovszky, Veermans, Hannula-Sormunen, & Lehtinen, 2016). This kind of narrow approach may fail to make use of children's spontaneous activities and play and could lead to little transfer of learning to novel contexts outside of formal mathematics lessons (Lehtinen & Hannula, 2006).

These negative views on the state of art in early childhood mathematical education on one hand, and lacking of support for the transfer of mathematical skills in school mathematics on the other hand, have been challenged by much more positive results in specific early math intervention studies. In a recent metaanalysis on early mathematical interventions, Wang, Firmender, Power, and Byrnes (2016) find a moderate-to-large effect size for the impacts of early childhood mathematics programs (Cohen's $d = 0.6$). More specifically, there was a tendency for programs to produce larger effects when they targeted a single content strand, presented content 120–150 min/week, designed programs for the prekindergarten environment, presented content individually to children, and used researcher-made mathematics assessments. Wang et al. (2016) concluded that interventions designed to improve the mathematics skills of young children can clearly have an effect and the effects are not limited to a single setting or particular program. Frye et al. (2013) reviewed 29 studies with experimental and quasi-experimental designs conducted in preschool, prekindergarten, or kindergarten in years from 1989 to 2011 and created recommendations for supporting early mathematical development. The five recommendations for teaching mathematics are as follows: (1) teach numbers and operations concepts; (2) teach geometry, measurement, and data analysis skills; (3) use progress monitoring tools; (4) teach students to view the world mathematically; and (5) integrate mathematics instruction throughout each school day. Frye et al. (2013) found moderate level of evidence for Recommendation 1 and a minimal level of evidence but positive effects for Recommendations 2–5.

However, despite evidence of positive effects from these interventions, further evidence suggests that many of these gained effects are not easily sustained once the interventions are over. This extends into primary school, as classroom interventions to improve specific mathematical skills may be sufficient for managing immediate change (e.g., Fuchs et al., 2013), but less successful in promoting sustained, transferable improvements in students' knowledge (De Corte, 1999). Indeed, recent metaanalyses have revealed that avoiding such fade-out of intervention effects, that is, truly promoting consistent change in learning trajectories, is hard to come by (Bailey et al., 2016). However, these authors suggest that more comprehensive environments, which work to fundamentally shift skills that would otherwise not be developed, are among the most promising to produce long-lasting, sustained effects (Bailey, Duncan, Odgers, & Yu, 2017).

Currently, there is great variability in the existing early math interventions. Some are computer game-based, narrowly targeted trainings, such as Grapho game or Number Race interventions (Räsänen, Salminen, Wilson, Aunio, & Dehaene, 2009). These interventions aim at promoting basic skills through more traditional pedagogical approaches, such as spaced practice and interleaving. Others are much more general and aim to support early mathematical development by producing a comprehensive mathematics curriculum with research-based learning trajectories and congruous professional development (e.g., Clements, Sarama, Spitler, Lange, & Wolfe, 2011). For example, Building Blocks program emphasizes teaching for understanding via learning trajectories and technology. Children are guided to extend and mathematize their everyday activities, from block building to art to songs to puzzles, through sequenced, explicit activities "throughout the day," off-computer and on-computer activities (Clements et al., 2011).

However, theoretical models or theory-based programs in mathematics education that make use of children's typical ways of learning in everyday surroundings seem to be rare, even though the need for them is recognized. A United States National Research Council report (Cross, Woods, & Schweingruber, 2009) emphasized the teacher's role in supporting children's ability to make meaning and mathematize the real world, as well as creating a nurturing math talk community (Fuson, Clements, & Sarama, 2015). This is in sharp contrast with many existing preschool and school mathematics materials, particularly for low-achieving children, which are often characterized by quite meaningless, rote-based learning tasks that neither make use of children's informal mathematical experiences nor support the transfer of learned numerical skills to other tasks or contexts.

This is even more typical for pedagogical models and interventions used for enhancing children's learning of fractions and decimal numbers.

External representations, such as pizza-models and recently also number line, are applied (Fuchs, Malone, Schumacher, Namkung, & Wang, 2017; Hamdan & Gunderson, 2017) but with minimal or no relation to various experiences children have with quantitative relations in their everyday life. Recent research has highlighted the difficulties many students face during the learning trajectory from natural numbers to fractions and decimals (McMullen, Laakkonen, Hannula-Sormunen, & Lehtinen, 2015) and insufficient ability of traditional mathematics teaching to support deeper understanding of rational numbers (Schneider & Siegler, 2010; Vosniadou, Vamvakoussi, & Skopeliti, 2008). Only a small part of students learn flexible and adaptive skills to apply rational number knowledge in varying situations (McMullen, Hannula-Sormunen, Lehtinen, & Siegler, 2020).

## Cultural activities and formal teaching in the development of numeracy

Training and interventions on (early) numerical skills have typically focused on directly enhancing particular numerical skills resulting in immediate outcomes, but which may not have sustainable effects. In this kind of approach, the child is an object of treatment with little agency or incentive to use the newly learnt skill beyond the training and testing context. Alternatively, supporting early mathematical development could consider that the development of mathematical thinking takes place not only during organized teaching-learning situations but also in many kinds of informal activities outside formal educational contexts. The large individual differences in children's mathematical skills when they enter formal schooling could be due to the qualitative and quantitative differences in these informal mathematical activities.

Fuson (1988), in her seminal book, described how children learn the meanings and uses of numbers in cardinal, ordinal, measurement, numerical, number sequence, and number system contexts in their early childhood. From an early age, children are involved in activities that are related to numbers and numerical relations. Before formal schooling, children are continuously dealing with magnitudes, for example, by comparing, grouping, and dividing various sets of objects. There are large interindividual differences between children in the intensity of engagement in these activities. However, basically all children have at least some experiences with magnitudes and numbers when they enter formal education (Fuson, 1988).

The role of these culturally mediated activities in early numerical development was furthered by Nunez (2017), who stated that the evidence of quantitative and numerical abilities suggests that.

> *biologically evolved preconditions for quantification do exist, but the emergence of number and arithmetic proper—absent in nonhuman animals—has materialized via cultural preoccupations and practices that are supported by language and symbolic reference—crucial dimensions that lie largely outside natural selection. (p. 409)*

This led him to conclude: "A distinction between quantical (e.g., quantity discrimination) and numerical (exact, symbolic) cognition is needed: quantical cognition provides biologically evolved preconditions for numerical cognition but it does not scale up to number and arithmetic, which require cultural mediation" (Nunez, 2017, p. 409). Nunez's model describes a different quantitative—nonsymbolic and numerical—symbolic view that resonates with our research on early numeracy development.

## Spontaneous focusing on numerosity and other numerical aspects

Since the first evidence on individual differences in Spontaneous Focusing on Numerosity (SFON; Hannula & Lehtinen, 2005), we have aimed to determine how early, low-level quantitative capacities are integrated with cultural practices of numeracy. To do so, we describe how children focus on exact numerosities and other mathematically relevant aspects of their surroundings and use their existing mathematical skills in their own, self-initiated actions. The transfer between an informal activities and formal mathematical knowledge and skills is at the core of this approach. In Hannula and Lehtinen (2005), we stated:

> *It is not only the sociocultural mediation of numerical cognition which develops the child's skill to focus on the aspect of numerosity and to utilize innate and cultural tools for enumeration, but also focusing on the aspect of numerosity develops the child's enumeration skills by activating the enumeration process and thus producing practice in enumeration. (p. 239)*

This practice, then, develops the child's quantifying skills in multiple ways. It not only broadens the counting range but also, along with more developed counting (quantifying) skills, a larger quantity may become a possible subject for counting. Furthermore, increasing knowledge about the use of quantifying means that different tasks will enhance the child's tendency to focus on numerosity in new, more demanding situations. As well, this practice is crucial for the integration of the necessary processes that lead to understanding the cardinality principle, as outlined by Gelman and Gallistel (1978).

This is in line with Sophian's theorizing about the iterative developmental processes of the conceptual knowledge of numbers and goal-directed numerical activities (Sophian, 1988). According to Saxe, Guberman, and Gearhart (1987), counting development is reciprocal in such way that socially structured goals of quantification change along with the development of skills, directing children's attention to different aspects of numbers and the ways they are used. Children seem to actively look for different situations where they can practice the phenomenon they want to master when their intention is focused on that specific phenomena (Lindahl & Samuelsson, 2001). Based on these accounts, it is clear that children's own spontaneous, self-initiated mathematical activities appear to be ripe for consideration in promoting early mathematical skills.

It has been 20 years since our first experimental task that assessed children's individual differences in SFON. In one of the first SFON tasks developed, the tester introduced a toy parrot "Vaakku" and his favorite berries to the child and said: "Watch carefully what I do, and then you do just like I did." Then the experimenter put two berries, one at a time into the parrot's mouth, and asked the child to do exactly like she had done. Some 3–5-year-old children notice, without any guidance or hints in the experimental setting, the exact number of berries the experimenter gave and systematically give the same amount of berries. However, some children of the same age do not pay attention to the exact number of berries, but instead focus on other aspects, such as the specific way in which the experimenter gave them to the bird. These children do not give the exact number of berries the experimenter give but many more, or just one berry to the parrot. These individual differences between children appear despite the fact that they do not differ from the other group on measures of attention or motivation to complete the task, and they are able to recognize and produce sets of two when asked. The basic idea behind these studies on SFON and other spontaneous mathematical focusing tendencies, investigated later by our group (for recent reviews, see Hannula-Sormunen, 2015; McMullen, Chan, Mazzocco, & Hannula-Sormunen, 2019), is that using number skills (such as exact number recognition) in natural surroundings is not completely automatic (Trick & Pylyshyn, 1994). Instead, this use is at least partly, a goal-directed, learned act. Furthermore, the amount of practice children acquire in using their early number skills in everyday settings differs substantially according to how frequently they focus on numerical properties (Hannula, 2005; Hannula & Lehtinen, 2005). Differentiating specific attentional processes that trigger the use of number skills from the skills themselves allows us to more exclusively capture the relevant subprocesses that are needed for this exact number recognition in everyday surroundings. When someone uses exact number recognition in their action, focusing on numerosity is an early step in this process. This recognition triggers an enumeration act based on subitizing or counting, which follows Gelman and Gallistel's (1978) counting principles, such as one-to-one correspondence, constant order, order irrelevance, abstraction, and the cardinality principle. Focusing on numerosity can happen spontaneously (i.e., without external guidance by others or task instructions in a certain situation) or in a guided way, when another person or task instructions help the person to focus their attention on numerosity. In this way, the term "spontaneous" does not refer to the developmental origins of this tendency, but rather the spontaneous recognition and use in a specific instance.

One major challenge in studying SFON is to develop testing procedures and activities that facilitated the measurement of focusing tendency strength and not just number skills. Our general aims when developing SFON assessments (Hannula, 2005) were as follows: (1) differentiating measures of spontaneous and guided focusing on numerical properties from each other (spontaneous, self-initiated enumeration vs. guided, prompted enumeration); (2) well controlled but meaningful tasks in which the probability of producing an accurate number (or order) of items or events unintentionally is very low, and the production of accurate numerosity (or order) thus indicates focusing on the number (or numerical order); (3) in the tasks, every child who has focused on the numerical aspect of the task should be able to proceed in the task accurately (i.e., nonsufficient enumeration, language, and procedural skills should not confound measures of SFON); and (4) no hints or guidance toward math or numerical issues is given by the experimenter or the task setting to allow measurements of spontaneous numerical behavior in the specific task context.

We have used the concept of SFON tendency to describe a child's general tendency across different tasks and time to focus on exact numerosity and use it in their actions (Hannula & Lehtinen, 2005). By comparing performance in guided and unguided numerical focusing tasks, we have shown that all children in the studies were capable of proceeding in the SFON tasks once their attention was focused on the exact numerosity involved in the task (Hannula & Lehtinen, 2005). These findings were confirmed in a recent study, where the response patterns showed that a SFON task does not measure enumeration accuracy but variability in numerical focusing (Nanu, McMullen, Munck, Pipari Study Group, & Hannula-Sormunen, 2018). Studies have also shown that SFON is not an artificial product of the specific measurement situations, but there is substantial correlation between children's SFON measures and the observed spontaneous use of numbers in everyday situations (Batchelor, 2014; Hannula, Mattinen, & Lehtinen, 2005; Rathé, Torbeyns, Hannula-Sormunen, De Smedt, & Verschaffel, 2016).

According to previous studies, learning to focus attention on the aspect of number in one's surroundings is a significant factor for successful early mathematical development, and it may provide children with the necessary practice in utilizing number recognition processes (e.g., Batchelor, 2014; Batchelor, Inglis, & Gilmore, 2015b; Edens & Potter, 2013; Hannula, 2005; Hannula & Lehtinen, 2005; Hannula, Lepola, & Lehtinen, 2010; Hannula, Räsänen, & Lehtinen, 2007; Hannula-Sormunen, 2015; Hannula-Sormunen, Lehtinen, & Räsänen, 2015; Kucian et al., 2012; McMullen, Hannula-Sormunen, & Lehtinen, 2015; Poltz et al., 2013; Rathé, Torbeyns, De Smedt, & Verschaffel, 2019; Sella, Berteletti, Lucangeli, & Zorzi, 2016). These studies revealed that some children's world may be full of opportunities to use their mathematical skills, while others seem to more often miss the mathematical features of the environment. Individual differences in SFON have been found to predict domain-specific differences in the development of mathematical skills throughout childhood and primary school (Hannula et al., 2010; Hannula-Sormunen et al., 2015; McMullen, Hannula-Sormunen, & Lehtinen, 2015; McMullen, Laakkonen, et al., 2015; Nanu et al., 2018). Furthermore, SFON tendency has been shown to be a stable component of early mathematical skills (Bojorque, Torbeyns, Hannula-Sormunen, Van Nijlen, & Verschaffel, 2017) that is not explained by motivational factors (Edens & Potter, 2013; Lepola & Hannula-Sormunen, 2019), general attentional skills, nonverbal IQ, or comprehension of instructions (Hannula et al., 2010; Hannula & Lehtinen, 2005). Across several such studies, the positive relation between SFON and mathematical skills remained significant, even after controlling for age and cognitive skills, including full-scale IQ (Nanu et al., 2018), verbal IQ (Poltz et al., 2013), or nonverbal IQ (Hannula et al., 2010; Hannula & Lehtinen, 2005; Hannula-Sormunen et al., 2015; Poltz et al., 2013); rapid serial naming (Hannula et al., 2010); working memory (Batchelor, Inglis, & Gilmore, 2015a; Nanu et al., 2018; Poltz et al., 2013); inhibition (Poltz et al., 2013), executive function skills, and vocabulary (Gray & Reeve, 2016); verbal comprehension (Hannula et al., 2010; Hannula &

Lehtinen, 2005); verbal production skills (Batchelor, Inglis, et al., 2015a); and spatial location detection (Hannula et al., 2010).

When children's number skills develop further, just focusing on the number of objects or events is not enough; they can start focusing on more complex mathematical phenomena. More recently, we have studied children's Spontaneous Focusing On quantitative Relations (SFOR), which is the spontaneous (i.e., unguided) focusing on quantitative relations and the use of these relations in situations that are not explicitly mathematical. Instead of focusing on a single numerical quantity, such as the total number of apples in a bowl (e.g. 6 apples in a bowl), SFOR involves focusing on the relation between two or more quantities (e.g., there are twice as many red apples as green apples; 1/3 of the apples are green). While SFOR tendency may often involve first focusing on number (as in the apples example), this may not always be the case, such as when dealing with continuous quantities. The relation between SFOR and SFOR tendency has yet to be explicitly examined.

By comparing students' performances in spontaneous and guided versions of the SFOR tasks, we have shown that individual differences in SFOR tendency are not entirely explained by the requisite skills needed to solve the tasks; however, there are substantial interindividual differences in young students' tendencies to spontaneously focus on quantitative relations (McMullen, Hannula-Sormunen, Laakkonen, & Lehtinen, 2016; McMullen, Hannula-Sormunen, & Lehtinen, 2014). Individual differences in SFOR tendency do predict domain-specific differences in the development of mathematical skills, particularly in rational number knowledge throughout childhood and primary school (McMullen et al., 2016; Van Hoof et al., 2016), even after controlling for nonverbal intelligence, arithmetical skills, and whole number estimation (McMullen, Hannula-Sormunen, & Lehtinen, 2015; McMullen, Laakkonen, et al., 2015), as well as general mathematical achievement (Van Hoof et al., 2016). SFOR tendency also has a long-term impact on mathematical development, stretching from primary into lower secondary school (McMullen, Hannula-Sormunen, & Lehtinen, 2017).

Spontaneous mathematical focusing tendencies are particularly interesting from the point of view of transfer in mathematical development because they seem to mediate between informal and formal mathematical aspects of mathematical knowledge. The iterative reciprocal development of numerical skills was first reported in Hannula and Lehtinen (2005), in which we showed that SFON tendency and enumeration skills are in a reciprocal relationship, with each predicting the other over time from age 3–6 years (Hannula & Lehtinen, 2005). In other studies, previously developed enumeration and subitizing skills have been found to support SFON tendency and, subsequently, enhanced SFON tendency has been a strong predictor of later mathematical development (Hannula-Sormunen et al., 2015). In our recent longitudinal study, we followed a sample of third to fifth graders over a 4-year period. The results showed that students' SFOR tendency and rational number conceptual knowledge were reciprocally related across the four time points (McMullen, Hannula-Sormunen, & Lehtinen, 2017).

## How to enhance SFON and SFOR?

Mirroring the well-known, and problematic, Mathew Effect (Mattinen, 2006) demonstrated that adults tend to provide more support for children who are more likely to notice number by providing these children with more and more numerical activities, while the children who do not attend to numerical features of their environment are often left without guidance on numerical activities. In response, we began to use the behaviors of those children with strong SFON tendencies to model the desired behaviors for children with lower SFON tendencies. A crucial element in applying relevant mathematical concepts to formal and informal settings is recognizing exactly when mathematical aspects are present and useful in reasoning (Lobato, 2012; Lobato, Rhodehamel, & Hohensee, 2012). In order to model the world mathematically, a child must first recognize that this can be done (McMullen & Resnick, 2018). In previous training studies aimed at supporting SFON tendency, the main goal was to make number a more direct target of focusing for children (e.g., Hannula et al., 2005; Mattinen, 2006). These programs explicitly highlighted and modeled when and how exact number recognition can be used in reasoning in and out of the original learning context throughout the day in preschool.

We based our first intervention study on previous research on theories of mathematical skill development (for a review, see Hannula, 2005) and sociocultural theories of adults' role in supporting early learning and attentional skills (Gauvain, 2001; Mattinen, 2006). Social interaction and the aid of more experienced members of society are crucial for the development of higher order processes in self-regulation, especially the controlling of attention (Gauvain, 2001; Ruff & Rothbart, 1996). According to Gauvain (2001), the social context affords children structured opportunities to practice, refine, and extend their cognitive skills. An important element in our

SFON interventions has been based on research showing how participation in shared activities requires and develops an understanding of the intentions of others, ways of cultural learning, motivation to share psychological states with others, and specific forms of cognitive representations for shared actions, which are all the bases of social life as human beings (Tommasello, Carpenter, Call, Behne, & Moll, 2005). For example, Gleissner, Meltzoff, and Bekkering (2000) demonstrated that 3-year-old children already code human behavior and imitate it on the basis of hierarchically organized goals, giving them the opportunity to focus on relevant and more abstract aspects of observed actions. The joint processes of children and adults helping to direct children's attention to relevant aspects of tasks enable children to acquire the culturally based numerical tools necessary for living in our society. Children learn to understand the purposes of the tasks and certain cognitive strategies embedded in a variety of everyday activities through these joint processes (Gauvain, 2001; Mix, 2002).

In her influential case study, Mix (2002) described a wide range of socially and linguistically mediated, numerically meaningful activities in everyday family life. Similarly, Lobato and her colleagues demonstrated that the social and physical aspects of a certain situation can support children's attention toward particular mathematical properties rather than others during mathematics lessons (Lobato, 2003; Lobato, Ellis, & Munoz, 2003). According to other studies, children's mathematical skills are related to parents' skills in modeling mathematical thinking in everyday contexts, as well as in contexts involving the usual preschool material (Anderson, 1997; Blevins-Knabe & Musun-Miller, 1996; Huntsinger, Jose, Larson, Krieg, & Shaligram, 2000; Linnell & Fluck, 2001; Saxe et al., 1987). Differences in the preschool and home learning environments are related to mathematics achievement up to age 10 (Melhuish, Sylva, Sammons, et al., 2008). Likewise, the amount and quality of parents' number-related talk has also been associated with early numerical development (Gunderson & Levine, 2011; Levine, Suriyakham, Rowe, Huttenlocher, & Gunderson, 2010).

In Hannula et al. (2005), we found that an intervention based on early educators' small-number-focused activities with 3-year-old children led to greater gains in the preschoolers' SFON tendencies, relative to a control group in which no such number-focused interactions were promoted. However, the effect of this intervention was apparent for only those preschoolers in the experimental group who had at least some measurable SFON tendency at baseline. This led us to wonder about the role of prior numerical skills and chances of triggering SFON for those who do not yet exhibit it.

In search of answers to this important open question, another set of studies was recently conducted. In this research on SFON enhancement, either 2.5–3-year-old children's SFON and small number recognition skills or 3–5-year-old children's SFON and counting skills were supported (Hannula-Sormunen et al., 2020). The programs aimed to promote noticing numerical features by embedding SFON baits around the daycare environment in addition to other numerical play and games. Intervention programs had several elements aimed at supporting transfer. First, all activities were conducted in small- and whole-group situations, as well as with individual children in suitable moments during the preschool or daycare day. In addition, when SFON or cardinality recognition skills were practiced with particular objects, children were always asked to look for other objects in their sight which they could enumerate. "What could you count here?" was a regular question asked of the children. The teachers and personnel were challenged to start seeing opportunities for guiding children's attention to exact numerosities during all daily activities so that eventually, numerosity would become a relevant focusing target and aspect of activities in the group.

The SFON baits were toys that were constructed by using identical objects and everyday life materials arranged in a manner that made the numerical features very salient against unanimous, typically single-color backgrounds. This often involved using several identical objects arranged in close proximity (e.g., two identical cups side-by-side), which increased the likelihood of counting behavior, as the items were more likely to be perceived as a set to be counted. If the child did not focus on the numerosity of the items in the SFON bait, the early educators were asked to explicitly guide the child's attention to numerosity by asking how many items there were or by taking away or adding items. Deliberate manipulation of numerosity has proven to be an efficient way of attracting children's attention toward numerosity of items in a set (Hannula et al., 2005). In contrast with previous SFON interventions that were effective only in children with some initial SFON tendency (Hannula et al., 2005), these recent interventions with SFON baits led to significant improvement in SFON tendency, as well as cardinality recognition and counting skills, particularly among those with the weakest SFON tendency and cardinality recognition skills at the start of the intervention. The active control groups of the studies also received special training in listening comprehension skills (Hannula-Sormunen et al., 2020).

Technology-based playful environments can also provide rich opportunities for triggering and enhancing children's tendency to focus on mathematical aspects of their environment. Recently, we found that 5-year-olds' arithmetic skills and SFON tendency developed as a result of playing the Fingu iPad game (Holgersson et al., 2016) integrated with SFON-based everyday activities (Hannula-Sormunen, Alanen, McMullen, & Lehtinen, 2016). The results showed a clear developmental advantage for the training group over the control group in arithmetic skills. Information technology was also used in the first intervention experiment aimed at enhancing students' SFOR tendency. Mobile technology was used as a tool to explore and identify quantitative relations in students' everyday environment outside of the classroom. The results showed that the intervention was successful in enhancing SFOR tendency (McMullen et al., 2019). In another recent study, classroom SFOR intervention was given during mathematics lessons that replaced part of the regular teaching of rational numbers (Määttä, Hannula-Sormunen, Halme, & McMullen, n.d.). In the control group, the same time was allotted for the direct teaching of rational numbers. The intervention resulted in immediate improvements in reasoning about multiplicative relations and leads to long-term improvements in number line estimation of rational numbers within the intervention group.

## Mathematical knowledge, transfer, and spontaneous mathematical focusing tendencies

In Lehtinen and Hannula (2006), we argued that the frequently observed situations of mathematics education in which children (or later adults) are not able to transfer their mathematical knowledge into new situations do not mean that the very nature of mathematical knowledge is situation-specific; there are several other reasons why people do not apply their (at least partly abstract) mathematical knowledge in these situations. We highlighted the importance of domain-specific attentional processes as essential parts of transfer and argued that one reason for transfer failures could be that the situations are not necessarily interpreted as mathematical or numerical by the child. This would mean that the child does not pay attention to the mathematically relevant aspects of the situation and is, thus, not able to carry out the expected mathematical operations.

Empirical findings also indicate that spontaneous mathematical focusing tendencies mediate formal mathematical knowledge and the practical application of numerosity, numerical order, and quantitative relations in non-explicitly mathematical situations. Indeed, explicit intervention aimed at increasing these tendencies may help create more opportunities to bridge this divide for more students.

Focusing attention on numerosity or other mathematically relevant aspects of the environment, which supports the utilization of mathematical knowledge, depends on the situational features, as modeled by McMullen and colleagues in their recent review (McMullen et al., 2019). For instance, Chan and Mazzocco (2017) found that by manipulating the relative salience of numbers (as a visual feature of the match options), the frequency of number-based matches could also be manipulated for children and adults during the task. As indicated by an analysis of children's behavioral patterns across spontaneous and guided versions of the same tasks, the existence of relevant mathematical knowledge is a necessary but not sufficient precondition for transfer (McMullen et al., 2014). Research on the early development of number skills shows that even young children have abstract mathematical concepts that are, in principle, transferable to different novel situations (Lehtinen & Hannula, 2006). The actual application depends on the child's spontaneous focusing on the quantitative or mathematical aspect of the situation, which, in turn, is dependent on the features of the situation and the child's experiences in the sociocultural environment. Failure of transfer may also be caused by failure of focusing, not just a failure of knowledge.

Beyond how a failure of transfer may also result from a failure of focusing, studies on spontaneous focusing on numerosity and quantitative relations have also highlighted the other direction of transfer, from practical experience to formal mathematical knowledge, which is almost completely neglected in traditional transfer research. The transfer from practical experience to formal mathematical knowledge cannot be described as the immediate application of previously learned knowledge in a new context but rather should be considered an instance of transfer in what Bransford and Schwartz (1999) called preparation for future learning. This might provide a clue as to why spontaneous mathematical focusing tendencies are such strong predictors of later mathematic learning.

The first hypothesized mechanism is based on the amount and intensity of practice. Mathematical proficiency requires extensive practice (e.g., Fuchs et al., 2010). Children with a strong tendency to spontaneously focus on mathematically relevant aspects both in and out of the classroom stand to gain much more intensive practice in mathematical thinking than children who focus on mathematics only when they are specifically guided by parents or teachers. This self-initiated practice is very similar with the early form of deliberate practice in music or sports, described in studies on expertise development (see, review, Lehtinen, Hannula-Sormunen, McMullen, & Gruber, 2017).

However, it is not only the amount and intensity, but also the quality, of practice that may matter. Practice in everyday life is qualitatively different from school math tasks because of the experience of variation (see Marton, 2006) children get in applying their mathematical thinking to somewhat "messy" situations in everyday life. One of the findings of studies of spontaneous mathematical focusing tendencies is that both SFON and SFOR predict the later conceptual understanding of rational numbers (McMullen, Hannula-Sormunen, & Lehtinen, 2015, 2017; McMullen, Laakkonen, et al., 2015). Spontaneous mathematical focusing may prepare children for learning rational numbers in the future because the experiences that children acquire when focusing on mathematical aspects of the environment outside of the formal educational context are different from what they would acquire while practicing with (clean and simplified) school math tasks (McMullen, Hannula-Sormunen, & Lehtinen, 2015; McMullen, Laakkonen, et al., 2015). For example, a higher SFON tendency might focus children's attention on situations where natural number is insufficient for dealing with a situation, such as sharing two apples between three people. A higher SFOR tendency would more often lead to situations where a mathematical relation would be best described using quantitative relations, as everyone would then get two-thirds of an apple. In both instances, these opportunities to describe and reason about the relational features of an everyday situation require first focusing on the mathematics embedded in the situation. Gaining more practice of this quality has the potential to transfer to stronger outcomes when learning about formal mathematics that relies on related concepts. This is important, because rational number knowledge is recognized as the hardest mathematical content to learn in primary and lower secondary schools, and a large part of the adult population has serious misconceptions about fractions, decimals, and percentages (Van Dooren, Lehtinen, & Verschaffel, 2015). An adequate conceptual understanding of rational numbers requires conceptual change in the beliefs about numbers (Obersteiner, Van Dooren, Van Hoof, & Verschaffel, 2013).

## Conclusions

Altogether, the studies on spontaneous mathematical focusing tendencies described in this chapter suggest a reciprocal relationship between the development of these tendencies and numerical skills, with numerical focusing tendencies having a positive impact on the development of early counting and enumeration skills and later conceptual understanding of rational numbers. These studies also indicate that the design of early childhood educational settings in which mathematics is not simply constrained to highly ritualized, decontextualized instructional techniques is a potential boon for the long-term development of children's mathematical knowledge and skills. Interventions designed to promote spontaneous mathematical focusing tendencies have the potential to provide long-lasting, sustained effects by aligning with recommendations from both national and research-based sources.

They do so by overcoming problematically minimal transfer effects of easily isolated drill-and-practice kinds of mathematical activities is an important goal for mathematics education (e.g., Lehtinen et al., 2017). First, promoting and supporting students' and children's spontaneous focusing tendencies, leading to self-initiated practice of newly learned mathematical skills, may help them begin to use these skills in their own activities, in addition to adult-guided mathematical exercises. This, in turn, may help children develop their mathematical understanding even further. Second, while designing mathematical learning environments and learning materials, educators should make sure their materials and training help students noticing the applicability of newly learnt mathematical skills in everyday contexts outside of mathematics lessons. In this way, they can help more children start seeing the mathematical world all around them.

## References

Anderson, A. (1997). Families and mathematics: A study of parent-child interactions. *Journal for Research in Mathematics Education*, *28*, 484–511.

Bailey, D. H., Duncan, G. J., Odgers, C. L., & Yu, W. (2017). Persistence and fadeout in the impacts of child and adolescent interventions. *Journal of Research on Educational Effectiveness.*. https://doi.org/10.1080/19345747.2016.1232459.

Bailey, D. H., Nguyen, T., Jenkins, J. M., Domina, T., Clements, D. H., & Sarama, J. S. (2016). Fadeout in an early mathematics intervention: Constraining content or preexisting differences? *Developmental Psychology*, *52*(9), 1457–1469. https://doi.org/10.1037/dev0000188.

Batchelor, S. (2014). *Dispositional factors affecting children's early numerical development.* Loughborough University.

Batchelor, S., Inglis, M., & Gilmore, C. (2015a). Spontaneous focusing on numerosity and the arithmetic advantage. *Learning and Instruction*, *40*, 79–88. https://doi.org/10.1016/j.learninstruc.2015.09.005.

Batchelor, S., Inglis, M., & Gilmore, C. (2015b). Spontaneous focusing on numerosity and the arithmetic advantage. *Learning and Instruction*, *40*, 116–135.

Björklund, C., & Barendregt, W. (2016). Teachers' pedagogical mathematical awareness in Swedish early childhood education teachers. *Scandinavian Journal of Educational Research*, *60*, 359–377.

Bojorque, G., Torbeyns, J., Hannula-Sormunen, M., Van Nijlen, D., & Verschaffel, L. (2017). Development of SFON in Ecuadorian Kindergartners. *European Journal of Psychology of Education, 32*(3), 449–462. https://doi.org/10.1007/s10212-016-0306-9.

Blevins-Knabe, B., Austin, A. B., Musun, L., Eddy, A., & Jones, R. M. (2000). Family home care providers' and parents' beliefs and practices concerning mathematics with young children. *Early Child Development and Care, 165*, 41–58. https://doi.org/10.1080/0300443001650104.

Blevins-Knabe, B., & Musun-Miller, L. (1996). Number use at home by children and their parents and its relationship to early mathematical performance. *Early Development and Parenting, 5*, 35–45.

Bransford, J. D., & Schwartz, D. L. (1999). Rethinking transfer: A simple proposal with multiple implications. In A. Iran-Nejad, & P. D. Pearson (Eds.), *Vol. 24. Review of research in education* (pp. 61–100). Washington, D.C.: American Educational Research Association.

Chan, J. Y.-C., & Mazzocco, M. M. M. (2017). Competing features influence children's attention to number. *Journal of Experimental Child Psychology, 156*, 62–81. https://doi.org/10.1016/j.jecp.2016.11.008.

Clements, D., Sarama, J., Spitler, M. E., Lange, A., & Wolfe, C. B. (2011). Mathematics learned by young children in an intervention based on learning trajectories: A large-scale cluster randomized trial. *Journal for Research in Mathematics Education, 42*(2), 127–166. https://doi.org/10.5951/jresematheduc.42.2.0127.

Cross, C. T., Woods, T. A., & Schweingruber, H. (Eds.). (2009). *Mathematics learning in early childhood: Paths towards excellence and equity*. Washington, D.C.: National Academy Press.

De Corte, E. (1999). On the road to transfer: An introduction. *International Journal of Educational Research, 31*, 555559.

Edens, K. M., & Potter, E. F. (2013). An exploratory look at the relationships among math skills, motivational factors and activity choice. *Early Childhood Education Journal, 41*(3), 235–243.

Freudenthal, H. (1991). *Revisiting mathematics education: China lectures*. Dordrecht: Kluwer.

Frye, D., Baroody, A. J., Burchinal, M., Carver, S. M., Jordan, N. C., & McDowell, J. (2013). *Teaching math to young children: A practice guide (NCEE 2014-4005)*. Washington, DC: National Center for education Evaluation and Regional Assistance (NCEE), Institute of Education Sciences, U.S. Department of Education. Retrieved from the NCEE website: http://whatworks.ed.gov.

Fuchs, L. S., Malone, A. S., Schumacher, R. F., Namkung, J., & Wang, A. (2017). Fraction intervention for students with mathematics difficulties: Lessons learned from five randomized controlled trials. *Journal of Learning Disabilities, 50*(6), 631–639. https://doi.org/10.1177/0022219416677249.

Fuchs, L. S., Powell, S. R., Seethaler, P. M., Cirino, P. T., Fletcher, J. M., Fuchs, D., et al. (2010). The effects of strategic counting instruction, with and without deliberate practice, on number combination skill among students with mathematics difficulties. *Learning and Individual Differences, 20*, 89–100.

Fuchs, L. S., Schumacher, R. F., Long, J., Namkung, J., Hamlett, C. L., Cirino, P. T., … Changas, P. (2013). Improving at-risk learners' understanding of fractions. Journal of Educational Psychology, 105(3), 683–700. https://doi.org/10.1037/a0032446.

Fuson, K. (1988). *Children's counting and concepts of number*. New York, NY: Springer Verlag.

Fuson, K. C., Clements, D. H., & Sarama, J. (2015). Making early math education work for all children. *Kappan, 97*, 63–68.

Gauvain, M. (2001). *The social context of cognitive development*. New York, NY: Guilford Press.

Gelman, R., & Gallistel, C. R. (1978). *The child's understanding of number*. Cambridge, MA: Harvard University Press.

Gleissner, B., Meltzoff, A. N., & Bekkering, H. (2000). Children's coding of human action: Cognitive factors influencing imitation in 3-year-olds. *Developmental Science, 3*, 405–414.

Gray, S, A, & Reeve, R, A. (2016). Number-specific and general cognitive markers of preschoolers' math ability profiles. *Journal of Experimental Child Psychology*, *147*, 1–21.

Gunderson, E. A., & Levine, S. C. (2011). Some types of parent number talk count more than others: Relations between parents' input and children's number knowledge. *Developmental Science*, *14*(5), 1021–1032.

Hachey, A. C. (2013). Teachers' beliefs count: Teacher beliefs and practice in early childhood mathematics education (ECME). *Dialog*, *16*, 77–85.

Hamdan, N., & Gunderson, E. A. (2017). The number line is a critical spatial-numerical representation: Evidence from a fraction intervention. *Developmental Psychology*, *53*(3), 587–596. https://doi.org/10.1037/dev0000252.

Hannula, M. M. (2005). *Spontaneous focusing on numerosity in the development of early mathematical skills (Doctoral dissertation)*. Finland, Painosalama: University of Turku.

Hannula, M. M., & Lehtinen, E. (2005). Spontaneous focusing on numerosity and mathematical skills of young children. *Learning and Instruction*, *15*(3), 237–256. https://doi.org/10.1016/j.learninstruc.2005.04.005.

Hannula, M. M., Lepola, J., & Lehtinen, E. (2010). Spontaneous focusing on numerosity as a domain-specific predictor of arithmetical skills. *Journal of Experimental Child Psychology*, *107*(4), 394–406. https://doi.org/10.1016/j.jecp.2010.06.004.

Hannula, M. M., Mattinen, A., & Lehtinen, E. (2005). Does social interaction influence 3-year-old children's tendency to focus on numerosity? A quasi-experimental study in day care. In L. Verschaffel, E. De Corte, G. Kanselaar, & M. Valcke (Eds.), *Powerful environments for promoting deep conceptual and strategic learning* (pp. 63–80). Leuven, Belgium: Leuven University Press.

Hannula, M. M., Räsänen, P., & Lehtinen, E. (2007). Development of counting skills: Role of spontaneous focusing on numerosity and subitizing-based enumeration. *Mathematical Thinking and Learning*, *9*, 51–57. https://doi.org/10.1207/s15327833mtl0901_4.

Hannula-Sormunen, M. M. (2015). Spontaneous focusing on numerosity and its relation to counting and arithmetic. In A. Dowker, & R. Cohen Kadosh (Eds.), *Oxford handbook of mathematical cognition* (pp. 275–290). Croydon: Oxford University Press.

Hannula-Sormunen, M., Alanen, A., McMullen, J., & Lehtinen, E. (2016). Integrating SFON enhancement with computerized arithmetical training—A pilot study. In *Poster presented in Workshop Domain-General and Domain-Specific Foundations of Numerical and Arithmetic Processing, University of Tubingen, Germany, 28–30 September.*

Hannula-Sormunen, M. M., Lehtinen, E., & Räsänen, P. (2015). Preschool children's spontaneous focusing on numerosity, subitizing and counting skills as predictors of their mathematical performance 7 years later at school a. *Mathematical Thinking and Learning*, *17*, 155–177.

Hannula-Sormunen, M. M., Nanu, C., Luomaniemi, K., Heinonen, M., Sorariutta, A., Södervik, I., et al. (2020). Promoting spontaneous focusing on numerosity and cardinality related skills at day care with one, two how many and count, how many programs. *Mathematical Thinking and Learning*, *22*(4), 312–331. https://doi.org/10.1080/10986065.2020.1818470.

Holgersson, I., Barendregt, W., Emanuelsson, J., Ottosson, T., Rietz, E., & Lindström, B. (2016). Fingu—A game to support children's development of arithmetic competence: Theory, design and empirical research. In P. S. Moyer-Packenham (Ed.), *International perspectives on teaching and learning mathematics with virtual manipulatives* (pp. 123–145). Switzerland: Springer International Publishing. https://doi.org/10.1007/978-3-319-32718-1_6.

Huntsinger, C. S., Jose, P. E., Larson, S. L., Krieg, D. B., & Shaligram, C. (2000). Mathematics, vocabulary, and reading development in Chinese American and European American children over the primary school years. *Journal of Educational Psychology*, *92*, 745–760.

Kucian, K., Kohn, J., Hannula-Sormunen, M. M., Richtmann, V., Grond, U., Käser, T., et al. (2012). Kinder mit Dyskalkulie fokussieren spontan weniger auf Anzahligkeit [children

with developmental dyscalculia focus spontaneously less on numerosities]. *Lernen Und Lernstörungen, 1*(4), 241–253.
Lehtinen, E., & Hannula, M. M. (2006). Attentional processes, abstraction and transfer in early mathematical development. In L. Verschaffel, F. Dochy, M. Boekaerts, & S. Vosniadou (Eds.), *Advances in Learning and Instruction Series. Instructional psychology: Past, present and future trends. Fifteen essays in honour of Erik De Corte* (pp. 39–54). Elsevier.
Lehtinen, E., Hannula-Sormunen, M. M., McMullen, J., & Gruber, H. (2017). Cultivating mathematical skills: From drill-and-practice to deliberate practice. *ZDM—Mathematics Education, 49*(4), 625–636.
Lepola, J., & Hannula-Sormunen, M. M. (2019). Spontaneous focusing on numerosity and motivational orientations as predictors of arithmetical skills from kindergarten to grade 2. *Educational Studies in Mathematics, 100*, 251–269.
Levine, S. C., Suriyakham, L. W., Rowe, M. L., Huttenlocher, J., & Gunderson, E. A. (2010). What counts in the development of young children's number knowledge? *Developmental Psychology, 46*(5), 1309–1319.
Lindahl, M., & Samuelsson, I. P. (2001). The effect of maternal support for counting and cardinal understanding in pre-school children. *Social Development, 10*, 202–220.
Linnell, M., & Fluck, M. (2001). The effect of maternal support for counting and cardinal understanding in pre-school children. *Social Development, 10*, 202–220.
Lobato, J. (2003). How design experiments can inform a rethinking of transfer and vice versa. *Educational Researcher, 32*, 17–20.
Lobato, J. (2012). The actor-oriented transfer perspective and its contributions to educational research and practice. *Educational Psychologist, 47*(3), 232–247. https://doi.org/10.1080/00461520.2012.693353.
Lobato, J., Ellis, A. B., & Munoz, R. (2003). How "focusing phenomena" in the instructional environment support individual students' generalizations. *Mathematical Thinking and Learning, 5*, 1–36.
Lobato, J., Rhodehamel, B., & Hohensee, C. (2012). "Noticing" as an alternative transfer of learning process. *Journal of the Learning Sciences, 21*(3), 433–482. https://doi.org/10.1080/10508406.2012.682189.
Marton, F. (2006). Sameness and difference in transfer. *Journal of the Learning Sciences, 15*(4), 499–535.
Määttä, S., Hannula-Sormunen, M., Halme, H., & McMullen, J. (n.d.). Guiding students' attention towards multiplicative relations around them: A classroom intervention. *Journal of Numerical Cognition*. Submitted for publication.
Mattinen, A. (2006). *Huomio lukumääriin: Tutkimus 3-vuotiaiden lasten matemaattisten taitojen tukemisesta päiväkodissa [Focus on numerosities: A study on supporting 3 year-old children's mathematical development in daycare] (Doctoral dissertation)*. Finland, Turku: University of Turku.
McMullen, J., Chan, J. Y., Mazzocco, M. M. M., & Hannula-Sormunen, M. M. (2019). Spontaneous mathematical focusing tendencies in mathematical development and education. In A. Norton, & M. W. Alibali (Eds.), *Constructing number: Merging perspectives from psychology and mathematics education* (pp. 69–86). Springer.
McMullen, J., Hannula-Sormunen, M. M., Laakkonen, E., & Lehtinen, E. (2016). Spontaneous focusing on quantitative relations as a predictor of the development of rational number conceptual knowledge. *Journal of Educational Psychology, 108*, 857–868.
McMullen, J., Hannula-Sormunen, M. M., & Lehtinen, E. (2014). Spontaneous focusing on quantitative relations in relation to children's mathematical skills. *Cognition and Instruction, 32*(2), 198–218.
McMullen, J., Hannula-Sormunen, M. M., & Lehtinen, E. (2015). Preschool spontaneous focusing on numerosity predicts rational number knowledge six years later. *ZDM—International Journal on Mathematics Education, 47*(5), 813–824.

McMullen, J., Hannula-Sormunen, M. M., & Lehtinen, E. (2017). Spontaneous focusing on quantitative relations as a predictor of rational number and algebra knowledge. *Contemporary Educational Psychology, 51*, 356–365.

McMullen, J., Hannula-Sormunen, M., Lehtinen, E., & Siegler, R. (2020). Distinguishing adaptive from routine expertise with rational number arithmetic. *Learning and Instruction, 68*. https://doi.org/10.1016/j.learninstruc.2020.101347.

McMullen, J., Laakkonen, E., Hannula-Sormunen, M. M., & Lehtinen, E. (2015). Modelling the developmental trajectories of rational number concept(s): A latent variable approach. *Learning and Instruction, 37*, 14–20.

McMullen, J., & Resnick, L. B. (2018). Linking informal and formal mathematical reasoning: Two directions across the same bridge? In E. Bergqvist, M. Österholm, C. Granberg, & L. Sumpter (Eds.), *Proceedings of the 42nd conference of the International Group for the Psychology of mathematics education*. Sweden: Umeå.

Melhuish, E. C., Sylva, K., Sammons, P., et al. (2008). Preschool influences on mathematics achievement. *Science, 321*, 1161–1162.

Mix, K. S. (2002). The construction of number concepts. *Cognitive Development, 17*, 1345–1363.

Nanu, C., McMullen, J., Munck, P., Pipari Study Group, & Hannula-Sormunen, M. (2018). Spontaneous focusing on numerosity in preschool as a predictor of mathematical skills and knowledge in the fifth grade. *Journal of Experimental Child Psychology, 169*, 42–58.

National Research Council. (2009). *Mathematics Learning in Early Childhood: Paths toward Excellence and Equity*. Washington, D.C: The National Academies Press.

Nunez, R. (2017). Is there really an evolved capacity for number? *Trends in Cognitive Sciences, 21*(6), 409–424. https://doi.org/10.1016/j.tics.2017.03.005.

Obersteiner, A., Van Dooren, W., Van Hoof, J., & Verschaffel, L. (2013). The natural number bias and magnitude representation in fraction comparison by expert mathematicians. *Learning ad Instruction, 28*, 64–72. https://doi.org/10.1016/j.learninstruc.2013.05.003.

Poltz, N., Quandte, S., Wyschkon, A., Hannula-Sormunen, M. M., von Aster, M., & Esser, G. (2013). Spontaneous focusing on numerosity and the development of mathematical skills in kindergarten. In *Paper presented at the meeting of ECDP, Lausanne, Switzerland*.

Pongsakdi, N., Brezovszky, B., Veermans, K., Hannula-Sormunen, M. M., & Lehtinen, E. (2016). A comparative analysis of word problems in selected Thai and Finnish textbooks. In *Proceedings of the 40th Conference of the International Group for the Psychology of Mathematics Education, Szegeb, August 3–7, 2016*.

Räsänen, P., Salminen, J., Wilson, A., Aunio, P., & Dehaene, S. (2009). Computer-assisted intervention for early number skills in children with low numeracy: A short review and a randomized controlled trial. *Cognitive Development, 24*, 450–472. https://doi.org/10.1016/j.cogdev.2009.09.003.

Rathé, S., Torbeyns, J., De Smedt, B., & Verschaffel, L. (2019). Spontaneous focusing on Arabic number symbols and its association with early mathematical competencies. *Early Childhood Research Quarterly, 48*, 111–121. https://doi.org/10.1016/j.ecresq.2019.01.011.

Rathé, S., Torbeyns, J., Hannula-Sormunen, M. M., De Smedt, B., & Verschaffel, L. (2016). Spontaneous focusing on numerosity: A review of recent research. *Mediterranean Journal for Research in Mathematics Education, 15*, 1–25.

Repo, L., Paananen, M., Eskelinen, M., Mattila, V., Lerkkanen, M.-K., Gammelgård, L., et al. (2019). In KARVI (Ed.). *Kansallisen Koulutuksen arviointikeskuksen julkaisuja: 15. Varhaiskasvatuksen laatu arjessa. Varhaiskasvatussuunnitelmien toteutuminen päiväkodeissa ja perhepäivähoidossa (Quality of early education. Fullfillment of early education curriculums in day care centers and family day care)*. Tampere: KARVI. https://karvi.fi/app/uploads/2018/10/KARVI_vaka_laadun-arvioinnin-perusteet-ja-suositukset.pdf.

Ruff, H. A., & Rothbart, M. K. (1996). *Attention in early development: Themes and variations*. New York, NY: Oxford University Press.
Saxe, G. B., Guberman, S. R., & Gearhart, M. (1987). Social processes in early number development. *Monographs of the Society for Research in Child Development*, *52*(216).
Schneider, M., & Siegler, R. S. (2010). Representations of the magnitudes of fractions. *Journal of Experimental Psychology: Human Perception and Performance*, *36*, 1227–1238.
Sella, F., Berteletti, I., Lucangeli, D., & Zorzi, M. (2016). Spontaneous non-verbal counting in toddlers. *Developmental Science*, *19*(2), 329–337. https://doi.org/10.1111/desc.12299.
Sophian, C. (1988). Limitations on preschool children's knowledge about counting: Using counting to compare sets. *Developmental Psychology*, *24*, 634–640.
Tommasello, M., Carpenter, M., Call, J., Behne, T., & Moll, H. (2005). Understanding and sharing intentions: The origins of cultural cognition. *Behavioral and Brain Sciences*, *28*(5), 675–735.
Trick, L., & Pylyshyn, Z. W. (1994). Why are small and large numbers enumerated differently? A limited capacity preattentive stage in vision. *Psychological Review*, 80–102.
Van Dooren, W., Lehtinen, E., & Verschaffel, L. (2015). Unraveling the gap between natural and rational numbers. *Learning and Instruction*, *37*, 1–4.
Van Hoof, J, Degrande, T, McMullen, J, Hannula-Sormunen, M, Lehtinen, E, Verschaffel, L, et al. (2016). The relation between learners' spontaneous focusing on quantitative relations and their rational number knowledge. *Studia Psychologica*, *58*, 156–170. https://doi.org/10.21909/sp.2016.02.714.
Vosniadou, S., Vamvakoussi, X., & Skopeliti, I. (2008). The framework theory approach to conceptual change. In S. Vosniadou (Ed.), *International handbook of research on conceptual change* (pp. 3–34). Mahwah, NJ: Erlbaum.
Wang, A. H., Firmender, J. M., Power, J. R., & Byrnes, J. P. (2016). Understanding the program effectiveness of early mathematics interventions for prekindergarten and kindergarten environments: A meta-analytic review. *Early Education and Development*, *27*(5), 692–713. https://doi.org/10.1080/10409289.2016.1116343.
Wyndhamn, J., & Säljö, R. (1997). Word problems and mathematical reasoning—A study of Children's mastery of reference and meaning in textual realities. *Learning and Instruction*, *7*, 361–382.

CHAPTER 10

# Developmental course of numerical learning problems in children and how to prevent dyscalculia: A summary of the longitudinal examination of children from kindergarten to secondary school

**Karin Kucian**
Center for MR-Research, University Children's Hospital Zurich, Zurich, Switzerland

## Introduction

Numerical competences are essential for many aspects of day-to-day living, and they are becoming even more crucial with the increasing role of technology in contemporary society. Importantly, profound difficulties with numeracy are very common with a prevalence rate around 6% (for review, please see Kucian, 2016). Affected children are usually diagnosed not before 2nd or 3rd grade, when their numerical learning problems get so severe that they can no more be compensated. By that time, affected children lack the understanding of basic numerical understanding, had often negative experiences with math-related issues, and often develop specific mathematical anxiety. These negative emotional factors in numerical cognition have negative effects on mathematical learning but can also have detrimental long-term consequences on school achievement and quality of life (for review please see Dowker, Sarkar, & Looi, 2016; Suárez-Pellicioni, Núñez-Peña, & Colomé, 2016). Therefore, a goal should be to identify children with math-learning problems as early as possible and support them effectively. To achieve this aim, it is essential to obtain a clearer picture of the developmental trajectory of children with poor numerical performance. In a comprehensive study, about 2900 German children were followed from

*Heterogeneous Contributions to Numerical Cognition*
https://doi.org/10.1016/B978-0-12-817414-2.00002-6

kindergarten to secondary school (SCHUES-Study 01 GJ1011, https://www.empirische-bildungsforschung-bmbf.de/de/440.php).

First, the study revealed important information about the predictive value of math-specific precursor skills as well as general precursor skills such as working memory and intelligence in kindergarten for future numerical and calculation achievement in 1st grade (Gallit et al., 2018).

Second, the study evaluated the long-term effects of a preschool training stimulating numerical competencies in children at risk for dyscalculia (Moraske et al., 2019).

Third, the persistence and the impact of numerical difficulties on mathematical performance and general school success up to secondary school were investigated (Schulz et al., 2018).

In the following sections, main findings of these three research objectives will be summarized in more detail. All findings are based on original articles which have been published in German in the peer-reviewed journal "Lernen und Lernstörungen" (for details, please see Gallit et al., 2018; Moraske et al., 2019; Schulz et al., 2018).

## How to predict which child will develop dyscalculia

Mathematical understanding does not just start with school entry but has its basics already in toddler and preschool age, where important prior knowledge and the necessary precursor skills develop for numerical and calculation knowledge. A distinction is made between specific and general precursor skills. As specific are such skills described that are mainly relevant for number processing and calculation, but also other numeracy-related school topics such as physics or chemistry. In contrast, general precursor skills include those that are important for different school achievements. In the present study, basic magnitude-number competences were examined as most important specific precursor skill for calculation and numerical abilities. Intelligence and working memory were investigated as relevant general precursor skills. Gallit et al. (2018) examined the development and reciprocal relation between specific and general precursor skills in kindergarten, and their predictive value for mathematical achievement in the 1st grade.

## Methods

A total of 1897 children in kindergarten participated in the present study. They were tested in kindergarten with a mean age of 5.3 years ($N$=1897 children). Nine months later, these children were retested at the end of kindergarten (mean age 6 years, $N = 1724$ children) and during the second semester of the 1st grade (mean age 7.1 years, $N = 1354$ children).

At both measurement time points in kindergarten, individual basic magnitude-number knowledge was tested by the standardized BUEVA-III test battery (Esser & Wyschkon, 2016). The BUEVA-III tests counting skills (forward and backward), size and magnitude estimation, reading numbers, as well as simple additions and subtractions, which predominantly can be solved by counting but also by mental calculation.

In addition, intelligence was assessed at both time points in kindergarten. Here, the mean of the subtest nonverbal and verbal intelligence of the BUEVA-III was calculated. In the subtest nonverbal intelligence, children are supposed to choose that image from several alternatives which does not match the others. The subtest verbal intelligence requires the child to complete a sentence, in which an analogy has to be formed.

Working memory was measured at both time points in kindergarten by a combination of three subtests: the central executive (auditory memory span test, Schmid, Zoelch, & Roebers, 2008), the visual–spatial sketchpad (6-Block-Tapping-Test, Roebers & Zoelch, 2005), and the phonological loop (SETK 3–5, Grimm, 2001). Individual working memory capacity was determined by the mean of these three standardized subtests.

Finally, in the 1st grade, calculation skills were assessed with the standardized test battery DEMAT 1+ (Krajewski, Küspert, & Schneider, 2002), which includes following subtests: quantity counting, number space, addition and subtraction, splitting numbers, complete numbers, part-whole, sequential calculations, equations, and text problems.

Only children were included in the final analyses when data were available from the 1st and 2nd and/or the 3rd assessment. Children were excluded when a low IQ was detected (IQ < 70, $N = 43$ children) at the 1st assessment, when they were still in kindergarten at the 3rd assessment ($N = 112$ children), or already in 2nd grade at the 3rd assessment time point ($N = 3$ children). This resulted in a final sample of 1624 children that could be included in the data analyses.

## Results

The analyses of the developmental trajectory corroborated that children improved their skills in all measured abilities significantly from the 1st to the 2nd assessment time point in kindergarten 9 months later (please see Table 1).

Moreover, all measured skills correlated significantly with each other on a level of $P < 0.001$ (please see Table 2).

**Table 1** Descriptive statistics.

| | Kindergarten t1 | | Kindergarten t2 | | | 1st grade | |
|---|---|---|---|---|---|---|---|
| | *N* | *T*-values mean (SD) | *N* | *T*-values mean (SD) | Improvement | *N* | *T*-values mean (SD) |
| *Basic magnitude-number knowledge* | 1624 | 49.28 (8.45) | 1567 | 49.94 (8.31) | $F(1, 1566) = 2638.4$, $P < 0.001$, $\eta^2 = 0.63$ | | |
| Phonological loop | 1589 | 49.81 (8.28) | 1519 | 49.72 (8.28) | $F(1, 1493) = 155.0$, $P < 0.001$, $\eta^2 = 0.09$ | | |
| Visual–spatial sketchpad | 1587 | 50.54 (9.78) | 1523 | 50.26 (9.66) | $F(1, 1494) = 318.7$, $P < 0.001$, $\eta^2 = 0.18$ | | |
| Central executive | 1581 | 50.76 (9.61) | 1521 | 50.51 (9.52) | $F(1, 488) = 396.5$, $P < 0.001$, $\eta^2 = 0.21$ | | |
| *Working memory* | 1579 | 50.39 (6.61) | 1517 | 50.17 (6.56) | | | |
| Nonverbal IQ | 1624 | 49.33 (8.25) | 1567 | 50.20 (8.33) | $F(1, 1566) = 1160.6$, $P < 0.001$, $\eta^2 = 0.43$ | | |
| Verbal IQ | 1624 | 49.46 (8.44) | 1567 | 50.30 (8.33) | $F(1, 1566) = 1261.2$, $P < 0.001$, $\eta^2 = 0.45$ | | |
| *Intelligence* | 1624 | 50.31 (9.30) | 1567 | 51.42 (9.24) | | | |
| *Calculation* | | | | | | 1077 | 50.15 (9.92) |

This table summarizes all assessed behavioral measurements at the 1st assessment time point in kindergarten t1 and 9 months later (Kindergarten t2) and in the 1st grade. The number of children (N), mean *T*-values, and standard deviation (SD) is listed of basic magnitude-number knowledge, working memory, which is the mean of phonological loop, visuospatial sketchpad, and central executive, as well as intelligence, which is the combination of nonverbal and verbal IQ. Children improved significantly in all skills over time, which is indicated in the column "Improvement." For the 1st grade, mean *T*-values and SD are specified for calculation abilities.

Adapted from Gallit, F., Wyschkon, A., Poltz, N., Moraske, S., Kucian, K., von Aster, M. G., & Esser, G. (2018). Henne oder Ei: Reziprozität mathematischer Vorläufer und Vorhersage des Rechnens. *Lernen und Lernstörungen,* 7(2), 81–92.

**Table 2** Intercorrelations.

| | *1* | *2* | *3* | *4* | *5* | *6* | *7* |
|---|---|---|---|---|---|---|---|
| *1. Basic magnitude-number knowledge t1* | – | 0.74*** | 0.57*** | 0.53*** | 0.50*** | 0.51*** | 0.58*** |
| *2. Basic magnitude-number knowledge t2* | | – | 0.52*** | 0.56*** | 0.45*** | 0.49*** | 0.59*** |
| *3. Intelligence t1* | | | – | 0.67*** | 0.45*** | 0.43*** | 0.40*** |
| *4. Intelligence t2* | | | | – | 0.43*** | 0.45*** | 0.44*** |
| *5. Working memory t1* | | | | | – | 0.56*** | 0.39*** |
| *6. Working memory t2* | | | | | | – | 0.44*** |
| *7. Calculation t3* | | | | | | | – |

***$P < 0.001$.

Adapted from Gallit, F., Wyschkon, A., Poltz, N., Moraske, S., Kucian, K., von Aster, M. G., & Esser, G. (2018). Henne oder Ei: Reziprozität mathematischer Vorläufer und Vorhersage des Rechnens. *Lernen und Lernstörungen,* 7(2), 81–92.

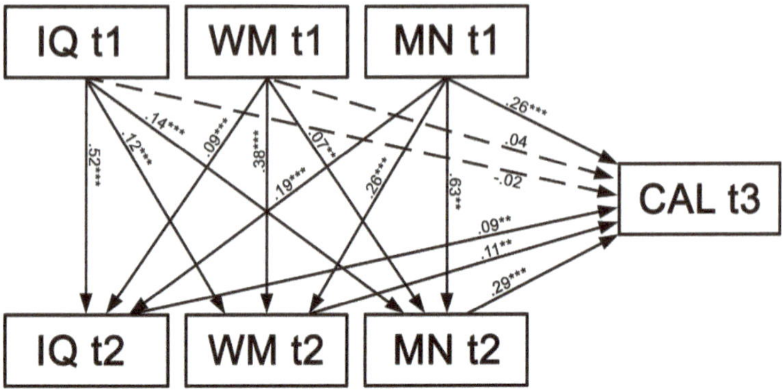


**Fig. 1** Predictive effects of specific and general precursor skills on calculation abilities in school. This figure illustrates predictive effects of precursor skills in kindergarten at t1 and t2 on later calculation abilities in 1st grade. *CAL*, calculation skills; *IQ*, Intelligence; *MN*, magnitude and number knowledge; *WM*, working memory; *t1*, first assessment time point in kindergarten; *t2*, second assessment time point in kindergarten 9 months later; *t3*, third assessment time point in 1st grade; $^{**}P < 0.01$; $^{***}P < 0.001$; *dashed lines* are not significant $P > 0.05$.

Of great interest is the possibility in such a longitudinal design to evaluate which factors predict later outcome. Fig. 1 illustrates how specific and general precursor skills in kindergarten predict calculation skills in the 1st grade. Most important findings here show that specific magnitude and number knowledge in kindergarten is the most important predictor for later calculation performance in school. In addition, also higher general precursor skills, such as intelligence and working memory capacities, forecast better calculation abilities in school.

In more detail, magnitude-number knowledge is the best predictor at both measurement time points in kindergarten for later calculation abilities in school. Only the general precursor skills, intelligence and working memory, assessed at the 2nd time point in kindergarten predicted calculation skills in 1st grade directly. In contrast, general precursor skills at the 1st assessment in kindergarten did not predict calculation skills in school but had an indirect effect over magnitude-number knowledge at t2 on later calculation abilities in grade 1. Interestingly, basic magnitude-number knowledge at t1 predicted also the general abilities, intelligence and working memory, 9 months later.

## Conclusions

These findings emphasize that specific precursor skills such as basic magnitude and number knowledge are the best predictor of mathematical achievement in 1st grade. This means that poor-performing children in kindergarten might still count to the poor math performers in school, although they show individual improvements. It can be concluded that numerical problems manifest early and lead to difficulties in calculation skills. Results also revealed that higher general cognitive abilities in kindergarten (higher intelligence and working memory capacities) promote magnitude and number knowledge before school entry. More gifted children with better general cognitive skills are probably in a better starting position for later numerical achievement in school. Interestingly, also magnitude and numerical knowledge promoted the establishment of intellectual and memory functions in kindergarten. The authors assume that superior developmental processes explain these effects. Children with higher numerical knowledge might expose themselves to more demanding calculation problems and train so in parallel higher cognitive functions. Based on these findings, we have to be aware that numerical problems develop early and that the developmental course of children with deficits in basic magnitude and number knowledge in kindergarten is rather negative, although they show individual improvements. However, many children with early numerical difficulties will not be able to catch up to the mathematical levels their peers with age-appropriate basic numerical skills in kindergarten will achieve in school. Therefore, particularly children with problems in specific numerical precursor skills are dependent on adequate support at an early age to prohibit further negative developments in mathematics. However, this raises the question if specific intervention in kindergarten for children with impaired numerical abilities can facilitate the acquisition of calculation skills in school and prevents them from further failure in mathematical topics.

## Is prevention of dyscalculia possible?

Empirical findings to predict and evaluate preventive options of basic numerical skills are rare. Only in recent years, longitudinal studies focus on the early phases of numerical competences and the evaluation of interventions that promote numerical precursor skills. What we know is that already in kindergarten, it is possible to predict future calculation skills and mathematical learning difficulties. Such numerical precursor abilities include, among others counting skills, reading Arabic digits, judging magnitude relations, and simple arithmetic operations. The option to identify math-specific developmental risks at an early stage gives hope to counteract with appropriate intervention programs and thereby reduce the risk of school problems in math. Hence, the goal of Moraske et al. (2019) was the evaluation of long-term effects of a specific preschool training in children at risk for the development of dyscalculia. It has to be noted that the authors put special attention to the training procedure, which was carried out under conditions that tended more to the kindergarten routine than to a scientific controlled condition as the authors were interested in the outcome of an intervention that was carried out under realistic conditions in daily kindergarten routine.

## Methods

From totally tested 1897 children in kindergarten, 88 children were identified as at risk for the development of dyscalculia and included in the present study (Moraske et al., 2019). A child was considered at risk for dyscalculia when her/his performance in the subtest counting and magnitude understanding of the BUEVA-III (Esser & Wyschkon, 2016) was in the lowest 10%. This tests and evaluates counting and magnitude skills by measuring individual forward and backward counting skills, size and magnitude estimation, reading numbers, as well as simple additions and subtractions, which predominantly can be solved not only by counting but also by mental calculation. Moreover, these children showed no risk for dyslexia, assessed by the subtest phonological awareness (BUEVA-III, Esser & Wyschkon, 2016), and had an intelligence level above 70. The allocation of kindergartens to the prevention or control group was dependent on the teachers' possibility and willingness to perform the prevention program. All kindergartens with at least one child, who has been identified as being at risk, received the opportunity to perform the prevention program. Fifteen out of 65 kindergartens decided to participate and to perform the prevention program. Resulting in half of the children at risk who underwent a specific intervention program and the other half of the children at risk for dyscalculia visited regular kindergarten. All children at risk were retested in 1st, 2nd, and 3rd grades. In each grade, individual calculation skills were measured by the standardized test batteries BUEGA (Esser, Wyschkon, & Ballaschk, 2008) and the DEMAT 1+ in 1st grade (Krajewski et al., 2002), DEMAT 2+ in 2nd (Krajewski, Liehm, & Schneider, 2004), and DEMAT 3+ in 3rd grade (Roick, Gölitz, & Hasselhorn, 2004). The assessed numerical skills in 1st grade are based on subtests magnitude numbers, number space, addition, subtraction, split or complement numbers, part-whole relation, chain problems (e.g. 4+8–3), inequations, and word problems. In 2nd grade, the test battery included addition, subtraction, doubling, and division in half. In 3rd grade, numerical skills are based on number line tasks, addition, subtraction, and multiplication. Additionally, intelligence was assessed at each grade level by the BUEGA test battery (Esser et al., 2008).

To evaluate the effectiveness of the intervention under everyday conditions in kindergarten, a minimum of 50% participation was required. The final sample of the intervention group involved 20 children at risk for dyscalculia, which participated in 18.5 training sessions of totally 22 sessions (SD = 3.5, Min = 12, Max = 22). The final control group included 29 children at risk for dyscalculia. Training and control group were balanced in terms of basic numerical precursor skills, intelligence, gender, age, socioeconomic status, and language skills.

The present intervention was based on the training programs "Mathematik im Vorschulalter" (Rademacher, Lehmann, Quaiser-Pohl, Günther, & Trautewig, 2009) and "Mengen, zählen, Zahlen" (Krajewski, Nieding, & Schneider, 2013). About three quarters of the tasks stem from the training "Mathematik im Vorschulalter" and included visual differentiation skills, spatial perception, magnitude estimation, concept of numbers, simple arithmetic problems, manipulation of symbols, recognition of abstract-logical associations, and causation. About one quarter was selected from the training "Mengen, zählen, Zahlen" and included counting, knowledge of Arabic digits up to 10, and the understanding of the quantity concept. The intervention lasted 11 weeks, with 2 sessions per week, each 30–40 min long. The training was carried out by kindergarten teachers who received a two-day schooling and a helpline was provided. In the middle of the training period, a supervision by a project worker was conducted. The training sessions were performed in small groups of three to six children.

## Results

The longitudinal effects of the prevention program conducted in kindergarten with children at risk for dyscalculia were evaluated in the 1st, 2nd, and 3rd school classes. Fig. 2 illustrates that both groups showed comparable calculation skills in 1st grade. However, average calculation performance of children in the training group in 1st grade was on average performance levels ($t$-values $> 40$), whereas children of the control group achieved only levels below average ($t$-values $< 40$). In 2nd and 3rd grades, children at risk for dyscalculia who accomplished the intervention in kindergarten achieved significantly higher calculation skills compared to children at risk for dyscalculia without preventive support. These findings are promising and provide evidence that a preschool training stimulating numerical competencies under practical conditions leads to long-term improvements of calculation skills.

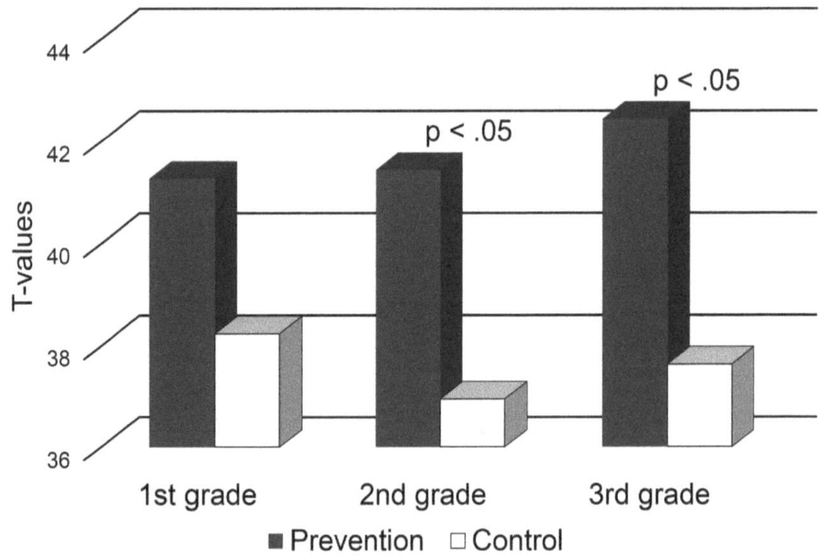


**Fig. 2** Positive long-term effects of mathematical intervention in kindergarten. This figure demonstrates positive long-term effects of a mathematical prevention program in kindergarten for children at risk for dyscalculia. Illustrated are long-term calculation skills in 1st, 2nd, and 3rd grades of children at risk for dyscalculia. Gray bars include children at risk who performed a number specific prevention program in kindergarten, whereas white bars include children at risk without intervention in kindergarten. Calculation skills are depicted as mean *T*-values of two standardized numerical test batteries BUEGA and DEMAT (Esser et al., 2008; Krajewski et al., 2002, 2004; Roick et al., 2004). *T*-values below 40 are classified as below average. *Based on Table 3 of Moraske, S., Wyschkon, A., Poltz, N., Kohn, J., Kucian, K., von Aster, M. G., & Esser, G. (2019). Indizierte Prävention von Rechenschwächen im Vorschulalter: Effekte bis Klasse 3. Lernen und Lernstörungen, 8(3), 141–153.*

In addition, the authors evaluated how many of the children at risk for dyscalculia in kindergarten really developed dyscalculia as defined by the clinical criteria including numerical and calculation performance one standard deviation below norm ($t$-values $<40$) and age-appropriate general cognitive skills. Calculation skills were evaluated by the mean of two standardized numerical test batteries BUEGA and DEMAT (Esser et al., 2008; Krajewski et al., 2002, 2004; Roick et al., 2004). Strikingly, the number of children with developmental dyscalculia was significantly lower in the group of children at risk who received the prevention in kindergarten compared to the untrained group. The effect sizes can be evaluated as big (1st grade: $w=0.46$, 2nd grade: $w=0.80$, 3rd grade: $w=0.55$). Fig. 3 illustrates that significantly less children of the training group developed severe mathematical learning problems in the sense of developmental dyscalculia on a long-term course.

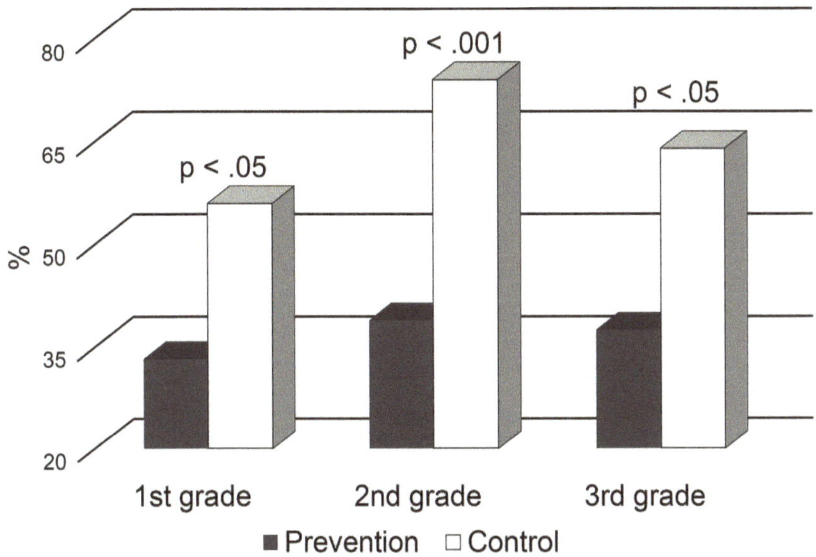


**Fig. 3** Preventive effect of numerical intervention in kindergarten to develop dyscalculia. This figure exhibits the preventive effect of a numerical training in kindergarten for children at risk for dyscalculia. Illustrated are long-term distributions of developmental dyscalculia in 1st, 2nd, and 3rd grades for children who were identified at risk for dyscalculia in kindergarten. Indicated are percentages of children, who fulfilled the clinical criteria for the diagnosis of developmental dyscalculia (mean calculation skills > *T*-value 40, age appropriate general cognitive skills). Gray bars include children at risk who performed a numerical specific prevention program in kindergarten, whereas white bars include children at risk without intervention in kindergarten. Significantly less children of the prevention group developed dyscalculia. *Based on Table 3 of Moraske, S., Wyschkon, A., Poltz, N., Kohn, J., Kucian, K., von Aster, M. G., & Esser, G. (2019). Indizierte Prävention von Rechenschwächen im Vorschulalter: Effekte bis Klasse 3. Lernen und Lernstörungen, 8(3), 141–153.*

## Conclusions

The present study demonstrates how in practice an early diagnosed risk of developing dyscalculia could effectively be counteracted. A relatively short intervention of 11 weeks, which was carried out under conditions corresponding to daily kindergarten routine, leads to significant improvements of early basic numerical understanding in children who were identified at risk for dyscalculia. Strikingly, children who received such an intervention in kindergarten showed long-term benefits, indicated by average calculation skills in 1st, 2nd, and 3rd grades. Moreover, preschool support of children at risk for dyscalculia has even the potential to diminish the probability to develop severe mathematical learning disorders in school. In sum, results confirm long-term efficacy of a preschool training stimulating numerical competencies and point to the potential of preventing children at risk to develop severe math problem. Accordingly, present long-term study demonstrates that a wide use of a preschool intervention to promote numerical and calculation precursor skills would be possible and encourages us to first identify children at risk for dyscalculia early and, second, offer them adequate support to facilitate school entry, school development and protect these children from failure, the development of severe math learning problems, and all related negative experiences.

## Persisting character of developmental dyscalculia

It is inevitable to conduct longitudinal studies to obtain a clearer picture of the developmental course of numerical learning disorders. Only longitudinal research allows to look at changes over time—how certain things change at different time points and explore why developmental shifts take place. However, longitudinal research is complex, requires enormous amounts of time, is expensive, and researchers have to deal with participants dropping out over time. Most likely, these reasons account for the small amount of existing longitudinal studies that examine the developmental trajectory of developmental dyscalculia. However, existing findings suggest that numerical learning disorders are not growing out overdevelopment. In contrast, they manifest overdevelopment (McCaskey et al., 2017; Shalev, Manor, Auerbach, & Gross-Tsur, 1998; Shalev, Manor, & Gross-Tsur, 2005), whereas the later a child is diagnosed with developmental dyscalculia, the higher is the persisting character (Kohn, Wyschkon, Ballaschk, Ihle, & Esser, 2013). For the first time, Schulz et al. (2018) examined in a large-scale evaluation the developmental course of numerical learning disorder from primary to secondary schools. During this time period, all children were assessed three times and allocated to one of four groups: children with either mathematical learning disorder, mathematical learning difficulties, control children with normal intelligence, or control children with IQ below average. Due to the fact that the definition of dyscalculia is not consistent and dependent on the clinical criteria used, the authors resigned to use the term dyscalculic. Instead, they preferred to label the groups as mathematical learning disorder cohort or mathematical learning difficulty subgroup. However, dependent on the clinical criteria applied, both subgroups fulfill the diagnostic premises of developmental dyscalculia. Hence, the goal of this study was to shed light on the persistence and the impact of numerical difficulties on later mathematical performance and school-related success.

## Methods

A total number of 2909 children could be included in the study of Schulz et al. (2018). Based on the first assessment, these children were divided into four different groups:

1. *Children with mathematical learning disorder:* 1.5 standard deviation below the average of class level in mathematics (*t*-value $<35$, percentile rank $<7$) AND 1.5 standard deviation discrepancy of mathematical to individual intelligence level. The number of children in this group was 40. Mean IQ was in the average range with mean *t*-values of 49.2 (4.4).
2. *Children with mathematical learning difficulty:* All children had to score below average performance in mathematics (*t*-value $<40$, percentile rank $<16$). In addition, one of the discrepancy criteria had to be met: Either 1.5 standard deviation below the average of class level in mathematics OR 1.5 standard deviation discrepancy of individual mathematical to intelligence level. The number of children in this group was 325. Mean IQ was in the average range with mean *t*-values of 41.6 (7.2).
3. *Control children with average IQ:* Children without mathematical difficulties and age-appropriate intelligence levels (mathematical and intellectual skills *t*-values $>40$). The number of children in this group was 2361. Mean IQ was in the average range with mean *t*-values of 54.0 (8.1).
4. *Control children with IQ below average:* Children without mathematical difficulties, but with intelligence levels below average (mathematical skills *t*-values $>40$, IQ *t*-values $30<t$-values $<39$). The number of children in this group was 183. Mean IQ was below average with mean *t*-values of 36.8 (2.4).

Individual intelligence levels were assessed my means of the standardized test battery BUEGA (Esser et al., 2008). Also, numerical performance was evaluated by the numerical subtest of the BUEGA for children up to the 5th grade including forward and backward counting skills, size and magnitude estimation, reading numbers, and simple additions and subtractions. For older children, a different age-appropriate test battery was chosen. Children from the 6th grade on performed the subtest calculation of the WISC-III (Wechsler, 1999) including mental calculation task which has to be solved without any utilities. Adolescents older than 16 years performed the subtest calculation from the WAIS-R (Tewes, 1991) in which verbal calculation problems are presented and have to be solved by mental calculation.

Children were assessed three times, whereas at the 1st assessment time point mean age was around 8 years, at the 2nd time point around 11 years, and at the 3rd time point around 14 years of age. Between the 1st and 2nd time points were approximately 3 years (mean 36.8 (8.1) months, min 23, max 59) and between the 2nd and the 3rd time points lay about 2.5 years (mean 30.5 (12.7) months, min 3, max 56).

Fig. 4 illustrates the study design and the allocation of children to the four different subgroups, age, and gender distributions.

School-related success at the 3rd assessment was defined as the combination of the mathematical grade, the language grade, and visiting school type. School types were categorized into 0 = gymnasium, 1 = comprehensive school types, 2 = school types for children with special needs. In doing so, lower values stand for higher school success. Totally 1144 adolescents could be included in the analysis of school success: Adolescents with dyscalculia $N=81$, control adolescents with average IQ $N=1017$, control adolescents with IQ below average $N=46$.

## Results

### *Prevalence*

Prevalence rates at the 1st assessment time point were dependent on the criteria for mathematical learning problems. Whereas 1.4% were classified as children with math learning disorder, 11.2% fulfilled the criteria of mathematical learning difficulties. Again, it has to be noted that dependent on the clinical criteria, both groups of this study could also be labeled as dyscalculic. In literature, prevalence rates for developmental dyscalculia vary between 3% and 6% (for review please see Kucian & von Aster, 2015). The differentiated analyses in the present study make clear that prevalence rates are highly dependent on the definition and criteria chosen and can therefore diverge. Including an additional discrepancy to intelligence results in a much lower prevalence rate with 1.4%. In contrast, present findings also demonstrate that the incidence of math learning difficulties most probably affects more pupils than expected with 11.2%. Regarding gender differences, more girls were affected than boys, whereas girls were three times more affected from math learning disorders than boys (3:1). For mathematical learning difficulties, the gender differences were a bit less prominent, but still girls were more often affected (1.3:1). These results indicate that girls seem to be more often affected from specific numerical learning problems and this holds particularly true for more severe forms of dyscalculia.

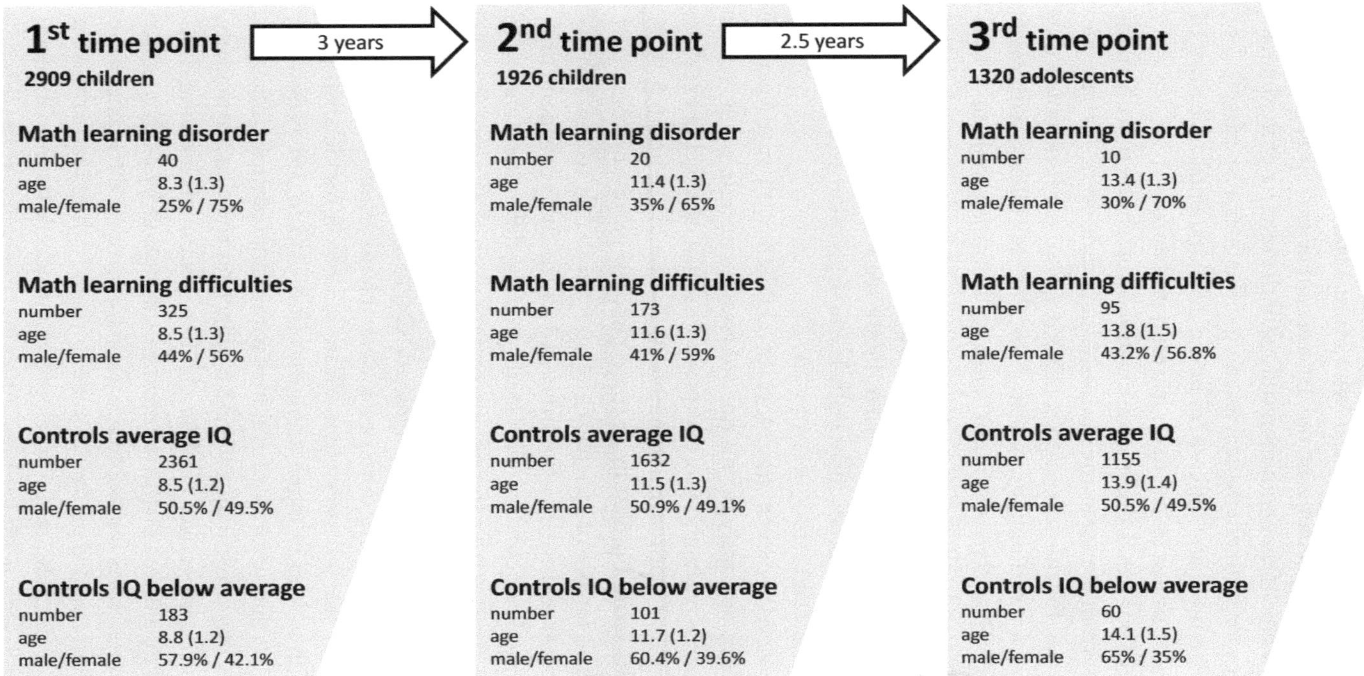


**Fig. 4** Longitudinal study design. Totally, 2909 children were followed longitudinally and assessed three times. Groups were defined as children with mathematical learning disorder (mathematical skills $t$-values $< 35$ and 1.5 SD discrepancy to IQ), mathematical learning difficulties (mathematical skills $t$-values $< 40$ OR 1.5 SD discrepancy to IQ), control children with average IQ (mathematical and IQ level $t$-values $> 40$), and control children with IQ below average (mathematical skills $t$-values $> 40$, IQ $t$-values $30 < t$-values $< 39$). *Based on Table 1 of Schulz, F., Wyschkon, A., Gallit, F., Poltz, N., Moraske, S., Kucian, K., … Esser, G. (2018). Rechenprobleme bei Grundschulkindern: Persistenz und Schulerfolg nach fünf Jahren. Lernen und Lernstörungen, 7(2), 67–80.*

### *Developmental course of diagnosis*

The question is, how these children who were identified with specific mathematical learning problems at the 1st assessment time point develop over time. How many of these children still fulfill the criteria of mathematical learning disorder or mathematical learning difficulties, respectively, 3 or 5.5 years later? Are there any differences in the developmental trajectory of children with more or less severe math learning problems?

First, the developmental course was analyzed for children with math learning disorder from the 1st to the 2nd assessment time point. No analyses could be performed for this subgroup at the 3rd assessment time point, due to the high dropout rate. Findings indicated that 75% of children with math learning disorder still suffered from mathematical learning problems 3 years later, whereas 30% still fulfilled the strong criteria of math learning disorder and 45% of math learning difficulties. Only 25% reached school-grade appropriate calculation skills 3 years later. Please see Fig. 5 for illustration. These percentages demonstrate that math learning disorder is a persisting learning problem, indeed.

Second, the developmental trajectory of children with mathematical learning difficulties was analyzed till the 3rd assessment time point, which was 5.5 years later. Results showed that more than half of affected children still showed profound math learning problems 3 years (54%) and 5.5 years (57%) later. Math-related problems of some children got even worse over time. This was true for 8% of affected children from the 1st to the 2nd assessment time point and for 5% from the 2nd to the 3rd assessment time point. Importantly, those children with math-learning difficulties who were able to catch up 3 years later to age-appropriate math skills were also the ones that showed average math performance 5.5 years later (please see also Fig. 5). Substantially only one-tenth of children who still had poor math performance at the 2nd time point were able to improve to an age-appropriate level at the 3rd assessment. Please see Fig. 5 for detailed percentages of children who changed categories from 2nd to 3rd assessment. This illustrates that the majority of children with math learning difficulties still show equally low math abilities 3 or 5.5 years later and that, particularly, the developmental perspectives of older children with math learning difficulties are bad. The authors further confirmed this notion by analyzing the developmental trajectories for children with math learning difficulties who were younger (2nd and 3rd grades) or older (4th and 5th grades) at the 1st assessment time point. The same developmental outlook became evident; the older a child is with math learning difficulties, the worse is her/his developmental perspective.

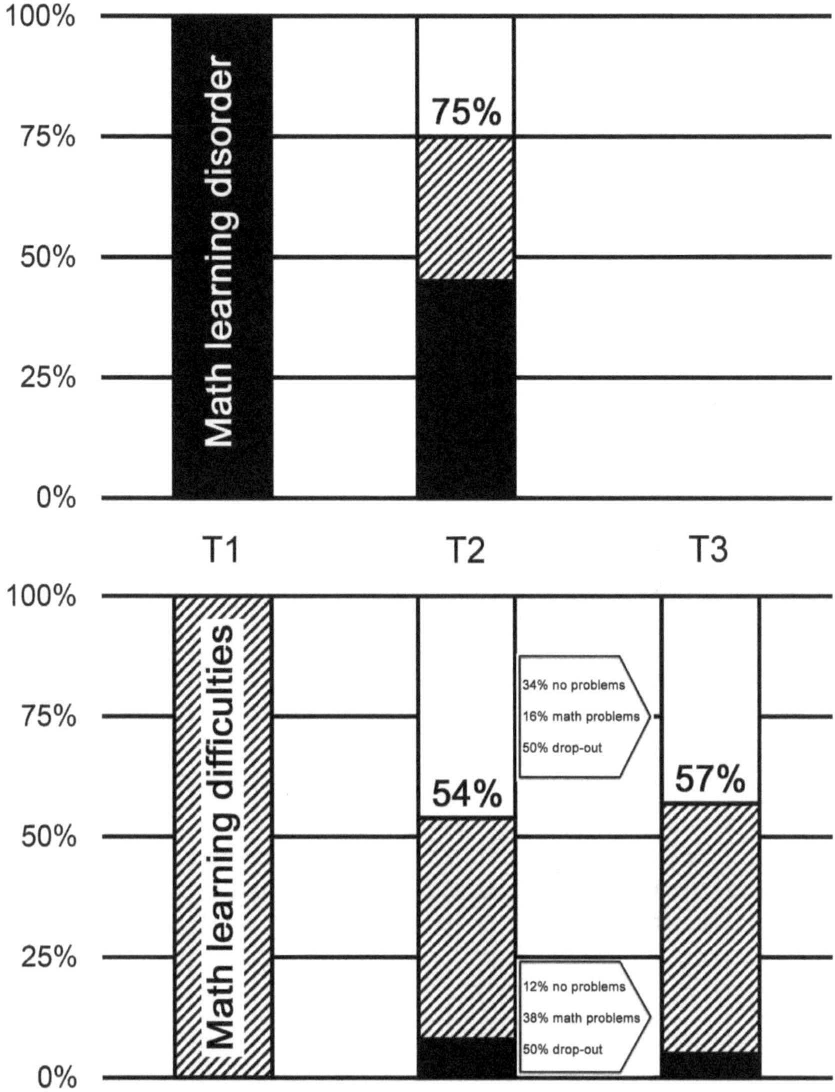


**Fig. 5** Developmental trajectory of math learning disorder and math learning difficulties. This figure illustrates the developmental course of children who were affected by math learning disorder (upper part, black bars) or math learning difficulties (lower part, striped bars) at the 1st assessment time point (t1). Seventy-five percent of children with math learning disorder still suffered from mathematical problems 3 years later at the 2nd assessment time point (t2). Of these children, 30% still fulfilled the strong criteria of math learning disorder and 45% of math learning difficulties. Only 25% reached age-appropriate math skills at t2 (white bars). No data analyses were performed at the 3rd assessment (t3) for children with math learning disorder due to the high dropout rate. The longitudinal course of children with math learning difficulties (lower part, striped bars) revealed that the majority of affected children still belong to the poor performers 3 years later (t2, 54%) or 5.5 years later (t3, 57%). Some of these children get even worse, 8% at t2 and 5% at t3. Black bars: Math learning disorder; striped bars: math learning difficulties; white bars: age-appropriate math skills; t1: 1st assessment time point; t2: 2nd assessment time point; t3: 3rd assessment time point; time between t1 and t2: 3 years; time between t2 and t3: 2.5 years. *Based on data of Schulz, F., Wyschkon, A., Gallit, F., Poltz, N., Moraske, S., Kucian, K., … Esser, G. (2018). Rechenprobleme bei Grundschulkindern: Persistenz und Schulerfolg nach fünf Jahren. Lernen und Lernstörungen, 7(2), 67–80.*

### *Developmental course of mathematical school performance*

This persistent character of math learning problems is also reflected in the sustained lower calculation skills of children with math learning difficulties over time. Children with math learning difficulties showed significantly lower mathematical performance in school compared to control children with average IQ or IQ levels below average. Throughout, highest mathematical performance scores in school were reached by control children with average IQ. This holds true for both assessment time points (t2 and t3: children with math learning difficulties have lower math performance in school compared to control children with average IQ $P<0.001$ and compared to control children with IQ below average $P<0.001$. Control children with average IQ have higher math performance in school than control children with IQ below average $P<0.001$).

### *Developmental course of general school-related success*

Next to the actual mathematical performance in school, also general school-related success was examined at the 3rd assessment. School-related success was defined as the combination of the mathematical grade, the language grade, and visiting school type. In terms of general school-related success, dyscalculic children achieved significantly lower school success compared to typically achieving peers with average IQ ($P<0.001$) (Fig. 6). Similarly, also children with intellectual deficits had significantly lower school-related success compared to control children with average IQ ($P<0.001$) (Fig. 6). Remarkably, children with dyscalculia did not differ in general school success from children with intellectual deficits (Fig. 6). As a result, dyscalculia affects school development just as unfavorably as below-average intelligence. However, authors point to the fact that such an interpretation has to be taken with care since the group of children with dyscalculia showed intellectual levels that were only marginally age appropriate (mean *t*-values 41.62 (7.16)) at the 1st assessment time point. Accordingly, also slightly lower IQ initially (next to reduced number abilities) could contribute to the general lower school success, which was comparable to one of the children with IQ below average. Moreover, due to the small sample size, it was not possible to split groups of children with math learning problems into subgroups of children who had additionally reading and writing problems. However, according to the known comorbidity of dyscalculia and dyslexia, it can be assumed that some of the children with math problems show also difficulties in reading and writing which will further diminish individual school-related success.

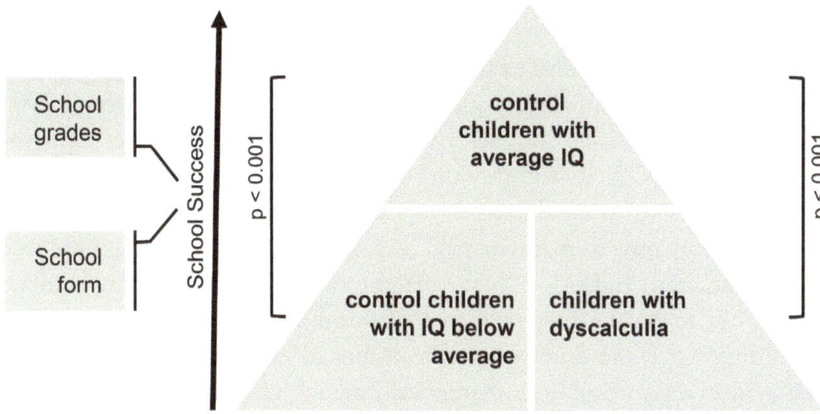


**Fig. 6** General school success of children with developmental dyscalculia. General school success was a composite score of the mathematical grade, the language grade, and school type. This figure illustrates that children with developmental dyscalculia have significantly lower school success compared to typically developing peers with average intellectual skills. Children with dyscalculia achieved similar school-related success to children with intellectual deficits. *Based on data of Schulz, F., Wyschkon, A., Gallit, F., Poltz, N., Moraske, S., Kucian, K., ... Esser, G. (2018). Rechenprobleme bei Grundschulkindern: Persistenz und Schulerfolg nach fünf Jahren. Lernen und Lernstörungen, 7(2), 67–80.*

## Conclusions

The present findings underscore first that developmental dyscalculia is prevalent learning disorder (1.4%–11.2%, higher in girls than boys) and second the negative developmental perspectives children with dyscalculia have. Particularly, children with more severe forms of dyscalculia (math learning disorder) will most probably not grow out of their learning disorder, showing that 75% of these children still suffer from mathematical learning problems 3 years later. Similarly, also the majority of children with milder forms of dyscalculia (math learning difficulties) will still have math learning problems 3 or 5.5 years later. Hence, this longitudinal study clearly corroborates that dyscalculia is a persisting learning disorder from primary to secondary school. This negative developmental course is characterized by lower math performance in school, and more importantly, also general lower school success, which is similar to children with intellectual abilities that are below average. The differentiated analyses of the present results could also highlight that the older a child is with dyscalculia, the worse are her/his developmental perspectives in terms of mathematical skills, general school success, as well as diagnosis of dyscalculia.

## Summary

The presented longitudinal examination of mathematical learning problems from kindergarten to primary and secondary school including about 2900 children offers valuable insights into the developmental course of math-related learning deficits and the long-term outcome of intervention. First, Gallit et al. (2018) were able to show that children at risk to develop later profound problems in mathematical learning could already be pinpointed in kindergarten. Individual basic magnitude and number competencies turned out to be the best predictor of future calculation skills in school. Accordingly, it is the domain-specific ability in kindergarten that provides best evidence for possible mathematical learning difficulties later in school. However, also low working memory and intelligence skills contributed to worse numerical abilities. These domain general competences put children probably in worse starting positions for later mathematical achievement in school. Interestingly, better magnitude and numerical knowledge promoted the establishment of intellectual and memory functions in kindergarten, which might be a result of superior learning processes. Children with higher numerical knowledge might expose themselves to more demanding calculation problems and train so in parallel higher cognitive functions. These findings make clear that numerical problems develop early and that the developmental trajectory of children with deficits in basic magnitude and number skills in kindergarten is rather pessimistic. Many of these children with early numerical difficulties will not be able to catch up in primary school to their peers with age-appropriate basic numerical skills in kindergarten, at least if early detection, prevention, and intervention will not take place.

As a matter of fact, children at risk for dyscalculia and children with mathematical learning deficits are dependent on adequate support and learning environment to interrupt the negative developmental trajectory. Accordingly, Moraske et al. (2019) evaluated an 11-week preschool intervention for children at risk for dyscalculia. Results were very promising and demonstrated that an adequate training in kindergarten for children at risk for dyscalculia has certainly the potential to prevent negative development of numerical skills and the development of dyscalculia. In more detail, while trained children at risk for dyscalculia improved their math competencies to an average level, mathematical performances of untrained children at risk for dyscalculia stayed below average in the 1st, 2nd, and 3rd grades. In addition, numerical preschool intervention diminished the probability to develop dyscalculia in school. These results confirm the long-term efficacy of a preschool prevention that stimulates numerical competences and should encourage kindergarten teachers to support children with numerical difficulties before entering school. Particularly, since the intervention was carried out under conditions tending to the kindergarten routine, the present study demonstrates that a wide use of such a preschool intervention would be possible and facilitates numerical learning at school entry, the development of numerical skills and prevents children at risk from failure, the development of dyscalculia and related negative experiences.

The long-term observation of children with mathematical learning problems until adolescents by Schulz et al. (2018) further corroborated that an early intervention should be aimed for. The longitudinal findings clearly indicated that the older a child is with dyscalculia, the worse are her/his developmental perspectives in terms of mathematical skills, general school success, as well as diagnosis of dyscalculia. Without adequate support, present findings once more demonstrate that dyscalculia is a persisting learning disorder and that affected children will most probably not grow out of their mathematical learning problems. Three-quarters of children with dyscalculia still suffered from severe mathematical learning difficulties 3 years later. Alike, the majority of children with milder forms of dyscalculia still had math learning problems 3 or 5.5 years later. The authors could demonstrate that this negative developmental course is marked by lower mathematical performance in school and more critically also by general lower school success. Notably, general school success of children with dyscalculia was comparable to general school achievement of children with intellectual deficits. This is alarming. It should not be the case that a specific learning disorder of calculation and numerical understanding has such a far-reaching negative effect on general school success. Moreover, the present findings affirm that dyscalculia is a popular learning disorder with 1.4%–11.2% of affected children. The prevalence rate varies depending on the clinical criteria that were considered. In general, more girls were affected particularly when looking at more severe forms of dyscalculia.

In conclusion, present longitudinal examination of numerical learning disabilities from kindergarten to secondary school substantiates that poor numerical achievement constitutes a prevalent and considerable developmental risk with far-reaching negative outcome. Notwithstanding, present findings also made clear that an early identification of children at risk for dyscalculia would be possible and that adequate preschool support would have the promising potential to intervene the negative developmental course of mathematical learning disorders.

## References


Dowker, A., Sarkar, A., & Looi, C. Y. (2016). Mathematics anxiety: What have we learned in 60 years? *Frontiers in Psychology*, 7, 508. https://doi.org/10.3389/fpsyg.2016.00508.

Esser, G., & Wyschkon, A. (2016). *BUEVA-III: Basisdiagnostik Umschriebener Entwicklungsstörungen im Vorschulalter - Version III.* Göttingen: Hogrefe.

Esser, G., Wyschkon, A., & Ballaschk, K. (2008). *BUEGA: Basisdiagnostik Umschriebener Entwicklungsstörungen im Grundschulalter.* Göttingen: Hogrefe.

Gallit, F., Wyschkon, A., Poltz, N., Moraske, S., Kucian, K., von Aster, M. G., et al. (2018). Henne oder Ei: Reziprozität mathematischer Vorläufer und Vorhersage des Rechnens. *Lernen und Lernstörungen*, 7(2), 81–92.

Grimm, H. (2001). *Sprachentwicklungstest für drei- bis fünfjährige Kinder (SETK 3–5).* Göttingen: Hogrefe.

Kohn, J., Wyschkon, A., Ballaschk, K., Ihle, W., & Esser, G. (2013). Verlauf von Umschriebenen Entwicklungsstörungen: Eine 30-Monats-Follow-up-Studie. *Lernen und Lernstörungen*, *2*(2), 77–89.

Krajewski, K., Küspert, P., & Schneider, W. (2002). *DEMAT 1+: Deutscher Mathematiktest für erste Klassen.* Beltz: Weinheim.

Krajewski, K., Liehm, S., & Schneider, W. (2004). *DEMAT 2+: Deutscher Mathematiktest für zweite Klassen.* Göttingen: Beltz.

Krajewski, K., Nieding, G., & Schneider, W. (2013). *Mengen, zählen, Zahlen: Die Welt der Mathematik verstehen.* Göttingen: Hogrefe.

Kucian, K. (2016). Developmental dyscalculia and the brain. In D. B. Berch, D. Geary, K. Mann Koepke, D. Geary, & K. Mann Koepke (Eds.), *Vol. 2. Development of mathematical cognition: Neural substrates and genetic influences* (pp. 165–193). Amsterdam: Elsevier.

Kucian, K., & von Aster, M. G. (2015). Developmental dyscalculia. *European Journal of Pediatrics*, *174*(1), 1–13. https://doi.org/10.1007/s00431-014-2455-7.

McCaskey, U., von Aster, M. G., Maurer, U., Martin, E., O'Gorman Tuura, R., & Kucian, K. (2017). Longitudinal brain development of numerical skills in typically developing children and children with developmental dyscalculia. *Frontiers in Human Neuroscience*, *11*, 629. https://doi.org/10.3389/fnhum.2017.00629.

Moraske, S., Wyschkon, A., Poltz, N., Kohn, J., Kucian, K., von Aster, M. G., et al. (2019). Indizierte Prävention von Rechenschwächen im Vorschulalter: Effekte bis Klasse 3. *Lernen und Lernstörungen*, *8*(3), 141–153.

Rademacher, J., Lehmann, W., Quaiser-Pohl, C., Günther, A., & Trautewig, N. (2009). *Mathematik im Vorschulalter.* Göttingen: Vandenhoeck & Ruprecht.

Roebers, C. M., & Zoelch, C. (2005). Erfassung und Struktur des phonologischen und visuell-räumlichen Arbeitsgedächtnisses bei 4-jährigen Kindern. *Zeitschrift für Entwicklungspsychologie und Pädagogische Psychologie*, *37*, 113–121.

Roick, T., Gölitz, D., & Hasselhorn, M. (2004). *DEMAT 3+: Deutscher Mathematiktest für dritte Klassen.* Göttingen: Beltz.

Schmid, C., Zoelch, C., & Roebers, C. M. (2008). Das Arbeitsgedächtnis von 4- bis 5-jährigen Kindern: Theoretische und empirische Analyse seiner Funktionen. *Zeitschrift für Entwicklungspsychologie und Pädagogische Psychologie*, *40*, 2–12.

Schulz, F., Wyschkon, A., Gallit, F., Poltz, N., Moraske, S., Kucian, K., et al. (2018). Rechenprobleme bei Grundschulkindern: Persistenz und Schulerfolg nach fünf Jahren. *Lernen und Lernstörungen*, 7(2), 67–80.

Shalev, R. S., Manor, O., Auerbach, J., & Gross-Tsur, V. (1998). Persistence of developmental dyscalculia: What counts? Results from a 3-year prospective follow-up study. *Journal of Pediatrics*, *133*(3), 358–362.

Shalev, R. S., Manor, O., & Gross-Tsur, V. (2005). Developmental dyscalculia: A prospective six-year follow-up. *Developmental Medicine and Child Neurology*, *47*(2), 121–125.

Suárez-Pellicioni, M., Núñez-Peña, M. I., & Colomé, À. (2016). Math anxiety: A review of its cognitive consequences, psychophysiological correlates, and brain bases. *Cognitive, Affective, & Behavioral Neuroscience, 16*(1), 3–22. https://doi.org/10.3758/s13415-015-0370-7.
Tewes, U. (1991). *WAIS-R: Wechsler Adult Intelligence Scale revised (German Version: HAWIE-R Hamburg-Wechsler-Intelligenz-Test für Erwachsene Revision)*. Bern: Huber.
Wechsler, D. (1999). *WISC-III Wechsler intelligence scale for children* (3rd ed.). Bern Göttingen Toronto Seattle: Hans Huber.

# CHAPTER 11

# Neurocognitive mechanisms of numerical intervention studies: The case of brain stimulation

**Nienke E.R. van Bueren[a,b], Evelyn H. Kroesbergen[a], and Roi Cohen Kadosh[b]**
[a]Behavioural Science Institute, Radboud University Nijmegen, Nijmegen, The Netherlands
[b]Department of Experimental Psychology, University of Oxford, Oxford, United Kingdom

## Introduction

Behavioral and neurocognitive understanding of mathematical skills is of high importance due to a strong relation to the success of one's future career and socioeconomic status (Ritchie & Bates, 2013). In addition, the process of developing skills needed for mathematical learning varies greatly between individuals. At the lower extreme of the continuum are children with developmental dyscalculia in which learning of mathematical skills is severely impaired, which affects about 3%–6% of school-aged children (Butterworth, Varma, & Laurillard, 2011; Kucian & von Aster, 2015; Rubinsten & Henik, 2009). In addition, approximately 20% of healthy children and adults suffer from low numeracy (Parsons & Bynner, 2005). Therefore it is imperative to investigate possible interventions to influence numerical skills to enhance performance. In this chapter, we focus on the possibilities to use brain stimulation as means to improve mathematical skills and related cognitive mechanisms. Such an approach could also progress our understanding of the causal mechanisms of mathematical skills. We start with describing these mathematical skills and underlying mechanisms.

## Basic and advanced mathematical skills

The development of numerical skills can be detected already at an early age using basic numerical skills. This includes, for example, non-symbolic (dots) and symbolic numbers (digits) comparison when two stimuli are compared for their quantities, or using number line estimation that requires an estimation of a given number (e.g., 23) between two anchors (e.g., 0 and 100) (Sasanguie, De Smedt, Defever, & Reynvoet, 2012). In preschool

*Heterogeneous Contributions to Numerical Cognition*
https://doi.org/10.1016/B978-0-12-817414-2.00014-2

and kindergarten, children learn the number words and counting and start with the first manipulations of quantities and number (Mullis, Martin, & Loveless, 2016). These form the basis for learning the first mathematical skills in the early school years, i.e., basic whole-number operations (i.e., simple addition, subtraction, multiplication, and division). Behavioral measures of these skills have been used to obtain more clarity about children's development of early mathematical skills. At school entry level, mathematical skills have been indicated as the strongest predictor for mathematical achievement later in life, as well as for other academic domains (Duncan et al., 2007; Toll, Van der Ven, Kroesbergen, & Van Luit, 2011). Problems with these basic skills could thus affect later development. Knowledge of the development of these skills could contribute to early screening and remediation of mathematical learning problems and to possible interventions. Once children have mastered (most of) the basic mathematical skills, they shift toward more advanced mathematical skills such as complex arithmetic problem solving, calculations with fractions, percentages, and geometry. Even if children have mastered the basic skills, it is not a guarantee for learning more complex skills (Huijsmans, Kleemans, van der Ven, & Kroesbergen, 2020). For a review about mathematics in the developing brain, see Peters and De Smedt (2018).

The mathematical processes that will be reviewed in this chapter entail rote memorization or calculation learning. Rote memorization learning is dependent on memory formation in order to recall previously learned mathematical problems and is strengthened by repetition of the fact (e.g., $3\times14=42$) (Delazer et al., 2005). In contrast, learning by calculation involves a basic understanding of mathematical rules and algorithms (e.g., $18+2-5=15$), as well as basic numerical understanding (e.g., understanding of cardinality and ordinality), and is thought to underlie deep-level cognitive processing. Both rote memorization and calculation seem to activate different brain regions. According to several functional magnetic resonance imaging (fMRI) studies, retrieving mathematical facts was paired with activation mainly in the left angular gyrus and calculating activated mainly the frontoparietal network (Delazer et al., 2003; Grabner et al., 2009; Zamarian, Ischebeck, & Delazer, 2009). This finding supports the notion that calculation requires more mental effort in comparison to rote memorization of mathematical facts and thus activates a broader network of brain areas.

In this chapter, we first give a short overview of the neurocognitive mechanisms underlying basic mathematical skills (i.e., basic whole-number operations). To elaborate, it has been argued that several domain-general neurocognitive processes are linked to mathematical skills, such as working memory (Raghubar & Barnes, 2016). Second, also domain-specific correlates such as non-symbolic and symbolic comparison skills are linked to numerical cognition (see De Smedt, Noël, Gilmore, & Ansari, 2013 for a review).

Cragg, Keeble, Richardson, Roome, and Gilmore (2017) propose a hierarchical multicomponent model of mathematics in which mathematical skills are supported by domain-specific factors (e.g., number tasks) that draw on domain-general factors (e.g., working memory). In short, we will discuss recent developments of the relation between these correlates and basic and advanced numerical skills (i.e., complex arithmetic problem solving) and summarize the underlying neurocognitive mechanisms.

## Domain-general cognitive factors

Working memory is the capacity to remember information while processing new information at the same time (Kail & Hall, 2001). Working memory skills are found to be linked to mathematical skill development (Friso-Van Den Bos, Van Der Ven, Kroesbergen, & Van Luit, 2013). More specifically, working memory can be divided into verbal and visuospatial working memory, which all can be linked to mathematical skills (Bull, Espy, & Wiebe, 2008; Rasmussen & Bisanz, 2005). A better domain-general working memory can help an individual to faster acquire skills needed for domain-specific procedures (e.g., solving mathematical problems) in comparison to individuals with a weaker domain-general working memory (Hambrick & Engle, 2002; Lee & Bull, 2015). A study that also included children (6–10 years) with mathematical learning disabilities (MLD) showed that children who performed low on either a visual-spatial working memory task also scored lower on a mathematical task. The same was found for number tasks (domain-specific factor). Children performing low on both working memory and number tasks showed the lowest mathematical skills. Working memory is also strongly related to mathematic skills later in primary education (Friso-Van Den Bos et al., 2013; Palmer & Baroody, 2011; Toll, Kroesbergen, & Van Luit, 2016; Van de Weijer-Bergsma, Kroesbergen, Prast, & Van Luit, 2015). In addition, working memory explains a higher variance in mathematics performance as opposed to measures of intelligence (for a review see Raghubar, Barnes, & Hecht, 2010). However, the role of intelligence in basic and advanced mathematical skills has yet to be fully explored. To tailor interventions that can influence domain-general cognitive factors, it is important to know which neuroanatomical correlates are involved during domain-general processing. The neuroanatomical correlates of working memory skills entail both the dorsolateral prefrontal cortex (DLPFC) and the posterior parietal cortex (PPC) since mathematics involves both storing and processing information. This is true for both basic and advanced mathematical skills (Andrews, Hoy, Enticott, Daskalakis, & Fitzgerald, 2011; Peng, Namkung, Barnes, & Sun, 2016; Wu et al., 2014).

Executive functions are needed when new information is presented to correctly respond in a goal-directed way. Three executive functions are distinguished: updating (monitoring and addition and/or deletion of working memory aspects), shifting (flexible adjusting), and inhibition (overwriting dominant processes) (Miyake & Friedman, 2012). Updating tasks, for example, contain a series of arithmetic equations together with a list of unrelated words whereby the participants must solve the equations and recall the words (Miyake et al., 2000). Shifting is usually measured with a dual-task task paradigm whereby the participants are instructed to shift from a set of rules or strategy to another, for example, by shifting between adding and multiplying numbers. A well-known task to measure inhibition is the Stroop-task, whereby the participants are presented with incongruent words printed in different colors (Stroop, 1992). Instead of naming the words, the participants are asked to name the respective colors in a certain timeframe.

Many studies previously investigated the association between executive functions and mathematical abilities due to the importance of these cognitive skills in learning (for a review, see Cragg & Gilmore, 2014). A longitudinal study using a large sample size of 211 children of 7–8-year olds showed that updating skills were strongly related to mathematical development, but inhibition and shifting were not (Lee & Bull, 2015). A meta-analysis also showed that out of these three different executive functions, updating was the most important predictor of mathematical skills in comparison to shifting and inhibition in primary school children (Friso-Van Den Bos et al., 2013). Children with better updating skills showed higher performance improvements on a number line estimation task, comparison task, and a marginal effect on a categorization task (e.g., cards with numbers had to be sorted in the correct category) (Kolkman, Hoijtink, Kroesbergen, & Leseman, 2013). Several other studies are in line with these findings concerning a strong relation between updating and early mathematical skills in children (Kroesbergen, Van Luit, Van Lieshout, Van Loosbroek, & Van De Rijt, 2009; Lee et al., 2012; St Clair-Thompson & Gathercole, 2006; Van der Ven, Kroesbergen, Boom, & Leseman, 2012). Executive functions are also more closely related to early mathematical skills than fluid intelligence is

(Friedman et al., 2006; Kroesbergen et al., 2009). Fluid intelligence, or problem solving and adaptive reasoning, is stronger related to executive functioning than crystalized intelligence or the ability to solve problems with existing knowledge (Van Aken, Kessels, Wingbermühle, Van Der Veld, & Egger, 2015). Tamnes et al. (2010) showed that the neuroanatomical correlates differ between updating, inhibition, and shifting. They measured performance on a keep track and letter memory task (updating), Antisaccade and Stroop (inhibition), and plus-minus and trail-making task (shifting) in children and adolescents in combination with an MRI. Interestingly, updating was associated with a thinner cortex bilaterally for clusters encompassing both frontal and parietal regions. Shifting showed mixed results: an age-independent negative association with cortical thickness in the central sulcus in the left hemisphere, and a positive association in the right occipital lobe. Inhibition was associated with thinner cortex bilaterally in regions encompassing occipital and parietal regions. Additionally, performance on inhibition tasks was furthermore associated to estimated rates of cortical maturation in posterior regions. Note that the prefrontal cortex thickness was not associated to inhibition (Tamnes et al., 2010). Even though research investigating the relation between executive function and basic or advanced mathematical skills is growing, there is still a necessity toward a better understanding of the development of this relation in different age groups. Also, research toward executive functions as a reliable target for future interventions is lacking.

## Domain-specific cognitive factors

Non-symbolic number tasks involve skills that are needed to perform numerical estimations or transformations on number representations that do not include symbols (e.g., estimating or comparing objects in one or several sets). With non-symbolic number tasks, we also refer to quantity manipulation skills in this chapter (i.e., addition and subtraction of numerosities). These quantity manipulation skills have been indicated to relate to mathematical achievement later in life (Geary & vanMarle, 2016). It should be noted that non-symbolic skills that are assessed by means of a magnitude comparison task are not as strongly related to later mathematical achievement as symbolic skills (Leibovich & Ansari, 2016; Schneider et al., 2017; Toll, Van Viersen, Kroesbergen, & Van Luit, 2015). In contrast, symbolic number skills refer to the understanding of the numerical meaning of numbers, and subsequently using this to count and manipulate numbers. Likewise, these skills acquired in preschool have also been linked to later mathematical achievement (Kolkman, Kroesbergen, & Leseman, 2013; Merkley & Ansari, 2016).

Interestingly, in some studies non-symbolic and symbolic number skills differ in terms of neural activation in the left and right hemisphere, particularly the intraparietal sulcus (IPS). In those studies, non-symbolic number representations mainly activate the right hemisphere, while symbolic number representations also activate the left hemisphere indicating a distinction in processing for non-symbolic and symbolic number skills (Chassy & Grodd, 2012; Emerson & Cantlon, 2015; Klein, Moeller, Nuerk, & Willmes, 2010). However, such differences were less prominent in other studies (Bulthé, De Smedt, & Op de Beeck, 2014; Cohen Kadosh et al., 2011; Eger et al., 2009; Piazza, Pinel, Le Bihan, & Dehaene, 2007). The neural activation in the left IPS but not the right increases over age and becomes more specialized over time for processing symbolic numerical comparisons (Vogel, Goffin, & Ansari, 2015). Not only does the left IPS show developmental changes as experience increases, but also individual differences between 8-year-old children in performance on a symbolic comparison task are shown in alterations of left IPS activation, which is related to future mathematical achievement (Bugden, Price, McLean, & Ansari, 2012). This hemispheric laterization seems to apply only to the IPS, since an fMRI study showed that functional connectivity in the frontoparietal network within both the left and right hemisphere in 4–11-year-old children was indicated to be math specific using a numerical mapping task (Emerson & Cantlon, 2012). With this, it was hypothesized that the IPS is involved during domain-specific activation in mathematics while the frontoparietal network is involved in more domain-general processes such as working memory, attentional, and cognitive control. In addition, neural activation of frontoparietal regions overlap with number-related regions that are generally modulated by response times. In the same study, no overlap in neural activation was found between number-specific IPS regions and regions modulated by reaction time (Emerson & Cantlon, 2012). It is worth to note that the level of activation (i.e., BOLD amplitude) in the frontoparietal areas is equal for children and adults. However, even when performance is the same between different age groups, adults exhibit greater functional connectivity in the left hemisphere. This difference is still present after controlling for verbal intelligence scores (Brown et al., 2005; Emerson & Cantlon, 2012). Summarized, the mentioned hemispheric asymmetry in neural activation of the IPS should be kept in mind when applying an intervention combined with non-symbolic and symbolic tasks. Thereby, neural activity related to numerical processing seems to differ over age in terms of left frontoparietal functional connectivity. It is important to contemplate these domain-specific and domain-general neural differences in numerical transcranial electrical stimulation (tES) interventions due to the diversity in montages in terms of stimulating a specific region or network and stimulating a given brain area unilaterally or bilaterally.

## Numerical interventions and non-invasive brain stimulation

Several approaches of numerical interventions have been studied ranging from behavioral to neuroimaging methods (Cohen Kadosh, Dowker, Heine, Kaufmann, & Kucian, 2013). The emphasis of this chapter is on the neurocognitive mechanisms of non-invasive brain stimulation (NIBS) as numerical intervention. NIBS techniques have gained a substantial amount of interest in cognitive research. This method allows to infer causality between altered neural activity and behavior and could provide a direct intervention to suboptimal brain functions in a clinical population, but also in the healthy population (Silvanto & Pascual-Leone, 2012). NIBS can be divided into two groups based on working mechanism, namely, tES and transcranial magnetic stimulation (TMS). During TMS, a pulse of current is sent through a coil that is placed on the head which generates a magnetic field that enters the brain. Hereby, an electrical current inhibits or excites the underlying neural activity and can modulate behavior (Hallett, 2000). For more information about the neurocognitive bases of numerical cognition from TMS studies, see Faye et al. (2019) for a review.

We primarily focus on tES since more enhancement research has been concluded regarding numerical cognition, and non-invasive tES provides an interesting solution to finding a suitable intervention to improve mathematical skills (Grabner, Rütsche, Ruff, & Hauser, 2015; Hauser, Rotzer, Grabner, Mérillat, & Jäncke, 2013; Schroeder et al., 2017; Simonsmeier, Grabner, Hein, Krenz, & Schneider, 2018; Snowball et al., 2013). In tES interventions, the researcher applies an electrical current (mostly 1–2 mA) through one or more electrodes that are placed on the scalp. The area that the researcher is interested in targeting is usually estimated using the international 10–20 electroencephalography (EEG) system. In contrast to TMS, the sensation from tES is minimal, is considered painless, and a safer method of brain stimulation. In contrast to TMS, the induced effect is not an action potential, but a more subtle modulation of the underlying neuronal population. Different types of tES are documented in relation to diverse cognitive interventions, such as transcranial random noise stimulation (tRNS), transcranial direct current stimulation (tDCS), and transcranial alternating current stimulation (tACS). tRNS makes use of applying alternating electrical current with randomly selected frequencies from a normal distribution of oscillations and enhances cortical excitability under both surface electrodes (Terney, Chaieb, Moliadze, Antal, & Paulus, 2008). Especially high-frequency tRNS (e.g., 101–640 Hz) has been shown to be most effective in influencing performance in the numerical cognition domain (e.g., Cappelletti et al., 2013; Pasqualotto, 2016; Popescu et al., 2016; Snowball et al., 2013). In contrast to tRNS, a direct current flows between the two electrodes when tDCS is administered. While the effects of tDCS depend on the strength and the duration of stimulation (Batsikadze, Moliadze, Paulus, Kuo, & Nitsche, 2013), it is assumed that the anode electrode excites the underlying neural population which leads to depolarization. Contrary, the cathode electrode inhibits the underlying area and leads to hyperpolarization (Nitsche & Paulus, 2000). As mentioned above, the current induced by tES is too low to directly induce action potentials at cellular level. Instead, it has been suggested that these methods induce a subthreshold environment that increases the possibility of spikes (Bikson et al., 2004; Ruffini et al., 2013). In contrast to both aforementioned techniques, it is thought that tACS is able to directly interfere with cortical oscillations whereby stimulation with one specific frequency leads to neural synchrony and several to desynchronization (Desmaisons et al., 2011; Helfrich et al., 2014). Neural oscillations are critically involved in the operational architecture of the brain (Uhlhaas, Roux, Rodriguez, Rotarska-Jagiela, & Singer, 2010), and tACS has been shown to influence this architecture. In short, oscillations are a basic principle for information processing in the brain whereby tACS is able to influence this process. For example, specific entrainment of oscillations by tACS has been related to a temporary improvement in working memory (Jaušovec & Jaušovec, 2014) and perception (Ambrus et al., 2015). Clinical applications of tACS are particularly important in brain disorders with abnormal oscillatory patterns (e.g., Parkinson's or schizophrenia: Brittain, Probert-Smith, Aziz, & Brown, 2013; Gonzalez-Burgos & Lewis, 2008) since tACS is able to modulate these patterns and related behavior. TACS has also yielded benefits in the case of tumor growth (Kirson et al., 2007) and optic nerve injury (Sabel et al., 2011). It should be noted that the influence and exact working mechanism of tACS on numerical cognition is understudied at this moment. Future numerical research should explore the potential of tACS as an intervention and the related neurocognitive mechanisms.

Most importantly, the main reason for applying tES as a numerical intervention is that the brain network involved during mathematical problem solving has been clearly pinpointed in recent neuroimaging studies. To elaborate, processing of mathematical operations has been attributed mainly to involvement of the human frontoparietal network (Grabner et al., 2009; Klein, Moeller, Glauche, Weiller, & Willmes, 2013; Matejko & Ansari, 2015; Nieder, 2016). A shift from the involvement of the prefrontal regions in this network to the parietal regions, mostly lateralized to the left hemisphere, has been observed with fMRI when expertise increases as consequence of a mathematical training paradigm (Ischebeck, Zamarian, Egger, Schocke, & Delazer, 2007; Kucian, von Aster, Loenneker, Dietrich, & Martin, 2008; Rivera, Reiss, Eckert, & Menon, 2005).

Not only does tES influence cognition in a healthy population, but studies showed that electrical stimulation can also have a therapeutic effect in a wide range of disorders (e.g., depression, Alzheimer's, Parkinson, and schizophrenia; see Kekic, Boysen, Campbell, and Schmidt (2016) for a systematic review). Increased popularity also depend on the low costs of tES, the easiness to use, and that they are considered safe (Poreisz, Boros, Antal, & Paulus, 2007; Riggall et al., 2015). Altogether, non-invasive tES appears to be a promising method to enhance numerical cognition in an efficient and harmless way. Not the least, interpreting results of this intervention should be considered with care due to the issue of individual differences such as the subject's brain state, hormonal levels, and age (Krause & Cohen Kadosh, 2014). It should be noted that the lack of stimulation efficacy due to individual differences begs for better and more sophisticated ways to apply tES in a clinical sample.

## The importance of transfer

There is a substantial amount of comprehensive reviews available that recently investigated the effects of non-invasive brain stimulation on numerical cognition in humans (see Schroeder et al., 2017 for a review). This section will primarily focus on studies that included a measurement of transfer of skill (henceforth, transfer). Transfer refers to an improvement in skills in another (similar) task that is not identical to the initial training task. Most studies exclusively studied whether a specific type of tES and stimulation parameters can increase mathematical or numerical performance during a learning task performed by healthy individuals. A measurement of transfer may occur as a result of numerical cognition training and is a crucial aspect to include, mainly due to the information it provides on the effects of brain stimulation. For example, transfer effects can specify if the applied current flow induces long-lasting effects that do not disappear directly after completing a learning task and could possibly be an indicator of neuroplasticity. In addition, examining the effects of transfer is of crucial practical importance for clinicians in educational settings in order to find methods that lead to a design that trains skills in the most generalizable way. Resources could be reduced this way since people need less time to train on a specific skill. One example of this type of measurement can be included in the form of a numerical transfer task performed after the initial learning task that includes presentation of new numerical or mathematical material.

Surprisingly, when reviewing studies that investigated the use of tES on numerical cognition, it became clear that the majority does not include any type of measurement in order to investigate transfer effects induced by brain stimulation (with exception of studies mentioned in Table 1, which will be discussed later). It seems that research in this field is currently focused on direct training effects, which might be because mathematics is a complex concept and involves several domain-related processes. We are not discarding tES studies without implementing transfer as ineffective. However, we argue that tES outcomes with transfer are more reliable and conclusive in terms of the behavioral effects found. Due to the neurocognitive complexity of this field, the next sections are organized in studies that discuss the contribution of tES as intervention and summarize core findings of each study in order to give the reader a thorough understanding of the most recent advances.

## Influencing basic mathematical skills with tES

Neurocognitive mechanisms of tES are dependent on the type of stimulation protocol that is used in combination with the behavioral (i.e., mathematical) task. For example, tES studies can apply an online or an offline approach. The "online" approach is when stimulation is applied during the performance of a task or neuroimaging that allows for quantification, interfering, and modulation of neuronal activity (Bergmann, Karabanov, Hartwigsen, Thielscher, & Siebner, 2016). The "offline" approach is when stimulation is applied before onset of a task or neuroimaging to induce (meta)plasticity to excite or inhibit neuronal activity by inducing long-term potentiation (LTP) or long-term depression (LTD) (Müller-Dahlhaus & Ziemann, 2015). The former is particularly important for examining aftereffects of tES and relates outcomes to a possible application in a clinical

**Table 1** Overview of basic and advanced mathematical skills' studies using tES and a measure of transfer.

| Study | tES design (type, current, duration, sample) | Montage | Task(s) | Measure of transfer | Outcome transfer | Effect size |
|---|---|---|---|---|---|---|
| *Basic mathematical skills* | | | | | | |
| Cappelletti et al. (2013) | tRNS; online → offline 1 mA, 20 min (5 days), between design, $N=10$ in each of four groups | Parietal lobe; P3–P4<br>Motor cortex; C3–C4 | Numerosity training | Retest of numerosity training up to 16 weeks posttraining and assessed transfer to other quantity judgments | Improvement (Weber fraction) in numerosity discrimination when training and tRNS were combined up to 16 weeks and no transfer in other domains | $d=1.12$–1.25 |
| Cohen Kadosh, Soskic, Iuculano, Kanai, and Walsh (2010) | tDCS; online → offline 1 mA, 20 min (6 days), between design, $N=5$ in each of three groups | Parietal lobe<br>Left anodal: P3–P4<br>Right anodal: P4–P3 | Learning paradigm of artificial digits | Performance measured after 6 months | Sustained performance after 6 months for the right anodal montage, and underperformance for the left anodal montage | $d=1.68$–2.32 |
| Looi et al. (2017) | tRNS; online, 0.75 mA, 20 min (twice a week for 5 weeks in total), between design, $N=6$ in tRNS group and $N=6$ in sham | Bilateral DLPFC; F3–F4 | Number line mapping in game format | Follow-up after 9 days | Increased accuracy, and steeper learning slope after tRNS | $d=1.54$–2.22 |

*Continued*

**Table 1** Overview of basic and advanced mathematical skills' studies using tES and a measure of transfer—cont'd

| Study | tES design (type, current, duration, sample) | Montage | Task(s) | Measure of transfer | Outcome transfer | Effect size |
|---|---|---|---|---|---|---|
| *Advanced mathematical skills* | | | | | | |
| Pasqualotto (2016) | tRNS; online 1 mA, 20 min, within-between design, $N=18$ in each of three groups | Frontal and Parietal lobe; F3–F4 and P3–P4 | Subtraction verification task | Follow-up after 7 days | Reduced response times and better accuracy | $d=0.59$–0.93 |
| Snowball et al. (2013) | tRNS; online 1 mA, 20 min (6 days), between design, $N=13$ in tRNS group and $N=12$ in sham | Frontal lobe; F3–F4 | Mathematical problem solving | Performance measured after 6 months | Steeper learning rate facilitation for calculation | $d=1.09$ |
| Popescu et al. (2016) | tRNS; online 1 mA, 20 min (5 days), between design, $N=16$ in each of two groups | Frontal lobe; F3–F4 → parietal lobe; P3–P4 | Mathematical problem solving (calculation and rote memorization) | New mathematical problems presented on the last day (day 5) | Improved calculation training effect | $d=1.46$ |
| Grabner et al. (2015) | tDCS; online 1.5 mA, 30 min, between design, $N=20$ in each of three groups | Parietal lobe Left anodal: P3–P4 Right anodal: P4–P3 | Multiplication and subtraction problems | Performance measured after 24 h | Performance worsened for cathodal tDCS, anodal tDCS showed improvement for subtraction learning | $d=0.55$ |
| Looi et al. (2016) | tDCS; online 1 mA, 30 min, (2 days), between design, $N=10$ in each of three groups | Frontal lobe F3–F4 | Body-tracking video game on fractions | Performance measured after 2 months | Sustained transfer after 2 months for both frontal anodal and cathodal stimulation | $d=0.63$–0.72 |

population. However, we will focus on both online and offline approaches in combination with the effects on numerical skill transfer.

Cappelletti et al. (2013) primarily focused on transfer effects when healthy adults performed a numerosity discrimination task (a type of a non-symbolic comparison task) combined with online tRNS (see Table 1). This study compared results with outcomes of three control groups. The first group received numerosity training + tRNS over the parietal lobes. This group was compared to (1) a group that received stimulation but no numerosity training, (2) a group only received training but no stimulation over the parietal lobe, and (3) a group received training + stimulation over the motor cortex which served as a control area. The group receiving parietal tRNS in combination with numerosity training showed significant performance improvement (~33%) in numerosity discrimination in comparison to other groups that did not receive tRNS. Interestingly, performance improvement of this group that combined training with parietal stimulation was still present up to 16 weeks after the initial training. This was not the case for the group that only performed the training task but did not receive parietal stimulation. It should be mentioned that this stimulation effect was not reflected in general cognitive improvement (e.g., executive functions), but only in quantity training which indicates a possible domain-specific stimulation effect. The lack of transfer was also present when subjects solved mathematical equations and is thought to be due to the absence of stimulating the DLPFC since mathematical abilities also involve this area together with parietal areas (Ischebeck et al., 2006). One suggestion for the underlying mechanisms of the effects of online tRNS and the successful transfer of similar cognitive tasks has been related to hemodynamic changes (Snowball et al., 2013). Another suggestion is that by increasing noise into the neural system by tRNS, the membrane is more likely to depolarize which induces changes in the sodium and calcium channels and this effect on performance is strengthened when it is coupled with a certain cognitive task (Fertonani, Pirulli, & Miniussi, 2011). However, these changes in the sodium and calcium channels are direct consequences of brain stimulation. A recent in-vitro study suggested that short-term random noise stimulation modulates the kinetics of activation and inactivation of the sodium channels (Remedios et al., 2019). Unfortunately, there is not much known about the long-term mechanisms that underlie transfer effects. One explanation that Cappelletti and colleagues point out is the involvement of NMDA receptor signaling due to their contribution to neuroplasticity and LTP (Forsyth, Bachman, Mathalon, Roach, & Asarnow, 2015). However, this theory has been challenged by evidence found that aftereffects of tRNS are not NMDA receptor dependent but are solely sodium channel and benzodiazepines (GABA agonist) dependent (Chaieb, Antal, & Paulus, 2015), and further results showing that tRNS reduced GAD 65/67, a precursor of GABA (Sánchez-León et al., 2020). These results contrast with the aftereffects of tDCS, which is NMDA receptor dependent, and implies different underlying mechanisms between tRNS and tDCS.

As mentioned earlier, it is believed that non-symbolic numerical skills involve the right IPS in contrast to symbolic numerical skills that is also related to the left IPS. In a similar vein, Cohen Kadosh et al. (2010) showed that only right anodal-left cathodal tDCS over the parietal cortex led to sustained performance on an artificial numerical learning paradigm and lasted for 6 months after initial performance. Oppositely, anodal stimulation to the left parietal lobe (and cathodal to the right parietal lobe) led to underperformance on the task. Cohen Kadosh et al. (2010) used a somewhat longer training paradigm (6 days instead of 5 days), and the behavioral task combined with stimulation was an artificial learning paradigm instead of a numerosity discrimination task. One reason of the effectiveness of parietal tES might be due to the combination with a basic mathematical task instead of an advanced mathematical task, since this task is more prone to automatic tendencies (Cohen Kadosh et al., 2007) and is in line with IPS involvement.

Whereas the previous studies included healthy subjects, the study by Looi et al. (2017) showed that tRNS over the bilateral DLPFC during a number line estimation training increased mathematical learning and performance for children (aged 8.5–10.9) with mathematical learning disabilities (MLD) (see Table 1). This study is the first to apply tRNS in children. Therefore half of the normally administered current in adults was used, namely, 0.75 mA. Interestingly, children were not tested in a lab setting but at their school to reflect real-world learning. Other domain-general processes such as working memory were not influenced by tRNS. The result shows improvement in number line estimation by tRNS, and an interaction between tRNS and number line estimation during the training when predicting improvement on a standardized mathematical achievement test. Although promising, we would like to highlight that the studies in this section had a small sample size and future tES research that are well powered in typical and atypical neurodevelopmental populations is vital.

## Basic mathematical skills and tES: Open questions

The results from the previously mentioned tES studies appear to indicate that parietal tES can facilitate transfer effects of enhanced basic mathematical performance (see Table 1). More tES research is needed that includes transfer to reliably induce long-lasting basic mathematical skill improvement. However, it should be noted that the effects of tES on these skills might differ in different population groups. For example, since working memory processes are strongly related to basic mathematical skills and activate the DLPFC, it is interesting to see if prefrontal stimulation is effective on a transfer task. Moreover, since a previous study also indicated the importance of functional connectivity strength within frontoparietal areas in a numerical mapping task (Emerson & Cantlon, 2012), tES studies should explore this finding further in combination with a basic mathematical task and a measurement of transfer. For example, one recent study indicated that tACS is a reliable tool to modulate large-scale corticocortical coupling that outlasts the stimulation period (Schwab, Misselhorn, & Engel, 2019). This finding shows that synchronization within the frontoparietal network can be strengthened by frequency-specific stimulation. However, behavioral effects due to synchronization induced by stimulation are still unclear. It would be interesting to exploit this idea further by targeting the frontoparietal network during mathematical performance. However, in contrast to tRNS and tDCS that have shown long-term effects, the effect of tACS decayed shortly (e.g., 120 s) after tACS ceased, and better tACS protocols that would allow long-lasting effects are needed.

Because basic mathematical skills are still developing and unstable in children, it is unclear if similar effects that were found in Cappelletti et al. (2013) and Cohen Kadosh et al. (2010), who tested healthy adults, will be found in a child population. tES research in this population group is extremely limited, mainly due to ethical concerns since a child might not be aware or able to report any side effects of brain stimulation (Davis, 2014). However, neuroethicists have advocated such research in atypical developing children (Maslen, Earp, Cohen Kadosh, & Savulescu, 2014). Even though the study from Looi et al. (2017) shows improved learning and performance in atypically developing children, more research into the exact working mechanisms of tES is needed in order to apply this method successfully to the atypical developing brain.

## Influencing advanced mathematical skills with tES

Cappelletti et al. (2013) only looked at training involving quantity discrimination (e.g., basic mathematical skills) and transfer effects that did not generalize to advanced mathematics (e.g., complex mathematical problem solving). Therefore it is meaningful to determine which aspects do relate to successful transfer effects in advanced mathematics. For instance, the underlying neural circuit that is involved in quantity training differs from complex mathematical training as has been indicated in the previous paragraphs. One study by Pasqualotto (2016) who included a 1-day training paradigm that was specifically aimed at targeting mathematical processes combined with online tRNS found increased performance in terms of reduced response times and increased accuracy 7 days after the initial training for both frontal and parietal stimulation (see Table 1). Mathematical skills were measured by using a subtraction verification task and compared with a word classification task. Because subjects were tested on unfamiliar subtraction 7 days later, this effect is a good example of transfer of a specific skill. As mentioned previously, one possible explanation from Cappelletti et al. (2013) for the lack of transfer effects of a quantity training task could be due to not stimulating the DLPFC. However, findings from Pasqualotto (2016) are in contrast to this assumption due to frontal stimulation. Thus it seems plausible that transfer effects are strongly dependent on the specific mathematical task.

Two studies that tested the effects of online tRNS on mathematical training are from Snowball et al. (2013) and Popescu et al. (2016) (see Table 1). Both studies use a design with longer consecutive training and stimulation than the previously mentioned tRNS studies. Whereas Snowball et al. (2013) stimulated frontal sites (and parietal sites in a control experiment), Popescu et al. (2016) moved stimulation from the frontal sites after 3 days to parietal sites on training days 4 and 5. The rational for this movement is explained with reference of the involvement of the frontoparietal network during mathematical problem solving. The DLPFC has a greater involvement during skill acquisition, and this involvement gradually declines and is taken over by more automatic processing of the parietal cortex. Facilitation of transfer effects was measured on the last testing day with both old material (trained mathematical material) and new presented material (untrained mathematical material). It was found that tRNS facilitated response times for old and new mathematical problems compared to sham (placebo) tRNS. Again, no transfer effect was present for other cognitive tasks such as working memory. The study design of Snowball et al. (2013) is somewhat similar to the study design used by Popescu and colleagues but they coupled tRNS and consecutive mathematical training with near-infrared spectroscopy (NIRS). NIRS provides insight into hemodynamic processes that are present in the prefrontal cortex during stimulation and training. In short, a more efficient neurovascular coupling in the left frontal cortex was found compared to sham tRNS. A long-term behavioral and hemodynamic transfer effect for calculation was also found after 6 months of the initial training when subjects completed both old and new material. According to Snowball et al. (2013), an effect of transfer for calculation and not rote memorization could be explained due to involvement of deep-level cognitive processing as compared to shallower rote memorization learning (Craik, 2002; Craik & Lockhart, 1972). Namely, transfer effects are reinforced by deep-level processing, which changes the neural system and can interfere with tRNS. It should be stated that Snowball et al. (2013) also tested the same study design involving parietal stimulation instead of DLPFC stimulation. However, no effect was found. Therefore stimulating a brain region that is known to be activated during a specific task does not always lead to performance improvement on the same task.

A study that used a mathematical (fractions) training paradigm in video game format showed that right anodal-left cathodal tDCS over the DLPFC improved the performance, which was especially clear in a task with increased difficulty (Looi et al., 2016) (see Table 1). Increased performance was still present in subjects who received active stimulation after 2 days and 2 months later. The tDCS group performed even better in comparison to their initial performance, which could be due to an increased protein synthesis after anodal tDCS as suggested consolidation mechanism (Karni et al., 1998). Interestingly, this study also found an improvement on verbal (but not visuospatial) working memory task that was not part of the initial training paradigm. This is explained by the intertwined link between mathematics and verbal working memory (Wilson & Swanson, 2001). However, this specific mathematical training paradigm involved fraction problems which might be more advanced than solving subtraction and multiplication problems. Another tDCS study that compared typical mathematical skill acquisition between anodal and cathodal parietal stimulation found an increased learning rate after 24 h for the left-anodal parietal group (Grabner et al., 2015). The opposite effect was observed for the cathodal group. Long-term transfer effects were not considered in this study.

## Advanced mathematical skills and tES: Open questions

Surprisingly, there are not many tDCS studies available in current literature that included short or long-term measurement of advanced mathematical transfer effects. This is remarkable, since tDCS research has increased tremendous over the last couple of years. The strength and modulation of tRNS on numerical cognition performance appears to be more convincing in comparison to tDCS. This might explain why there is less research done investigating transfer effects after applying tDCS. Moreover, results of some of these studies that combined tDCS and mathematical training are inconsistent. For example, Hauser et al. (2013) found a positive outcome on a mental mathematical task for a group of subjects that received left-anodal frontal stimulation, and not for the right anodal or bilateral montage which contradicts the results of Cohen Kadosh et al. (2010). Both studies indicate that this effect is due to involvement of either the left or right DLPFC in mathematical processing. It is further argued that the discrepancy might be due to differences in the study design; learning of number-symbol associations versus mental mathematics, online versus offline stimulation, and assessment of different aspects of complex mathematical processing. Further research should include both short and long-term transfer effect measurements when applying left-anodal frontal tDCS. Other factor that could cause a discrepancy in the facilitation effects of tDCS in mathematical problem solving is the size of the operand. The effects of tDCS and oscillatory activity depend on whether a person is solving one-digit (small) or two-digit (large) mathematical operations (Rütsche, Hauser, Jäncke, & Grabner, 2015). More specifically, anodal tDCS to the left parietal cortex had a positive effect on response latencies in large mathematical facts, but a negative effect in small mathematical operations. Despite the need for more tDCS research into advanced mathematical transfer effects, there is marginal evidence that anodal tDCS can facilitate mathematical and numerical skills on a behavioral level after stimulation. This facilitation supports the view that tDCS can affect cellular and molecular mechanism such as LTP (Forsyth et al., 2015). In addition, an extensive neurobiological review of the after effects of NIBS indicated that tDCS modulates the cellular background and thereby directs synaptic plasticity (Cirillo et al., 2017).

As can be clear from Table 1, studies that included a measurement of transfer into their study design differ in terms of stimulation parameters. For example, both frontal, parietal, and a shift from frontal to parietal stimulation proved to be effective in relation to steeper learning rates. Nevertheless, there is some controversy of the effectiveness of stimulating the parietal areas which should be further investigated.

## Conclusion and future directions

Previous NIBS literature mainly focuses on the immediate effect on basic or advanced mathematical skills. These studies have shown substantial evidence that performance on a mathematical task can be enhanced, both offline and online (for a review, see Schroeder et al., 2017). However, there are not many studies that also examined either short-term (a couple of weeks) or long-term (several months) transfer effects of tES. These transfer effects could provide information about the facilitation of a certain skill and could indicate neuroplasticity. Especially long-term facilitation by tES on performance is essential to study, to further evaluate its adequacy in the future in a clinical or educational setting.

In continuation to the potential practice of tES in a clinical setting, there is currently only one single study on children with mathematical learning difficulties. The effects of tES in this subpopulation are not yet evident due to several neuroethical questions raised (Cohen Kadosh, Levy, O'Shea, Shea, & Savulescu, 2012). While current neurostimulation parameters appear to be safe for minors (Krishnan, Santos, Peterson, & Ehinger, 2015; Palm et al., 2016), the exact behavioral and neurobiological working mechanisms of tES are still unclear. It is therefore important to be agile and consider any important risk and pitfalls associated with stimulating the developing brain (Davis, 2014; Krause & Cohen Kadosh, 2013). Likewise, all tES studies discussed in this chapter tested healthy subjects instead of using tES as a potential intervention for learning disabilities such as dyscalculia. One study using tDCS in two adults with dyscalculia showed increased numerical learning when left-anodal and right-cathodal parietal stimulation was administered (Iuculano & Cohen Kadosh, 2014). This is in contrast with the finding that right-anodal and left-cathodal tDCS improved performance in healthy subjects (Cohen Kadosh et al., 2010). So, tES can have different effects in healthy and clinical populations that beg for more research.

It is furthermore important to point out some weaknesses with the current literature. First, there is a lack of replication studies. Second, the sample sizes used in most of the studied are small. Third, there is a lack of preregistration of studies in this field, although such movement has only begun more recently (Van 't Veer & Giner-Sorolla, 2016). Although most studies have appeared before awareness of this issue, future studies should consider improving these weaknesses in their prospective studies. Interestingly, it has recently been argued that there is only a weak publication bias effect of the effects of tES on performance and learning (Simonsmeier et al., 2018). In addition, the longevity of the tRNS and tDCS effects are largely unknown. Nonetheless, at least two tRNS and two tDCS studies suggest that there is still a behavioral improvement present induced by tES after a few months (Cappelletti et al., 2013; Cohen Kadosh et al., 2010; Looi et al., 2016; Snowball et al., 2013). Even though most of these studies made use of a relatively long training design (5–6 sessions), it is remarkable that this result is still present after that time period since no training or stimulation was applied until (and during) the follow-up. Notably, studies using tRNS and tDCS in other cognitive domains yielded similar results (Frank, Harty, Kluge, & Cohen Kadosh, 2018; Herpich et al., 2019; Reis et al., 2009). Moreover, the transfer effects presented in this chapter were limited to the trained material and did not generalize to other domain-general processes with exception of the study from Looi et al. (2016) who used a more complex mathematical training paradigm entailing fraction problems. Despite the specificity to trained material, there is still the possibility present that tES induced other concealed cognitive costs or gains (Brem, Fried, Horvath, Robertson, & Pascual-Leone, 2014; Iuculano & Cohen Kadosh, 2013; Sarkar, Dowker, & Cohen Kadosh, 2014).

An important component that we would like to highlight is the effect of individual differences in this line of research. Multiple factors can influence the response of non-invasive tES, such as age, hormonal level, and brain state (Krause & Cohen Kadosh, 2014; Krause, Márquez-Ruiz, & Kadosh, 2013). For example, neural oscillations are dependent on different mathematical problems based on problem size (Rütsche et al., 2015). To illustrate, lower-alpha desynchronization is increased in two-digit mathematical operations in contrast to theta synchronization, which is increased in one-digit mathematical operations. Moreover, a recent event-related potential (ERP) study highlighted the importance of individual differences in mathematical abilities since both the neural and behavioral correlates differ between people with high and low mathematical baseline performance (Artemenko, Soltanlou, Dresler, Ehlis, & Nuerk, 2018). In continuation, tES has shown to benefit more those with lower baseline abilities (Furuya, Klaus, Nitsche, Paulus, & Altenmüller, 2014; Looi et al., 2016; Sarkar et al., 2014) and has shown even impairment in those with very strong mathematical abilities (Krause, Dresler, Looi, Sarkar, & Cohen Kadosh, 2019).

More recent studies that investigated the beneficial effect of tES on skill learning imply a dependency on both behavioral and neuronal moderators (Sheffield, Raz, Sella, & Cohen Kadosh, 2020; Van Bueren, Francesco, Sheffield, Kroesbergen, & Cohen Kadosh, 2021). These findings imply that the effect of tRNS on arithmetic skill acquisition depends on the baseline ability and neurophysiology of the subject. Namely, subjects with a suboptimal arithmetic baseline ability and attentional electrocortical marker (theta/beta ratio) who received tRNS during arithmetic learning displayed enhanced performance as opposed to subjects with more optimal values.

Thus, the effect of tRNS was associated with domain-general components, in this case sustained attention, rather than domain-specific components. This conclusion fits also nicely with the recent finding that tRNS applied over the regions that improve mathematical learning and cognition and also benefited the executive functions training in children with attentional deficits (Berger, Dakwar-Kawar, Grossman, Nahum, & Kadosh, 2019).

It has also been argued that the type of study and task difficulty used in tES research is an important factor (Bieck, Artemenko, Moeller, & Klein, 2018; Popescu et al., 2016; Van Bueren et al., 2021). Additionally, tRNS might have a stronger effect on numerical learning and consolidation instead of modulating immediate performance which is in line with a recent meta-analysis from Simonsmeier et al. (2018) who showed tES improves learning more than performance. All these neurocognitive aspects of numerical cognition influence the consequences of tES on behavioral outcomes. This warrants more caution when using tES as a cognitive enhancement method and could explain the mixed findings in the field of tES.

In conclusion, tES research highlights that basic and advanced mathematical skills depend on a variety of different neurocognitive factors. However, the relative lack of systematic research, replication, and large sample sizes prevents this field from strong causal understanding of the multifactual nature of the underlying mechanisms for optimal intervention. However, given that this line of research is relatively new, we suggest that improving these caveats could, at least partly, allow better basic understanding with potential translational implications.

## References

Ambrus, G. G., Pisoni, A., Primaßin, A., Turi, Z., Paulus, W., & Antal, A. (2015). Bifrontal transcranial alternating current stimulation in the ripple range reduced overnight forgetting. *Frontiers in Cellular Neuroscience*, *9*, 374. https://doi.org/10.3389/fncel.2015.00374.

Andrews, S. C., Hoy, K. E., Enticott, P. G., Daskalakis, Z. J., & Fitzgerald, P. B. (2011). Improving working memory: The effect of combining cognitive activity and anodal transcranial direct current stimulation to the left dorsolateral prefrontal cortex. *Brain Stimulation*, *4*(2), 84–89. https://doi.org/10.1016/J.BRS.2010.06.004.

Artemenko, C., Soltanlou, M., Dresler, T., Ehlis, A.-C., & Nuerk, H.-C. (2018). The neural correlates of arithmetic difficulty depend on mathematical ability: Evidence from combined fNIRS and ERP. *Brain Structure and Function*, *223*(6), 2561–2574. https://doi.org/10.1007/s00429-018-1618-0.

Batsikadze, G., Moliadze, V., Paulus, W., Kuo, M.-F., & Nitsche, M. A. (2013). Partially non-linear stimulation intensity-dependent effects of direct current stimulation on motor cortex excitability in humans. *The Journal of Physiology*, *591*(7), 1987–2000. https://doi.org/10.1113/jphysiol.2012.249730.

Berger, I., Dakwar-Kawar, O., Grossman, E. S., Nahum, M., & Kadosh, R. C. (2019). Scaffolding the attention-deficit/hyperactivity disorder brain using transcranial direct current and random noise stimulation. *MedRxiv*. https://doi.org/10.1101/19005983, 19005983.

Bergmann, T. O., Karabanov, A., Hartwigsen, G., Thielscher, A., & Siebner, H. R. (2016). Combining non-invasive transcranial brain stimulation with neuroimaging and electrophysiology: Current approaches and future perspectives. *NeuroImage*, *140*, 4–19. https://doi.org/10.1016/J.NEUROIMAGE.2016.02.012.

Bieck, S. M., Artemenko, C., Moeller, K., & Klein, E. (2018). Low to no effect: Application of tRNS during two-digit addition. *Frontiers in Neuroscience*, *12*, 176. https://doi.org/10.3389/fnins.2018.00176.

Bikson, M., Inoue, M., Akiyama, H., Deans, J. K., Fox, J. E., Miyakawa, H., & Jefferys, J. G. R. (2004). Effects of uniform extracellular DC electric fields on excitability in rat hippocampal slices *in vitro*. *The Journal of Physiology*, *557*(1), 175–190. https://doi.org/10.1113/jphysiol.2003.055772.

Brem, A.-K., Fried, P. J., Horvath, J. C., Robertson, E. M., & Pascual-Leone, A. (2014). Is neuroenhancement by noninvasive brain stimulation a net zero-sum proposition? *NeuroImage*, *85*, 1058–1068. https://doi.org/10.1016/J.NEUROIMAGE.2013.07.038.

Brittain, J.-S., Probert-Smith, P., Aziz, T. Z., & Brown, P. (2013). Tremor suppression by rhythmic transcranial current stimulation. *Current Biology*, *23*(5), 436–440. https://doi.org/10.1016/J.CUB.2013.01.068.

Brown, T. T., Lugar, H. M., Coalson, R. S., Miezin, F. M., Petersen, S. E., & Schlaggar, B. L. (2005). Developmental changes in human cerebral functional organization for word generation. *Cerebral Cortex*, *15*(3), 275–290. https://doi.org/10.1093/cercor/bhh129.

Bugden, S., Price, G. R., McLean, D. A., & Ansari, D. (2012). The role of the left intraparietal sulcus in the relationship between symbolic number processing and children's arithmetic competence. *Developmental Cognitive Neuroscience*, *2*(4), 448–457. https://doi.org/10.1016/J.DCN.2012.04.001.

Bull, R., Espy, K. A., & Wiebe, S. A. (2008). Short-term memory, working memory, and executive functioning in preschoolers: Longitudinal predictors of mathematical achievement at age 7 years. *Developmental Neuropsychology*, *33*(3), 205–228. https://doi.org/10.1080/87565640801982312.

Bulthé, J., De Smedt, B., & Op de Beeck, H. P. (2014). Format-dependent representations of symbolic and non-symbolic numbers in the human cortex as revealed by multi-voxel pattern analyses. *NeuroImage*, *87*, 311–322. https://doi.org/10.1016/j.neuroimage.2013.10.049.

Butterworth, B., Varma, S., & Laurillard, D. (2011). Dyscalculia: From brain to education. *Science*, *332*(6033), 1049–1053. https://science.sciencemag.org/.

Cappelletti, M., Gessaroli, E., Hithersay, R., Mitolo, M., Didino, D., Kanai, R., … Walsh, V. (2013). Transfer of cognitive training across magnitude dimensions achieved with concurrent brain stimulation of the parietal lobe. *Journal of Neuroscience*, *33*(37). https://www.jneurosci.org.proxy.library.uu.nl/content/33/37/14899.short.

Chaieb, L., Antal, A., & Paulus, W. (2015). Transcranial random noise stimulation-induced plasticity is NMDA-receptor independent but sodium-channel blocker and benzodiazepines sensitive. *Frontiers in Neuroscience*, *9*, 125. https://doi.org/10.3389/fnins.2015.00125.

Chassy, P., & Grodd, W. (2012). Comparison of quantities: Core and format-dependent regions as revealed by fMRI. *Cerebral Cortex*, *22*(6), 1420–1430. https://doi.org/10.1093/cercor/bhr219.

Cirillo, G., Di Pino, G., Capone, F., Ranieri, F., Florio, L., Todisco, V., … Di Lazzaro, V. (2017). Neurobiological after-effects of non-invasive brain stimulation. *Brain Stimulation*, *10*(1), 1–18. https://doi.org/10.1016/j.brs.2016.11.009.

Cohen Kadosh, R., Cohen Kadosh, K., Schuhmann, T., Kaas, A., Goebel, R., Henik, A., & Sack, A. T. (2007). Virtual dyscalculia induced by parietal-lobe TMS impairs automatic magnitude processing. *Current Biology*, *17*(8), 689–693. https://doi.org/10.1016/j.cub.2007.02.056.

Cohen Kadosh, R., Dowker, A., Heine, A., Kaufmann, L., & Kucian, K. (2013). Interventions for improving numerical abilities: Present and future. *Trends in Neuroscience and Education*, *2*(2), 85–93. https://doi.org/10.1016/J.TINE.2013.04.001.

Cohen Kadosh, R., Soskic, S., Iuculano, T., Kanai, R., & Walsh, V. (2010). Modulating neuronal activity produces specific and long-lasting changes in numerical competence. *Current Biology: CB*, *20*(22), 2016–2020. https://doi.org/10.1016/j.cub.2010.10.007.

Cohen Kadosh, R. C., Bahrami, B., Walsh, V., Butterworth, B., Popescu, T., & Price, C. J. (2011). Specialization in the human brain: The case of numbers. *Frontiers in Human Neuroscience*, *5*, 62. https://doi.org/10.3389/fnhum.2011.00062.

Cohen Kadosh, R. C., Levy, N., O'Shea, J., Shea, N., & Savulescu, J. (2012). The neuroethics of non-invasive brain stimulation. *Current Biology*, *22*(4), R108–R111. https://doi.org/10.1016/J.CUB.2012.01.013.

Cragg, L., & Gilmore, C. (2014). Skills underlying mathematics: The role of executive function in the development of mathematics proficiency. *Trends in Neuroscience and Education*, *3*(2), 63–68. https://doi.org/10.1016/J.TINE.2013.12.001.

Cragg, L., Keeble, S., Richardson, S., Roome, H. E., & Gilmore, C. (2017). Direct and indirect influences of executive functions on mathematics achievement. *Cognition*, *162*, 12–26. https://doi.org/10.1016/j.cognition.2017.01.014.

Craik, F. I. M. (2002). Levels of processing: Past, present.. and future? *Memory*, *10*(5–6), 305–318. https://doi.org/10.1080/09658210244000135.

Craik, F. I. M., & Lockhart, R. S. (1972). Levels of processing: A framework for memory research. *Journal of Verbal Learning and Verbal Behavior*, *11*(6), 671–684. https://doi.org/10.1016/S0022-5371(72)80001-X.

Davis, N. J. (2014). Transcranial stimulation of the developing brain: A plea for extreme caution. *Frontiers in Human Neuroscience*, *8*, 600. https://doi.org/10.3389/fnhum.2014.00600.

De Smedt, B., Noël, M. P., Gilmore, C., & Ansari, D. (2013). How do symbolic and non-symbolic numerical magnitude processing skills relate to individual differences in children's mathematical skills? A review of evidence from brain and behavior. *Trends in Neuroscience and Education*, *2*(2), 48–55. https://doi.org/10.1016/j.tine.2013.06.001.

Delazer, M., Domahs, F., Bartha, L., Brenneis, C., Lochy, A., Trieb, T., & Benke, T. (2003). Learning complex arithmetic—An fMRI study. *Cognitive Brain Research*, *18*(1), 76–88. https://doi.org/10.1016/j.cogbrainres.2003.09.005.

Delazer, M., Ischebeck, A., Domahs, F., Zamarian, L., Koppelstaetter, F., Siedentopf, C. M., … Felber, S. (2005). Learning by strategies and learning by drill—Evidence from an fMRI study. *NeuroImage*, *25*(3), 838–849. https://doi.org/10.1016/j.neuroimage.2004.12.009.

Desmaisons, D., Vincent, J., Lledo, P., Del Testa, M., Rossi, A., & Rossi, S. (2011). Control of action potential timing by intrinsic subthreshold oscillations in olfactory bulb output neurons. *Journal of Neuroscience*, *19*(24), 10727–10737. https://doi.org/10.1523/jneurosci.0978-11.2011.

Duncan, G. J., Dowsett, C. J., Claessens, A., Magnuson, K., Huston, A. C., Klebanov, P., … Zill, N. (2007). School readiness and later achievement. *Developmental Psychology*, *43*(6), 1428. https://doi.org/10.1037/0012-1649.43.6.1428.supp.

Eger, E., Michel, V., Thirion, B., Amadon, A., Dehaene, S., & Kleinschmidt, A. (2009). Deciphering cortical number coding from human brain activity patterns. *Current Biology*, *19*(19), 1608–1615. https://doi.org/10.1016/j.cub.2009.08.047.

Emerson, R. W., & Cantlon, J. F. (2012). Early math achievement and functional connectivity in the fronto-parietal network. *Developmental Cognitive Neuroscience*, *2*, S139–S151. https://doi.org/10.1016/J.DCN.2011.11.003.

Emerson, R. W., & Cantlon, J. F. (2015). Continuity and change in children's longitudinal neural responses to numbers. *Developmental Science, 18*(2), 314–326. https://doi.org/10.1111/desc.12215.

Faye, A., Jacquin-Courtois, S., Reynaud, E., Lesourd, M., Besnard, J., & Osiurak, F. (2019). *Numerical cognition: A meta-analysis of neuroimaging, transcranial magnetic stimulation and brain-damaged patients studies* (p. 102053). NeuroImage: Clinical. https://doi.org/10.1016/j.nicl.2019.102053.

Fertonani, A., Pirulli, C., & Miniussi, C. (2011). Random noise stimulation improves neuroplasticity in perceptual learning. *Journal of Neuroscience, 31*(43), 15416–15423. https://www.jneurosci.org.proxy.library.uu.nl/content/31/43/15416.short.

Forsyth, J. K., Bachman, P., Mathalon, D. H., Roach, B. J., & Asarnow, R. F. (2015). Augmenting NMDA receptor signaling boosts experience-dependent neuroplasticity in the adult human brain. *Proceedings of the National Academy of Sciences of the United States of America, 112*(50), 15331–15336. https://doi.org/10.1073/pnas.1509262112.

Frank, B., Harty, S., Kluge, A., & Cohen Kadosh, R. (2018). Learning while multitasking: Short and long-term benefits of brain stimulation. *Ergonomics, 61*(11), 1454–1463. https://doi.org/10.1080/00140139.2018.1563722.

Friedman, N. P., Miyake, A., Corley, R. P., Young, S. E., DeFries, J. C., & Hewitt, J. K. (2006). Not all executive functions are related to intelligence. *Psychological Science, 17*(2), 172–179. https://doi.org/10.1111/j.1467-9280.2006.01681.x.

Friso-Van Den Bos, I., Van Der Ven, S. H. G., Kroesbergen, E. H., & Van Luit, J. E. H. (2013). Working memory and mathematics in primary school children: A meta-analysis. *Educational Research Review, 10*, 29–44. https://doi.org/10.1016/j.edurev.2013.05.003.

Furuya, S., Klaus, M., Nitsche, M. A., Paulus, W., & Altenmüller, E. (2014). Ceiling effects prevent further improvement of transcranial stimulation in skilled musicians. *Journal of Neuroscience, 34*(41), 13834–13839. https://doi.org/10.1523/JNEUROSCI.1170-14.2014.

Geary, D. C., & vanMarle, K. (2016). Young children's core symbolic and nonsymbolic quantitative knowledge in the prediction of later mathematics achievement. *Developmental Psychology, 52*(12), 2130–2144. https://doi.org/10.1037/dev0000214.

Gonzalez-Burgos, G., & Lewis, D. A. (2008). GABA neurons and the mechanisms of network oscillations: Implications for understanding cortical dysfunction in schizophrenia. *Schizophrenia Bulletin, 34*(5), 944–961. https://doi.org/10.1093/schbul/sbn070.

Grabner, R. H., Ansari, D., Koschutnig, K., Reishofer, G., Ebner, F., & Neuper, C. (2009). To retrieve or to calculate? Left angular gyrus mediates the retrieval of arithmetic facts during problem solving. *Neuropsychologia, 47*(2), 604–608. https://doi.org/10.1016/j.neuropsychologia.2008.10.013.

Grabner, R. H., Rütsche, B., Ruff, C. C., & Hauser, T. U. (2015). Transcranial direct current stimulation of the posterior parietal cortex modulates arithmetic learning. *European Journal of Neuroscience, 42*(1), 1667–1674. https://doi.org/10.1111/ejn.12947.

Hallett, M. (2000). Transcranial magnetic stimulation and the human brain. *Nature, 406*(6792), 147–150. https://doi.org/10.1038/35018000.

Hambrick, D. Z., & Engle, R. W. (2002). Effects of domain knowledge, working memory capacity, and age on cognitive performance: An investigation of the knowledge-is-power hypothesis. *Cognitive Psychology, 44*(4), 339–387. https://doi.org/10.1006/COGP.2001.0769.

Hauser, T. U., Rotzer, S., Grabner, R. H., Mérillat, S., & Jäncke, L. (2013). Enhancing performance in numerical magnitude processing and mental arithmetic using transcranial Direct Current Stimulation (tDCS). *Frontiers in Human Neuroscience, 7*, 244. https://doi.org/10.3389/fnhum.2013.00244.

Helfrich, R. F., Schneider, T. R., Rach, S., Trautmann-Lengsfeld, S. A., Engel, A. K., & Herrmann, C. S. (2014). Report entrainment of brain oscillations by transcranial alternating current stimulation. *Current Biology, 24*, 333–339. https://doi.org/10.1016/j.cub.2013.12.041.

Herpich, F., Melnick, M. D., Agosta, S., Huxlin, K. R., Tadin, D., & Battelli, L. (2019). Boosting learning efficacy with noninvasive brain stimulation in intact and brain-damaged humans. *Journal of Neuroscience*, *39*(28), 5551–5561. https://doi.org/10.1523/JNEUROSCI.3248-18.2019.

Huijsmans, M. D., Kleemans, T., van der Ven, S. H., & Kroesbergen, E. H. (2020). The relevance of subtyping children with mathematical learning disabilities. *Research in Developmental Disabilities*, *104*, 103704. https://doi.org/10.1016/j.ridd.2020.103704.

Ischebeck, A., Zamarian, L., Egger, K., Schocke, M., & Delazer, M. (2007). Imaging early practice effects in arithmetic. *NeuroImage*, *36*(3), 993–1003. https://doi.org/10.1016/j.neuroimage.2007.03.051.

Ischebeck, A., Zamarian, L., Siedentopf, C., Koppelstätter, F., Benke, T., Felber, S., & Delazer, M. (2006). How specifically do we learn? Imaging the learning of multiplication and subtraction. *NeuroImage*, *30*(4), 1365–1375. https://doi.org/10.1016/J.NEUROIMAGE.2005.11.016.

Iuculano, T., & Cohen Kadosh, R. (2013). The mental cost of cognitive enhancement. *The Journal of Neuroscience: The Official Journal of the Society for Neuroscience*, *33*(10), 4482–4486. https://doi.org/10.1523/JNEUROSCI.4927-12.2013.

Iuculano, T., & Cohen Kadosh, R. (2014). Preliminary evidence for performance enhancement following parietal lobe stimulation in developmental dyscalculia. *Frontiers in Human Neuroscience*, *8*, 38. https://doi.org/10.3389/fnhum.2014.00038.

Jaušovec, N., & Jaušovec, K. (2014). Increasing working memory capacity with theta transcranial alternating current stimulation (tACS). *Biological Psychology*, *96*, 42–47. https://doi.org/10.1016/J.BIOPSYCHO.2013.11.006.

Kail, R., & Hall, L. K. (2001). Distinguishing short-term memory from working memory. *Memory & Cognition*, *29*(1), 1–9. https://doi.org/10.3758/BF03195735.

Karni, A., Meyer, G., Rey-Hipolito, C., Jezzard, P., Adams, M. M., Turner, R., … Krakauer, J. W. (1998). The acquisition of skilled motor performance: Fast and slow experience-driven changes in primary motor cortex. *Proceedings of the National Academy of Sciences of the United States of America*, *95*(3), 861–868. https://doi.org/10.1073/pnas.95.3.861.

Kekic, M., Boysen, E., Campbell, I. C., & Schmidt, U. (2016). A systematic review of the clinical efficacy of transcranial direct current stimulation (tDCS) in psychiatric disorders. *Journal of Psychiatric Research*, *74*, 70–86. https://doi.org/10.1016/J.JPSYCHIRES.2015.12.018.

Kirson, E. D., Dbalý, V., Tovarys, F., Vymazal, J., Soustiel, J. F., Itzhaki, A., … Palti, Y. (2007). Alternating electric fields arrest cell proliferation in animal tumor models and human brain tumors. *Proceedings of the National Academy of Sciences of the United States of America*, *104*(24), 10152–10157. https://doi.org/10.1073/pnas.0702916104.

Klein, E., Moeller, K., Glauche, V., Weiller, C., & Willmes, K. (2013). Processing pathways in mental arithmetic—Evidence from probabilistic fiber tracking. *PLoS One*, *8*(1). https://doi.org/10.1371/journal.pone.0055455, e55455.

Klein, E., Moeller, K., Nuerk, H.-C., & Willmes, K. (2010). On the neuro-cognitive foundations of basic auditory number processing: An fMRI study. *Behavioral and Brain Functions*, *6*(1), 42. https://doi.org/10.1186/1744-9081-6-42.

Kolkman, M. E., Hoijtink, H. J. A., Kroesbergen, E. H., & Leseman, P. P. M. (2013). The role of executive functions in numerical magnitude skills. *Learning and Individual Differences*, *24*, 145–151. https://doi.org/10.1016/J.LINDIF.2013.01.004.

Kolkman, M. E., Kroesbergen, E. H., & Leseman, P. P. M. (2013). Early numerical development and the role of non-symbolic and symbolic skills. *Learning and Instruction*, *25*, 95–103. https://doi.org/10.1016/J.LEARNINSTRUC.2012.12.001.

Krause, B., & Cohen Kadosh, R. (2013). Can transcranial electrical stimulation improve learning difficulties in atypical brain development? A future possibility for cognitive training. *Developmental Cognitive Neuroscience*, *6*, 176–194. https://doi.org/10.1016/j.dcn.2013.04.001.

Krause, B., & Cohen Kadosh, R. (2014). Not all brains are created equal: The relevance of individual differences in responsiveness to transcranial electrical stimulation. *Frontiers in Systems Neuroscience*, *8*, 25. https://doi.org/10.3389/fnsys.2014.00025.

Krause, B., Dresler, M., Looi, C.Y., Sarkar, A., & Cohen Kadosh, R. (2019). Neuroenhancement of high-level cognition: Evidence for homeostatic constraints of non-invasive brain stimulation. *Journal of Cognitive Enhancement*, *3*(4), 388–395. https://doi.org/10.1007/s41465-019-00126-7.

Krause, B., Márquez-Ruiz, J., & Kadosh, R. C. (2013). The effect of transcranial direct current stimulation: A role for cortical excitation/inhibition balance? *Frontiers in Human Neuroscience*, 7, 602. https://doi.org/10.3389/fnhum.2013.00602.

Krishnan, C., Santos, L., Peterson, M. D., & Ehinger, M. (2015). Safety of noninvasive brain stimulation in children and adolescents. *Brain Stimulation*, *8*(1), 76–87. https://doi.org/10.1016/j.brs.2014.10.012.

Kroesbergen, E. H., Van Luit, J. E. H., Van Lieshout, E. C. D. M., Van Loosbroek, E., & Van De Rijt, B. A. M. (2009). Individual differences in early numeracy: The role of executive functions and subitizing. *Journal of Psychoeducational Assessment*, *27*(3), 226–236. https://doi.org/10.1177/0734282908330586.

Kucian, K., & von Aster, M. (2015). Developmental dyscalculia. *European Journal of Pediatrics*, *174*(1), 1–13. https://doi.org/10.1007/s00431-014-2455-7.

Kucian, K., von Aster, M., Loenneker, T., Dietrich, T., & Martin, E. (2008). Development of neural networks for exact and approximate calculation: A fMRI study. *Developmental Neuropsychology*, *33*(4), 447–473. https://doi.org/10.1080/87565640802101474.

Lee, K., & Bull, R. (2015). Developmental changes in working memory, updating, and math achievement. *Journal of Educational Psychology*, *108*(6), 869–882. https://doi.org/10.1037/edu0000090.

Lee, K., Ng, S. F., Pe, M. L., Ang, S. Y., Hasshim, M. N. A. M., & Bull, R. (2012). The cognitive underpinnings of emerging mathematical skills: Executive functioning, patterns, numeracy, and arithmetic. *British Journal of Educational Psychology*, *82*(1), 82–99. https://doi.org/10.1111/j.2044-8279.2010.02016.x.

Leibovich, T., & Ansari, D. (2016). The symbol-grounding problem in numerical cognition: A review of theory, evidence, and outstanding questions. *Canadian Journal of Experimental Psychology/Revue Canadienne de Psychologie Expérimentale*, *70*(1), 12–23. https://doi.org/10.1037/cep0000070.

Looi, C. Y., Duta, M., Brem, A.-K., Huber, S., Nuerk, H.-C., & Cohen Kadosh, R. (2016). Combining brain stimulation and video game to promote long-term transfer of learning and cognitive enhancement. *Scientific Reports*, *6*(1), 22003. https://doi.org/10.1038/srep22003.

Looi, C. Y., Lim, J., Sella, F., Lolliot, S., Duta, M., Avramenko, A. A., & Cohen Kadosh, R. (2017). Transcranial random noise stimulation and cognitive training to improve learning and cognition of the atypically developing brain: A pilot study. *Scientific Reports*, 7(1), 4633. https://doi.org/10.1038/s41598-017-04649-x.

Maslen, H., Earp, B. D., Cohen Kadosh, R., & Savulescu, J. (2014). Brain stimulation for treatment and enhancement in children: An ethical analysis. *Frontiers in Human Neuroscience*, *8*. https://doi.org/10.3389/fnhum.2014.00953.

Matejko, A. A., & Ansari, D. (2015). Drawing connections between white matter and numerical and mathematical cognition: A literature review. *Neuroscience & Biobehavioral Reviews*, *48*, 35–52. https://doi.org/10.1016/j.neubiorev.2014.11.006.

Merkley, R., & Ansari, D. (2016). Why numerical symbols count in the development of mathematical skills: Evidence from brain and behavior. *Current Opinion in Behavioral Sciences*, *10*, 14–20. https://doi.org/10.1016/J.COBEHA.2016.04.006.

Miyake, A., & Friedman, N. P. (2012). The nature and organization of individual differences in executive functions. *Current Directions in Psychological Science*, *21*(1), 8–14. https://doi.org/10.1177/0963721411429458.

Miyake, A., Friedman, N. P., Emerson, M. J., Witzki, A. H., Howerter, A., & Wager, T. D. (2000). The unity and diversity of executive functions and their contributions to complex "frontal lobe" tasks: A latent variable analysis. *Cognitive Psychology*, *41*(1), 49–100.

Müller-Dahlhaus, F., & Ziemann, U. (2015). Metaplasticity in human cortex. *The Neuroscientist*, *21*(2), 185–202. https://doi.org/10.1177/1073858414526645.

Mullis, I.V. S., Martin, M. O., & Loveless, T. (2016). *International trends in mathematics and science achievement, curriculum, and instruction.* TIMSS & PIRLS International Study Center.

Nieder, A. (2016). The neuronal code for number. *Nature Reviews Neuroscience*, *17*(6), 366. https://doi.org/10.1038/nrn.2016.40.

Nitsche, M. A., & Paulus, W. (2000). Excitability changes induced in the human motor cortex by weak transcranial direct current stimulation. *The Journal of Physiology*, *527*(Pt 3), 633–639. https://doi.org/10.1111/j.1469-7793.2000.t01-1-00633.x.

Palm, U., Segmiller, F. M., Epple, A. N., Freisleder, F.-J., Koutsouleris, N., Schulte-Körne, G., & Padberg, F. (2016). Transcranial direct current stimulation in children and adolescents: A comprehensive review. *Journal of Neural Transmission*, *123*(10), 1219–1234. https://doi.org/10.1007/s00702-016-1572-z.

Palmer, A., & Baroody, A. J. (2011). Blake's development of the number words "one," "two," and "three". *Cognition and Instruction*, *29*(3), 265–296. https://doi.org/10.1080/07370008.2011.583370.

Parsons, S., & Bynner, B. (2005). *Does numeracy matter more?*. National Research and Development Centre for Adult Literacy and Numeracy. https://dera.ioe.ac.uk/22317/1/doc_2979.pdf.

Pasqualotto, A. (2016). Transcranial random noise stimulation benefits arithmetic skills. *Neurobiology of Learning and Memory*, *133*, 7–12. https://doi.org/10.1016/j.nlm.2016.05.004.

Peng, P., Namkung, J., Barnes, M., & Sun, C. (2016). A meta-analysis of mathematics and working memory: Moderating effects of working memory domain, type of mathematics skill, and sample characteristics. *Journal of Educational Psychology*, *108*(4), 455–473. https://doi.org/10.1037/edu0000079.

Peters, L., & De Smedt, B. (2018). Arithmetic in the developing brain: A review of brain imaging studies. *Developmental Cognitive Neuroscience*, *30*, 265–279. https://doi.org/10.1016/j.dcn.2017.05.002.

Piazza, M., Pinel, P., Le Bihan, D., & Dehaene, S. (2007). A magnitude code common to numerosities and number symbols in human intraparietal cortex. *Neuron*, *53*(2), 293–305. https://doi.org/10.1016/j.neuron.2006.11.022.

Popescu, T., Krause, B., Terhune, D. B., Twose, O., Page, T., Humphreys, G., & Cohen Kadosh, R. (2016). Transcranial random noise stimulation mitigates increased difficulty in an arithmetic learning task. *Neuropsychologia*, *81*, 255–264. https://doi.org/10.1016/j.neuropsychologia.2015.12.028.

Poreisz, C., Boros, K., Antal, A., & Paulus, W. (2007). Safety aspects of transcranial direct current stimulation concerning healthy subjects and patients. *Brain Research Bulletin*, *72*(4–6), 208–214. https://doi.org/10.1016/j.brainresbull.2007.01.004.

Raghubar, K. P., & Barnes, M. A. (2016). Early numeracy skills in preschool-aged children: A review of neurocognitive findings and implications for assessment and intervention. *The Clinical Neuropsychologist*, *31*(2), 329–351. https://doi.org/10.1080/13854046.2016.1259387.

Raghubar, K. P., Barnes, M. A., & Hecht, S. A. (2010). Working memory and mathematics: A review of developmental, individual difference, and cognitive approaches. *Learning and Individual Differences*, *20*(2), 110–122. https://doi.org/10.1016/j.lindif.2009.10.005.

Rasmussen, C., & Bisanz, J. (2005). Representation and working memory in early arithmetic. *Journal of Experimental Child Psychology*, *91*(2), 137–157. https://doi.org/10.1016/J.JECP.2005.01.004.

Reis, J., Schambra, H. M., Cohen, L. G., Buch, E. R., Fritsch, B., Zarahn, E., … Krakauer, J. W. (2009). Noninvasive cortical stimulation enhances motor skill acquisition over multiple days through an effect on consolidation. *Proceedings of the National Academy of Sciences*, *106*(5), 1590–1595. https://doi.org/10.1073/pnas.0805413106.

Remedios, L., Mabil, P., Flores-Hernández, J., Torres-Ramírez, O., Huidobro, N., Castro, G., … Manjarrez, E. (2019). Effects of short-term random noise electrical stimulation on dissociated pyramidal neurons from the cerebral cortex. *Neuroscience*, *404*, 371–386. https://doi.org/10.1016/J.NEUROSCIENCE.2019.01.035.

Riggall, K., Forlini, C., Carter, A., Hall, W., Weier, M., Partridge, B., & Meinzer, M. (2015). Researchers' perspectives on scientific and ethical issues with transcranial direct current stimulation: An international survey. *Scientific Reports*, *5*(1), 10618. https://doi.org/10.1038/srep10618.

Ritchie, S. J., & Bates, T. C. (2013). Enduring links from childhood mathematics and reading achievement to adult socioeconomic status. *Psychological Science*, *24*(7), 1301–1308. https://doi.org/10.1177/0956797612466268.

Rivera, S. M., Reiss, A. L., Eckert, M. A., & Menon, V. (2005). Developmental changes in mental arithmetic: Evidence for increased functional specialization in the left inferior parietal cortex. *Cerebral Cortex*, *15*(11), 1779–1790. https://doi.org/10.1093/cercor/bhi055.

Rubinsten, O., & Henik, A. (2009). Developmental dyscalculia: Heterogeneity might not mean different mechanisms. *Trends in Cognitive Sciences*, *13*(2), 92–99. https://doi.org/10.1016/J.TICS.2008.11.002.

Ruffini, G., Wendling, F., Merlet, I., Molaee-Ardekani, B., Mekonnen, A., Salvador, R., … Miranda, P. C. (2013). Transcranial current brain stimulation (tCS): Models and technologies. *IEEE Transactions on Neural Systems and Rehabilitation Engineering*, *21*(3), 333–345. https://doi.org/10.1109/TNSRE.2012.2200046.

Rütsche, B., Hauser, T. U., Jäncke, L., & Grabner, R. H. (2015). When problem size matters: Differential effects of brain stimulation on arithmetic problem solving and neural oscillations. *PLoS One*, *10*(3). https://doi.org/10.1371/journal.pone.0120665, e0120665.

Sabel, B. A., Fedorov, A. B., Naue, N., Borrmann, A., Herrmann, C., & Gall, C. (2011). Non-invasive alternating current stimulation improves vision in optic neuropathy. *Restorative Neurology and Neuroscience*, *29*(6), 493–505. https://doi.org/10.3233/RNN-2011-0624.

Sánchez-León, C. A., Sánchez-López, A., Cordones, I., Gómez-Climent, M. A., Cohen Kadosh, R., & Márquez-Ruiz, J. (2020). Impact of chronic transcranial Random-Noise Stimulation (tRNS) on prefrontal cortex excitation-inhibition balance in juvenile mice. *bioRxiv*. https://doi.org/10.1101/2020.09.04.282889.

Sarkar, A., Dowker, A., & Cohen Kadosh, R. (2014). Cognitive enhancement or cognitive cost: Trait-specific outcomes of brain stimulation in the case of mathematics anxiety. *The Journal of Neuroscience: The Official Journal of the Society for Neuroscience*, *34*(50), 16605–16610. https://doi.org/10.1523/JNEUROSCI.3129-14.2014.

Sasanguie, D., De Smedt, B., Defever, E., & Reynvoet, B. (2012). Association between basic numerical abilities and mathematics achievement. *British Journal of Developmental Psychology*, *30*(2), 344–357. https://doi.org/10.1111/j.2044-835X.2011.02048.x.

Schneider, M., Beeres, K., Coban, L., Merz, S., Susan Schmidt, S., Stricker, J., & De Smedt, B. (2017). Associations of non-symbolic and symbolic numerical magnitude processing with mathematical competence: A meta-analysis. *Developmental Science*, *20*(3). https://doi.org/10.1111/desc.12372, e12372.

Schroeder, P. A., Dresler, T., Bahnmueller, J., Artemenko, C., Cohen Kadosh, R., & Nuerk, H.-C. (2017). Cognitive enhancement of numerical and arithmetic capabilities: A mini-review of available transcranial electric stimulation studies. *Journal of Cognitive Enhancement*, *1*(1), 39–47. https://doi.org/10.1007/s41465-016-0006-z.

Schwab, B. C., Misselhorn, J., & Engel, A. K. (2019). Modulation of large-scale cortical coupling by transcranial alternating current stimulation. *Brain Stimulation*, *12*(5), 1187–1196. https://doi.org/10.1101/484014.
Sheffield, J. G., Raz, G., Sella, F., & Cohen Kadosh, R. (2020). How can noise alter neurophysiology in order to improve human behaviour? A combined transcranial random noise stimulation and electroencephalography study. *BioRxiv*, *2020*. https://doi.org/10.1101/2020.01.09.900118, 900118.
Silvanto, J., & Pascual-Leone, A. (2012). Why the assessment of causality in brain–behavior relations requires brain stimulation. *Journal of Cognitive Neuroscience*, *24*(4), 775–777. https://doi.org/10.1162/jocn_a_00193.
Simonsmeier, B. A., Grabner, R. H., Hein, J., Krenz, U., & Schneider, M. (2018). Electrical brain stimulation (tES) improves learning more than performance: A meta-analysis. *Neuroscience & Biobehavioral Reviews*, *84*, 171–181. https://doi.org/10.1016/J.NEUBIOREV.2017.11.001.
Snowball, A., Tachtsidis, I., Popescu, T., Thompson, J., Delazer, M., Zamarian, L., … Cohen Kadosh, R. (2013). Long-term enhancement of brain function and cognition using cognitive training and brain stimulation. *Current Biology*, *23*(11), 987–992. https://doi.org/10.1016/j.cub.2013.04.045.
St Clair-Thompson, H. L., & Gathercole, S. E. (2006). Executive functions and achievements in school: Shifting, updating, inhibition, and working memory. *Quarterly Journal of Experimental Psychology*, *59*(4), 745–759. https://doi.org/10.1080/17470210500162854.
Stroop, J. R. (1992). Studies of interference in serial verbal reactions. *Journal of Experimental Psychology*, *121*(1), 15–23. https://doi.org/10.1037/0096-3445.121.1.15.
Tamnes, C. K., Østby, Y., Walhovd, K. B., Westlye, L. T., Due-Tønnessen, P., & Fjell, A. M. (2010). Neuroanatomical correlates of executive functions in children and adolescents: A magnetic resonance imaging (MRI) study of cortical thickness. *Neuropsychologia*, *48*(9), 2496–2508. https://doi.org/10.1016/J.NEUROPSYCHOLOGIA.2010.04.024.
Terney, D., Chaieb, L., Moliadze, V., Antal, A., & Paulus, W. (2008). Increasing human brain excitability by transcranial high-frequency random noise stimulation. *Journal of Neuroscience*, *28*(52). http://www.jneurosci.org.proxy.library.uu.nl/content/28/52/14147.short.
Toll, S. W. M., Kroesbergen, E. H., & Van Luit, J. E. H. (2016). Visual working memory and number sense: Testing the double deficit hypothesis in mathematics. *British Journal of Educational Psychology*, *86*(3), 429–445. https://doi.org/10.1111/bjep.12116.
Toll, S. W. M., Van der Ven, S. H. G., Kroesbergen, E. H., & Van Luit, J. E. H. (2011). Executive functions as predictors of math learning disabilities. *Journal of Learning Disabilities*, *44*(6), 521–532. https://doi.org/10.1177/0022219410387302.
Toll, S. W. M., Van Viersen, S., Kroesbergen, E. H., & Van Luit, J. E. H. (2015). The development of (non-)symbolic comparison skills throughout kindergarten and their relations with basic mathematical skills. *Learning and Individual Differences*, *38*, 10–17. https://doi.org/10.1016/J.LINDIF.2014.12.006.
Uhlhaas, P. J., Roux, F., Rodriguez, E., Rotarska-Jagiela, A., & Singer, W. (2010). Neural synchrony and the development of cortical networks. *Trends in Cognitive Sciences*, *14*(2), 72–80. https://doi.org/10.1016/j.tics.2009.12.002.
Van Aken, L., Kessels, R. P. C., Wingbermühle, E., Van Der Veld, W. M., & Egger, J. I. M. (2015). Fluid intelligence and executive functioning more alike than different? *Acta Neuropsychiatrica*, *28*(1), 31–37. https://doi.org/10.1017/neu.2015.46.
Van Bueren, N. E. R., Francesco, S., Sheffield, J. G., Kroesbergen, E. H., & Cohen Kadosh, R. (2021). *Suboptimal behavior and neurophysiology determines the benefit of noise on cognitive learning*. In preparation.
Van de Weijer-Bergsma, E., Kroesbergen, E. H., Prast, E. J., & Van Luit, J. E. H. (2015). Validity and reliability of an online visual–spatial working memory task for self-reliant administration in school-aged children. *Behavior Research Methods*, *47*(3), 708–719. https://doi.org/10.3758/s13428-014-0469-8.

Van der Ven, S. H. G., Kroesbergen, E. H., Boom, J., & Leseman, P. P. M. (2012). The development of executive functions and early mathematics: A dynamic relationship. *British Journal of Educational Psychology*, *82*(1), 100–119. https://doi.org/10.1111/j.2044-8279.2011.02035.x.

Van 't Veer, A. E., & Giner-Sorolla, R. (2016). Pre-registration in social psychology—A discussion and suggested template. *Journal of Experimental Social Psychology*, *67*, 2–12. https://doi.org/10.1016/j.jesp.2016.03.004.

Vogel, S. E., Goffin, C., & Ansari, D. (2015). Developmental specialization of the left parietal cortex for the semantic representation of Arabic numerals: An fMR-adaptation study. *Developmental Cognitive Neuroscience*, *12*, 61–73. https://doi.org/10.1016/J.DCN.2014.12.001.

Wilson, K. M., & Swanson, H. L. (2001). Are mathematics disabilities due to a domain-general or a domain-specific working memory deficit? *Journal of Learning Disabilities*, *34*(3), 237–248. https://doi.org/10.1177/002221940103400304.

Wu, Y.-J., Tseng, P., Chang, C.-F., Pai, M.-C., Hsu, K.-S., Lin, C.-C., & Juan, C.-H. (2014). Modulating the interference effect on spatial working memory by applying transcranial direct current stimulation over the right dorsolateral prefrontal cortex. *Brain and Cognition*, *91*, 87–94. https://doi.org/10.1016/J.BANDC.2014.09.002.

Zamarian, L., Ischebeck, A., & Delazer, M. (2009). Neuroscience of learning arithmetic—Evidence from brain imaging studies. *Neuroscience & Biobehavioral Reviews*, *33*(6), 909–925. https://doi.org/10.1016/J.NEUBIOREV.2009.03.005.

CHAPTER 12

# Intervention studies in math: A metareview

**Bert Reynvoet[a], Stefanie Vanbecelaere[b,c], Fien Depaepe[b,c], and Delphine Sasanguie[d]**
[a]KU Leuven, Faculty of Psychology and Educational Sciences, Brain & Cognition, Leuven, Belgium
[b]KU Leuven, Faculty of Psychology and Educational Sciences, Centre for Instructional Psychology and Technology, Leuven, Belgium
[c]KU Leuven, IMEC Research Group, ITEC, Leuven, Belgium
[d]HOGENT, Research Centre for Learning in Diversity, Ghent, Belgium

## Introduction

In this contribution, we aim to provide a comprehensive overview of interventions targeting early numeracy or more advanced mathematics, focusing on both typical and atypical (or at risk) achievers in preschool, primary, or secondary education. Math interventions differ in format and intensiveness. With respect to *format*, Dowker (2017) already highlighted that in addition to individual and small-group interventions, the last couple of years, there has been an increasing interest in whole-class approaches, mainly because several countries have become increasingly concerned with improving numeracy standards—and it is believed that whole-class approaches can offer solace in this regard. In fact, this tendency corresponds with the augmenting application of the "Response to Instruction" (or "Response to Intervention," RTI) approach, first in the United States (e.g., Fuchs, Fuchs, & Prentice, 2004), and later also in Europe, like in The Netherlands (e.g., Schölvinck & Jansen, 2014) and in Belgium (Prodia, 2016). The RTI approach in turn refers to the *differences in intensiveness* of interventions that can be noticed among math intervention studies (see also the other chapters in this book). RTI is an approach that aims to prevent that math failure occurs via early identification and screening, early intervention, frequent monitoring, and applying increasingly intensive evidence-based interventions to those learners who keep having difficulties with the math learning content. More specific, RTI provides math interventions to children with different levels of intensiveness, dependent on their response to these interventions. The progress of the children is monitored and the interventions are adjusted accordingly. Concrete, the

*Heterogeneous Contributions to Numerical Cognition*
https://doi.org/10.1016/B978-0-12-817414-2.00012-9

RTI process is usually described in three categories of interventions, described as "waves of support" in the United Kingdom, "tiers of support" in the United States (Dowker, 2017), and "rounds of support" in Belgium (Prodia, 2016). With each "tier of support" (from here onward, the term "tier" will be used), the interventions become more intensive. Tier 1 refers to "thoughtful math instruction" or whole-class teaching, designed to be suitable for and stimulating engagement of all children. Tier 1 interventions are research-based practices during the core instructional time with their teacher. The aim is that approximately 80%–85% of the learners make progress in math without further help after Tier 1. Tier 2 involves interventions in small groups, specifically targeting mild-to-moderate math difficulties. After Tier 2, the aim is that only 3%–6% of the learners keep on experiencing (severe) math difficulties. Those children will then receive a final, third tier of support (Tier 3), which is the most intensive (often one-on-one) intervention in regular education (see Fig. 1).

Based on this introduction, it becomes clear that the RTI approach is widely supported by researchers and practitioners to foster number and math and that interventions are very diverse in terms of format and intensiveness.

## Exponential increase of math interventions: Some facts and numbers

To have an idea about the number of intervention studies targeting mathematics learning, the following search string was inserted in the electronic database Web of Science (all collection): (intervention OR training) AND (math OR early numeracy) AND (kindergarten OR preschool OR "primary school" OR "secondary school"). This search resulted in 544 hits (date of search: October 28, 2019) with an exponential increase of math

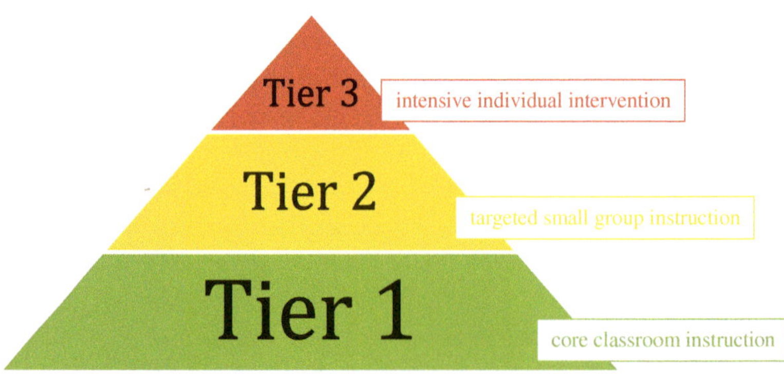


**Fig. 1** The three waves/tiers/rounds of support.

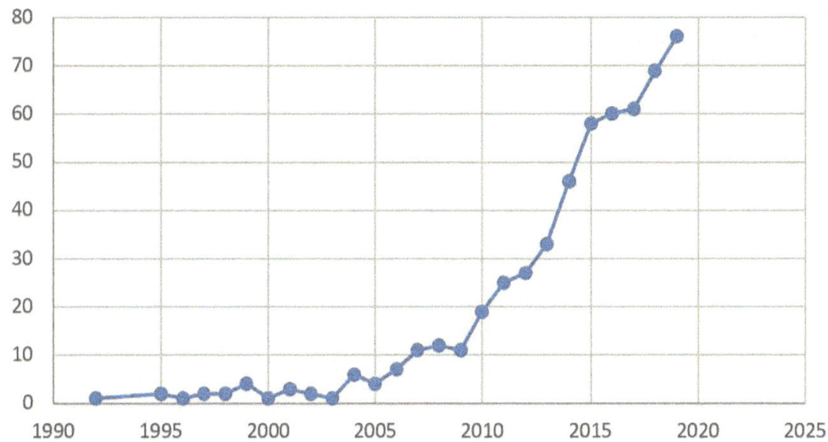


**Fig. 2** Number of math intervention studies.

interventions in recent years (Fig. 2). Most studies were conducted in the area of educational research ($N=252$), followed by developmental psychology ($N=143$), educational psychology ($N=112$), special education ($N=47$), and multidisciplinary psychology ($N=33$). These numbers reveal that math interventions are often investigated from both an educational and psychological perspective.

As it is hard to derive general conclusions from one single study, a lot of researchers have synthesized the results of multiple intervention studies. The importance of these syntheses has been emphasized before: for instance, Fischer, Moeller, Cress, and Nuerk (2013) mention that "it is essential to compare intervention studies with one another to be able to choose the most appropriate treatment for a child. To this end, research syntheses such as systematic reviews and metaanalyses can provide such an overview" (p. 90). Systematic reviews and metaanalyses are two well-known methodologies to accomplish this; the first is a systematic and explicit methodology to identify, select, and critically evaluate the studies included (Bettany-Saltikov, 2010a), while the latter is a statistical tool for estimating the mean and variance of underlying population effects from a collection of empirical studies addressing ostensibly the same research question (Field & Gillett, 2010). Although these are the best known methodologies, also other types of reviews (e.g., narrative review) are used but less frequently (Grant & Booth, 2009). As in this contribution we are interested in enlarging our understanding about interventions in general, we searched the literature to identify studies that **synthesized intervention studies** by now inserting

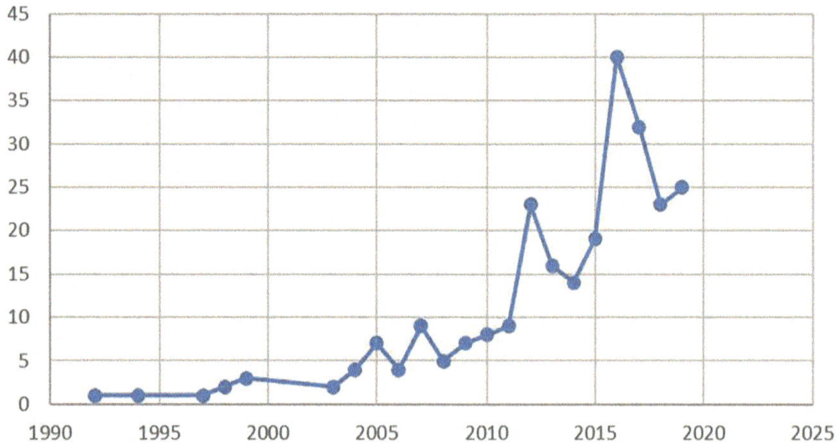


**Fig. 3** Number of systematic reviews/metaanalyses on the effectiveness of math interventions.

the following string in Web of Science: (review OR metaanalysis) AND (intervention OR training) AND (math OR "early numeracy"). This search revealed 169 hits (date: August 26, 2019). The results show a strong increase of hits recently, indicating that the number of conducted systematic reviews and metaanalyses on mathematics interventions over the past years has increased exponentially (Fig. 3).

This large number of hits on systematic reviews and metaanalyses based on one single search in one single electronic database is striking and motivated us to conduct a systematic review of reviews and metaanalyses, a so-called metareview (Blackwood, 2016). Indeed, we believe that a synthesis of the findings of previous systematic reviews and metaanalyses can shed additional light on the understanding of the effectiveness of math interventions in general. This way, on the one hand, we hope to provide practitioners with evidence-based guidelines that can be applied in the classroom. On the other hand, we aim to inform researchers about the current state-of-the-art, gaps in the literature, and challenges for future research.

## Methodology

### Metareview

A metareview, which is a review of the results described in previously published systematic reviews and metaanalyses, was conducted. Given the multiplicity and heterogeneity with respect to samples, instructional methods, and content among the previous reviews and metaanalyses on this topic, we were particularly interested in extracting the general, recurring conclusions on the effectiveness of math interventions. We conducted the metareview following the guidelines to conduct a review of Bettany-Saltikov (2010a, 2010b). Below, we elaborate on each conducted step.

The metareview started with a nonsystematic review to get familiar with the specific focuses of reviews and metaanalyses investigating intervention studies in math. This review and search indicated that a lot of reviews and metaanalyses were conducted in the past years. Among these reviews and metaanalyses, we found very diverse research questions with respect to sample characteristics, instructional methods, and content that was trained. This initial search was important to determine the search strategy.

In a next step, the overarching review question was defined: "What is the effectiveness of intervention studies in math?." Based on the nonsystematic review, we identified important terms in the field which we used to explore different search strings. We chose for the following search string (see also introduction): (review OR metaanalysis) AND (intervention OR training) AND (math OR "early numeracy"). Next, inclusion and exclusion criteria were described. Studies were *included* when (1) the study dealt with the effectiveness of interventions aiming to foster numeracy or mathematics; (2) effects on children's learning were examined; (3) the sample consisted of normal achievers and/or learners with difficulties/disorders; (4) the participants were in preschool, primary, or secondary education; and (5) the study was a review or metaanalysis. Studies were *excluded* when the study (1) did not address the effectivity of an intervention (e.g., theoretical papers), (2) did not consider the effects on children or students (but on teachers for example), (3) did not concern a math intervention, (4) took place in informal settings (e.g., home numeracy interventions), (5) examined the effect of specific characteristics of an intervention (e.g., feedback), (6) did not concern a review or a metaanalysis, and (7) concerned STEM or general academic performance instead of pure math.

Next, the search string was entered in the electronic database Web of Science (all collection, date: August 26, 2019). We chose this database because of its importance in the field of educational and psychological research. The search resulted in 169 records. All titles, abstracts, and references were exported to a word document and were systematically screened by the second author. Each article was subjected to the inclusion and exclusion criteria. All articles were coded: (a) definitely included, (b) maybe included, and (c) definitely rejected. Full texts were checked in case the article received tag b. In case of doubt, the first two authors discussed the issues until full agreement was achieved. Finally, 29 articles were selected for inclusion. In addition, reference lists and citations of these papers were checked which resulted in 12 additional articles. In total, 41 studies were included for further investigation (see Appendix).

## Description of the analyses

The full texts of each included article were read and basic information was extracted from each study: authors, year of publication, type of study (review or metaanalysis), age of participants, trained content, sample characteristics (normal achiever, difficulties), and instructional method (see Appendix). This information was extracted by either the first or second author, and in case of doubts, the first and second author discussed the issue to arrive to an agreement.

In addition, the results (quantitative/qualitative arguments/conclusions) of each study were summarized. Based on these summaries, the authors held several meetings to discuss the results in order to identify different themes the results referred to. In the end, the following five themes were derived:

**(1)** *when* ("How effective are intervention studies in math for participants in a particular age category?");

**(2)** *what and how long* ("How effective are intervention studies in math addressing particular math content?" and/or "Does the effectivity of the intervention study in math depend on the length of the intervention?");

**(3)** *how* ("How effective are intervention studies in math making use of a particular instructional method?");

**(4)** *who* ("How effective are intervention studies in math for a specific group of participants?"); and.

**(5)** *quality* ("How is the effectiveness of the intervention studies in math related to the quality of the design in the study?").

The first two authors independently coded the results of each study in terms of the five categories. This means that all quantitative/qualitative arguments/conclusions were labeled to one of the five categories (i.e., when, how, what, who, quality). To report the results, within each category, we searched for similarities, differences, and ambivalence among the results of the different studies. It is important to mention that these identified categories are not completely independent factors: for instance, the targeted age (= when) will evidently also define its content (= what). In several sections, we will also highlight their interrelatedness.

# Results

## Descriptive statistics

The characteristics of the included studies are presented in Appendix. First, the studies were ordered by publication date. We observed that 35 out of 41 studies were conducted since 2010.

The second column describes the methodology that was used. Each study was labeled as a metaanalysis, systematic review, or another type of synthesis (e.g., a narrative review). Almost half of the studies ($n=18$) followed a quantitative approach by conducting a metaanalysis. Another large part of the studies conducted a systematic review ($n=19$). The remaining studies consisted of a critical analysis ($n=1$), a review (but not a systematic one) ($n=2$), and a narrative review ($n=1$).

Third, the table provides information about the age category that was targeted in the study, either preschool children, elementary school, secondary school, or not specified. It is apparent from the table that half of the studies ($n=20$) did not target a specific age category and wanted to generalize over different ages (from kindergarten to grade 12, often referred to as K-12 education), while other studies specifically chose to report the results of studies targeting a specific group: preschool ($n=3$), elementary school ($n=8$), or secondary school ($n=2$). Some studies investigated intervention effects on preschool and elementary school children ($n=5$) or on elementary and secondary school children ($n=3$).

In the fourth column, studies are labeled by the content that was targeted in the math intervention. We opted for the following categories: early numeracy skills, not specified, and specific content. Most studies ($n=27$) did not focus on one particular aspect of math (=not specified). Another large part of the studies focused on specific content, such as word problem solving ($n=4$), computational skills ($n=3$), fractions ($n=2$), or executive functions ($n=1$). Four studies focused on studies training early numeracy skills (such as counting, digit comparison, number line estimation).

Appendix also contains the sample characteristics per study. This fifth column dissociates between studies that did not specify the sample (i.e., the review or metaanalysis dealt with different samples), studies that specifically focused on at-risk learners (which have difficulties with math or have a learning disorder), or studies that included another specific group of learners (for instance, children with an autism spectrum syndrome (ASS) or emotional disorders). Generally, Appendix shows that many studies focused on children-at-risk, learners with math difficulties or disorders (e.g., dyscalculia) ($n=21$).

Furthermore, a lot of studies ($n = 13$) did not specify the sample characteristics of the studies they included, suggesting that possibly both typical achievers and learners with difficulties or other problems might have been included. Four reviews examined only studies that solely included children with ASS, two reviews focused solely on children with emotional disorders, and one review included only minority students. In general, these latter studies focusing were reviews and metaanalyses of single-case studies, whereas reviews and metaanalyses which do not specify the sample often are based on group comparison design studies.

Finally, a sixth column provides information about the instructional methods under investigation in each study. Most studies did not target a specific instructional method ($n = 29$) and compared a range of instructional techniques. Some studies ($n = 7$) included only studies that fostered math using digital technologies ranging from technology-based instruction to digital game-based learning. The last category "other" consisted of studies that targeted studies who used a specific instructional method (e.g., peer and cross age tutoring, cognitive and metacognitive instruction).

## Results of the metaanalyses and reviews

Below an overview is given of the results of the metaanalyses and reviews included in this metareview. The results are reported for each category described in the section "Description of the analyses" (i.e., when, how, who, what and quality).

### *When?*

A first question that we wanted to resolve with our metareview is *at what age* are interventions most effective? Our systematic analysis made very clear that interventions are set up for children with different ages and that in all these age categories positive effects are reported (see also Dowker, 2017 for a similar conclusion). Although there is a tendency that less mathematical interventions are administered in secondary school, the metaanalysis of Stevens, Rodgers, and Powell (2018) investigating the effectiveness of interventions in children at-risk or with mathematical problems showed that grade did not significantly moderated the intervention effectiveness. This observation was confirmed in another metaanalysis a few years earlier by Chodura, Kuhn, and Holling (2015). Also, for digital-based interventions, a similar conclusion was reached: a systematic review of Kiru, Doabler, Sorrells, and Cooc (2018) concluded that interventions are effective at all ages throughout primary school. Cheung and Slavin (2013) took age into account as a possible moderator in their metaanalysis on the effectiveness of technology-based interventions, but they also found comparable size effects for both primary and secondary school interventions.

In contrast with interventions targeting children at risk for dyscalculia which mostly focus on primary school, King, Lemons, and Davidson (2016) concluded that math interventions for children with autism are as common in primary as in secondary school. Moreover, 71% of these interventions were effective and no clear differences were reported between interventions targeting primary or secondary school children (King et al., 2016). Whereas several metaanalyses and reviews do not report an effect of age on the effectiveness of interventions, some do report such differences. For instance, Nelson and McMaster (2018) found in their metaanalysis of early numeracy interventions that interventions in younger children were more effective. This conclusion was confirmed in a metaanalysis on interventions targeting at-risk children from Haberstroh and Schulte-Körne (2019). Also, Methe, Kilgus, Neiman, and Riley-Tillman (2012) reached this conclusion in their metaanalysis on interventions training of addition and subtraction skills in primary school. However, an opposite conclusion was suggested by Dennis et al. (2016): in their metaanalysis, larger effect sizes were observed for interventions conducted in primary school compared to interventions in preschool children.

From these reviews and metaanalyses, it is clear that the effect of age on the effectiveness of interventions is nonconsistent, and if there is one, it is most likely very small. Yet, a frequently heard claim in several of the studies in our metareview was that interventions should be implemented at an early age in typical developing children and children at risk in order to boost numeracy (e.g., Cohen Kadosh, Dowker, Heine, Kaufmann, & Kucian, 2013; Day-Hess & Clements, 2017; Haberstroh & Schulte-Körne, 2019). In part, this claim is also motivated by the fact that boosting early numeracy can have secondary advantages such as the increased acceptance of math (Day-Hess & Clements, 2017) and less mathematics anxiety (Dowker, 2017).

### What and how long?

A second question we wanted to answer is *what content* is most effective for interventions and *how long* should an intervention last? As for *content*, our metareview made clear that the content targeted in interventions is very diverse, ranging from early numeracy (e.g., Nelson & McMaster, 2018), over arithmetic (e.g., Stocker Jr & Kubina Jr, 2017), fractions (e.g., Roesslein & Codding, 2019), and word problem solving (e.g., Hwang & Riccomini, 2016) to functional content, e.g., learning to work with money (Barnett & Cleary, 2015). Most importantly, all these interventions addressing different contents were found to be effective.

Some reviews and metaanalyses on numerical interventions reported differences in intervention effectivity depending on the specific content. For instance, in a metaanalysis on interventions on early numeracy, Nelson and McMaster (2018) observed that the presence of math vocabulary instruction (e.g., including ordinal and position words such as under, above, before, and after) in an early intervention results in a higher effect size. In a systematic review, De Smedt, Noël, Gilmore, and Ansari (2013) suggested that early interventions should primarily target symbolic number skills (and not nonsymbolic number skills). For interventions targeting more advanced numerical skills, Stevens et al. (2018) noticed that interventions with fractions are more effective than interventions in which the focus is on basic math skills (i.e., addition, multiplication). In contrast with Stevens et al. (2018), Kroesbergen and Van Luit (2003) found that interventions targeting basic math skills have the largest effect size. In a metaanalysis of Higgins, Huscroft-D'Angelo, and Crawford (2019) on the effects of technology in math education, it was found that studies targeting numbers and operations, geometry or multiple content areas showed a stronger influence on achievement than studies that addressed algebra. Barnett and Cleary (2015) concluded that in children with ASS, often low-level mathematical content is effectively trained and that studies investigating the effects of training of higher level mathematical content such as problem solving are largely missing. Finally, in a metaanalysis of numerical interventions in children with emotional disorders, Losinski, Ennis, Sanders, and Nelson (2019) claimed that the largest effects were found for early numeracy interventions, followed by fraction and word problem interventions.

Another recurring theme in metaanalyses and reviews is the question whether interventions should be focused one particular skill or should rather consist of a general and broad selection of different numerical skills. All included reviews and metaanalyses that addressed this question suggested that preference should be given to an intervention that tackles the specific problem area (e.g., Dowker, 2017, Fischer et al., 2013); Haberstroh & Schulte-Körne, 2019; Kroeger, Brown, & O'Brien, 2012). A related issue is whether interventions should focus on domain-general skills such as memory or executive functions. In their reviews, both Raghubar and Barnes (2017) and Dowker (2017) discussed the effectivity of interventions aiming at improving domain-general skills and concluded that there is currently very little evidence for a positive effect on mathematics from such trainings. Raghubar and Barnes (2017) came to the same conclusion when language skills were addressed in the interventions with the aim to improve numerical skills. In another review, Day-Hess and Clements (2017) do emphasize the importance of executive functions in early math skills. However, their suggestion is not to train domain-general executive skills, but to train those early math skills that also heavily rely on executive functioning with the aim to improve both math skills and executive functioning.

Inconsistent claims are mentioned in the reviewed papers with respect to *how long* an intervention should last. Based on the results of a metaanalysis of interventions in primary school children at risk or with mathematical problems, Chodura et al. (2015) concluded that there was no effect of intervention intensity. This was confirmed in a systematic review of early numeracy interventions in at-risk children by Mononen, Aunio, Koponen, and Aro (2014). In some of the reviews and metaanalyses, a plea is made for longer, more intensive interventions (Dowker, 2017; Haberstroh & Schulte-Körne, 2019; Xin & Jitendra, 1999). However, other studies observed that longer interventions in fact yielded smaller effect sizes (Nelson & McMaster, 2018). Also, Kroesbergen and Van Luit (2003) found that the duration of the interventions in children at risk negatively correlated with effect size. It should be noted though that conclusions with respect to the duration are possibly biased by the content of the intervention that was discussed before. Interventions addressing a broad range of skills evidently last longer than more focused interventions. We already mentioned that multicomponential interventions are less effective than focused interventions, which could also explain the counterintuitive finding that shorter, but focused interventions are more effective (Fischer et al., 2013).

### How?

A third section of this metareview deals with the question *which instructional methods are the most effective?* Analyzing the reviews and metaanalyses made clear that different instructional methods have been examined and proven to be effective: drill and practice, direct instruction, self-management techniques, peer tutoring, and technology-assisted instruction. For instance, in a systematic review on early numeracy interventions, Mononen et al. (2014) concluded that computer-assisted instruction, direct instruction, gameplay, and the use of "concrete → representational → abstract" (CRA) instructions all led to improvement in math performance. Also, Chodura et al. (2015) concluded that all instructional types lead to positive effects. Finally, on the basis of a metaanalysis of word problem interventions in children with emotional disorders, Losinski et al. (2019) concluded that different types of instructions (e.g., heuristics, drill and practice, direct instruction, and peer-assisted learning) were effective.

Some of the studies were set up to directly compare the effect sizes of different instructional methods. For instance, for interventions targeting arithmetic problems in primary school children at risk for mathematical problems, Methe et al. (2012) found that fluency-based interventions (i.e., drill and practice) are more effective than acquisition-based interventions, which aim to enhance performance through explanation, modeling, and feedback (i.e., concrete → representational → abstract). A similar observation emerged from the metaanalysis on arithmetic intervention in primary school children by Codding, Burns, and Lukito (2011). They compared four different instructional methods (practice with modeling, practice without modeling, drill, and self-management) and did find the strongest effect sizes for drill and practice with modeling, whereas self-management techniques only yielded moderate effects. In a metaanalysis considering a broader range of interventions in primary school children (i.e., not only interventions targeting arithmetic but also other skills such as algebra), Gersten et al. (2009) observed that most instructions (e.g., explicit instruction, use of heuristics, student verbalizations of their math reasoning, using visual representations) resulted in significant effect sizes. Only two components failed to reveal a significant effect: goal-setting strategies and peer-assisted learning. When all these components were entered in a regression analysis to find out which components lead to unique contributions, only teaching of heuristics and the use of direct/explicit instruction turned out to have a positive effect on intervention outcome. The effectivity of direct instruction as an instructional principle can also be found in many other studies that were included in our metareview. For instance, Kroesbergen and Van Luit (2003) observed that direct instruction appears to be more effective than other instructional approach (i.e., mediated/assisted instruction and realistic math education). In a very recent metaanalysis, Stevens et al. (2018) observed that explicit instruction is indeed the most frequently used instruction principle for interventions in at-risk elementary school children. The integration of explicit instruction into digital-based learning environments is also reported to lead to higher effect sizes (Kiru et al., 2018; Shin & Bryant, 2015). Finally, direct instruction as a principle is also advised for interventions targeting children with autism, although behavioral strategies such as visual prompting and self-management are also effective in this sample (Hughes & Yakubova, 2019).

The emergence of digital technologies has offered many new opportunities to train numeracy and mathematics. In the past decade, particular features such as interactivity, immediate feedback, adaptivity, and choice have been identified as powerful characteristics of digital technologies (National Academies of Sciences Engineering and Medicine, 2018). This also resulted in several metaanalyses and reviews focusing exclusively on the effectiveness of digital (game)-based learning/interventions (e.g., Byun & Joung, 2018; Chauhan, 2017; Cheung & Slavin, 2013; Hughes & Yakubova, 2019; Kiru et al., 2018; Lämsä, Hämäläinen, Aro, Koskimaa, & Äyrämö, 2018; Higgins et al., 2019). Early metaanalyses failed to demonstrate positive effects of digital-based interventions. For instance, in their metaanalysis on interventions in special need education, Kroesbergen and Van Luit (2003) concluded that computer-assisted interventions produced lower effect sizes than traditional approaches. This observation was echoed by the metaanalysis of Cheung and Slavin (2013) a few years later, who found that computer-assisted interventions generally produced positive, though modest effects. The possibilities of technology have increased enormously recently, and a more positive picture of digital interventions is provided by recent studies. For instance, Chodura et al. (2015) reported that digital interventions were just as effective as nondigital interventions. Also, Byun and Joung (2018) and Kiru et al. (2018) observed an overall positive effect size for digital-based interventions in children at-risk and typically developing children (although they did not compare with nondigital interventions). Recently, Higgins et al. (2019) investigated the effects of digital-based math interventions on both cognitive outcomes and learners' motivation and attitudes toward math. The results yielded a significant overall positive impact on math achievement and a moderate effect of using technology on motivation and attitudes. This effect was moderated by treatment type (i.e., studies using a traditional control group showed less effects compared to studies that employed multiple technologies as treatments or a one group pretest-posttest design), targeted content area, and intervention duration.

Here too, it is clear that instructional type cannot be considered as an intervention feature independent from the other characteristics. First, a rather evident observation is that instructions depend on content ("what"). Drill and practice is a very efficient instruction in interventions targeting simple arithmetic, whereas teaching heuristics is more popular in interventions targeting word problem solving. Second, there are indications that instructions should differ depending on the targeted sample ("who"). For instance, in a discussion on the effectivity of peer tutoring, Dowker (2017) suggested that this particular form of instruction may be effective in children with a mild retardation of mathematical skills, but not in children from special needs education. Moreover, the instructional principles for children at risk for dyscalculia cannot be copy-pasted to interventions that target the mathematical problems of children with autism or other emotional disorders (see Gevarter et al., 2016; Hughes & Yakubova, 2019; King et al., 2016 for reviews or metaanalyses on mathematical interventions in these samples).

Finally, in several reviews and metaanalyses included in this metareview, advice was found on who should be involved in the interventions that are set up. Kroesbergen and Van Luit (2003) see a very important role for the teacher, but also the involvement of trained specialists (Haberstroh & Schulte-Körne, 2019) and researchers (Dennis et al., 2016) has been mentioned as a requirement for effective interventions. The importance of the home environment was stressed in the review of Day-Hess and Clements (2017) and in the metaanalysis of Chauhan (2017). For instance, in their study examining the effectiveness of digital-based interventions, Chauhan (2017) observed a higher effect size for informal interventions in the home environment than for formal interventions in a school context. These observations make clear that different actors (teachers, researchers, parents) can contribute to the success of interventions, an observation that nicely fits the argumentation of Codding et al. (2011) who claimed that interventions in which different partners collaborate are the most effective.

### Who?

Our final question is in *which sample* interventions are most effective? As can be seen in Table A.1, the metareview contained studies that covered all tiers of the RTI framework that were described in the beginning of this chapter: studies reviewing interventions for normal achievers (TIER 1), interventions for children at risk or children with difficulties (TIER 2), and interventions for children with severe learning disorders or emotional disorders (TIER 3). The general conclusion is that all these interventions are effective. Some reviews and metaanalyses have compared interventions in different samples but consistent results remained absent. For instance, in a metaanalysis on interventions targeting early numeracy skills, Nelson and McMaster (2018) found higher effect sizes for interventions in children with some risk for later math difficulties when compared to children with no risk. However, interventions in students with high risks for math difficulty and with a low socio-economic status were less effective when compared to interventions in children with no risk. Similarly, Kroesbergen and Van Luit (2003) concluded that interventions are more effective in children with a mild retardation in math than in children with more severe learning difficulties (see also Xin & Jitendra, 1999). In contrast with these findings, Dennis et al. (2016) observed better intervention outcomes in children with greater difficulties in mathematics. Finally, Chodura et al. (2015) observed no differences in effectiveness between children at risk for mathematical problems and children with severe learning difficulties.

### *Quality of math interventions*

A reoccurring commentary in the metaanalyses and reviews on numerical interventions is the observation that there is a very large heterogeneity between the effect sizes of the studies (e.g., Baker, Gersten, & Lee, 2002; Stevens et al., 2018). Although a substantial part of this heterogeneity can be explained by the above factors (when, what, how, who), methodological issues also have contributed to the diverse outcomes of intervention studies. It is commonly recognized that (math) intervention studies have a lot of methodological problems (Dennis et al., 2016; King et al., 2016; Losinski et al., 2019; Mulcahy, Krezmien, & Travers, 2016; Nelson & McMaster, 2018). In what follows, we have categorized the methodological problems we encountered into measurement issues, design issues, and replication issues.

First, with respect to *measurement issues*, several metaanalyses and systematic reviews have indicated that the outcome measure used to evaluate the effectiveness of interventions seriously moderates the effect size. For instance, in their metaanalysis on early numeracy interventions, Nelson and McMaster (2018) observed larger effect sizes in studies using proximal measures (i.e., outcome measures closely related to what is being trained, also in some papers referred to as "near transfer") than in studies using norm-referenced measures, which are broader and tap more into far transfer. Similar observations and claims can be found in the studies by Roesslein and Codding (2019), Gersten et al. (2009), and Mononen et al. (2014). Moreover, several authors argue that it is currently unclear whether numerical interventions result in any far transfer whatsoever, like improvements on standardized mathematic tests (Stevens et al., 2018) or to real-world problems (Hwang & Riccomini, 2016). Another underinvestigated aspect in evaluating the effectivity of interventions is the presence of sustainable effects (Barnett & Cleary, 2015). For instance, Nelson and McMaster (2018) mentioned that only 10 studies (out of 34) examined such long-lasting effects. Evidently, the effect size of these delayed assessments is much smaller than the effect size of measures obtained directly after the intervention and as such contributes to the heterogenic pattern of results. Finally, the meta-analysis of Higgins et al. (2019) is the only study in our metareview that investigated the effect of numerical interventions on motivation and attitudes toward math, which could indirectly have a positive effect on children's math learning outcomes (see also Dowker, 2017). According to these authors, future intervention work should also include noncognitive measures such as self-efficacy, student engagement, motivation, or attitudes.

Second, several studies in this metareview also mentioned problems with several aspects of the *design* of intervention studies. A first problem that was echoed in several studies was the operationalization of a control condition, needed to evaluate the effect of an intervention (e.g., Szűcs & Myers, 2017). Some interventions have opted for a passive control condition, in which students continue with their business (in class) as usual. The progress (if any) on one or more mathematic outcomes made by these children is then compared with the progress made by children who participated in an intervention. Other researchers have compared the outcomes of an intervention with the outcomes of children in an active control condition. Such an active control condition is operationalized in many different ways, ranging from a control condition in which children are trained on non-related skills (e.g., a reading exercise), in which children are trained on the same skill but with another method (e.g., pen-and-paper vs. digital based) or on another numerical skill. It is evident that the specificity of the control condition has a serious impact on the effect size. As noted by Fischer et al. (2013), studies that compared the experimental (= intervention) group with an active control group produced smaller effect sizes. This was echoed in a systematic review of technology-based interventions by Lämsä et al. (2018), who concluded that the operationalization of the control condition has a huge impact on effect sizes. In addition to the type of activities in the control condition, also the composition of the control group can affect the effect size. For instance, Fischer et al. (2013) mentioned that studies that do not make use of a performance-matched control group, reported in general larger effect sizes.

A third problem that was mentioned in several studies is the need for replication. Both Stevens et al. (2018) and Szücs and Myers (2017) recognized the problem that a lot of interventions have very small sample sizes and, as a consequence, are low-powered studies. More problematic is that a metaanalysis on technology-based interventions by Cheung and Slavin (2013) showed that studies with small sample sizes produce, on average, twice the effect sizes of those with large sample sizes. The latter observation suggests that there is a large publication bias for effective numerical interventions, a problem also mentioned by Barnett and Cleary (2015) and Xin and Jitendra (1999). Publication bias is not only a problem for interventions targeting number skills in kindergartners or children at risk (Tier 1 and 2) but also in Tier 3 interventions targeting children with severe mathematical difficulties. For instance, Losinski et al. (2019) explicitly mentioned the need for replications of numerical interventions in children with emotional disorders.

Concluding, several measurement, design, and replication issues have been described, all having an enormous impact on the reported effect size of numerical interventions. All these methodological problems also probably have led to one of the major findings of the metaanalysis by Kroesbergen and Van Luit (2003) who concluded that the most important predictor of effect sizes of numerical interventions is the quality of the research design (see also Methe et al., 2012 for a similar conclusion).

## Conclusion

This metareview aimed to synthesize metaanalyses and reviews on math interventions. The synthesis suggested four questions that are regularly addressed in metaanalyses and reviews on four questions: at what age interventions should be delivered (*when*)? Which skills should be trained and for how long (*what*)? What is the type of instruction that is recommended (*how*)? Who should be targeted in interventions (*who*)? By doing so, we wanted to derive some general conclusions for practitioners and researchers about the current state-of-the-art that should also inform future intervention work.

The metareview made clear that there is a large heterogeneity in the focus of reviews and metaanalyses on numerical interventions in terms of samples (e.g., children with no, mild, or severe difficulties, children in preschool, elementary, or secondary school), instructional methods (e.g., traditional vs. digital interventions, drill vs. deeper acquisition methods), and mathematical skills that are trained (e.g., multiple or one specific skill such as early numeracy, arithmetic fluency, word problem solving). Despite these different foci, some general conclusions can be derived. **First**, in terms of the learners' age, the results reveal that interventions are effective at any age, but that it is recommended to start early as this is especially beneficial for enhancing children's attitudes toward numbers and math. **Second**, interventions focused on one specific skill are more effective than interventions targeting multiple skills. In addition, convincing evidence is lacking for the training of domain-general skills (memory, executive functions) to enhance math skills. **Third**, various instructional methods can be used in interventions and all have been shown to be effective. Particularly direct instruction and drill and practice have been demonstrated as effective instructional methods (although this may depend on the content that needs to be learned). The emergence of digital technologies has resulted in many new evolutions such as educational games, intelligent tutoring systems, computer-assisted instruction, and blended learning. These methods can be as effective as traditional teaching approaches and may also have a positive effect on learners' motivation. **Fourth**, the results revealed that all children, ranging from normal achievers to children with mild or severe difficulties, benefit from an intervention, although it will be difficult to bring children with difficulties or at-risk to the same level of their normal developing peers.

Based on this metaanalysis, the advice that can be given to practitioners is that the emphasis should be on early detection and prevention, by fostering early numeracy skills, one at the time and making use of direct instructions, if possible combined with digital technologies to make learning more enjoyable. This does certainly not mean that all other types of interventions should be abandoned. The metareview demonstrated that all types of interventions result in moderate-to-large positive effects on the numerical performance of children. In other words, this metareview also subscribes Dowker's (2017) statement that "any help is effective."

The most important advice for researchers is that the overall methodological quality of the intervention studies should be scaled up. Several methodological problems were raised in the available metaanalyses and reviews including, measurement issues, design insufficiencies and a lack of replications. These methodological problems may have led (in part) to overly positive conclusions concerning the effects of interventions. Evidence-based practice has become the golden standard in many domains such as medicine and education. There is however still a lot of room for improvement when it comes to the quality of the evidence, or in other words the quality of numerical intervention studies. Preregistration of studies and/or multilab collaborations, two practices that are increasingly more applied in the field of psychology and education, may help in this respect.

## Appendix

See Table A.1.

**Table A.1** Characteristics of the included studies.

| Studies | Methodology | Age | Content | Sample characteristics | Instructional method |
|---|---|---|---|---|---|
| Higgins et al. (2019) | MA ($k=24$) | Not specified (K-12 students) | Not specified | Not specified | Technology-based |
| Nelson and McMaster (2018) | MA ($k=34$) | Preschool (to first grade) | Early numeracy | Not specified | Not specified |
| Hughes and Yakubova (2019) | Systematic review ($k=11$) | Not specified | Not specified | Other (autism spectrum disorder) | Technology-based (video-based instruction) |
| Roesslein and Codding (2019) | Systematic review ($k=12$) | Elementary school | Specific content (fractions) | At-risk (struggling math learners) | Not specified |
| Haberstroh and Schulte-Körne (2019) | MA ($k=$not clear from the text) | Not specified | Not specified | At-risk (learners with dyscalculia, math disorder) | Not specified |
| Losinski et al. (2019) | MA ($k=17$) | Not specified | Not specified | Other (students with emotional disorders) | Not specified |
| Stevens et al. (2018) | MA ($k=24$) | Preschool/Elementary school (grades 4–12) | Not specified | At-risk (students with difficulties or disability) | Not specified |
| Lämsä et al. (2018) | Systematic review | Not specified | Not specified (reading and math) | At-risk (learning disabilities) | Technology-based (games) |
| Kiru et al. (2018) | Systematic review ($k=19$) | Not specified (K-12 students) | Not specified | At-risk (students with or at risk for mathematics learning disabilities) | Technology-based (technology-mediated mathematics (TMM)) |

*Continued*

**Table A.1** Characteristics of the included studies.—cont'd

| Studies | Methodology | Age | Content | Sample characteristics | Instructional method |
|---|---|---|---|---|---|
| Byun and Joung (2018) | MA ($k=17$) | Not specified (K-12 students) | Not specified | Not specified | Technology-based (DGBL) |
| Day-Hess and Clements (2017) | Systematic review | Preschool | Specific content (EF, math) | Not specified | Not specified |
| Dowker (2017) | Systematic review (book chapter) | Elementary school | Not specified | At-risk (children with difficulties with math) | Not specified |
| Raghubar and Barnes (2017) | Systematic review | Preschool | Not specified | At-risk (children with dyscalculia, high-risk or low math abilities) | Not specified |
| Stocker Jr and Kubina Jr (2017) | Systematic review | Not specified | Specific content (computational skills) | At-risk (students with disabilities/math deficits) | Other (CCC: cover-copy and compare) |
| Chauhan (2017) | MA ($k=41$) | Elementary school | Not specified (all subjects: including math) | Not specified | Technology-based |
| Szücs and Myers (2017) | Other (critical analysis) | Not specified | Early numeracy (train ANS to improve symbolic arithmetic) | Not specified | Not specified |
| Gevarter et al. (2016) | Systematic review | Not specified | Not specified | Other (autism spectrum disorder) | Not specified |
| Dennis et al. (2016) | MA ($k=25$) | Elementary school | Not specified | At-risk (students with math difficulties) | Not specified |

| | | | | | |
|---|---|---|---|---|---|
| Hwang and Riccomini (2016) | Systematic review ($k = 17$) | Secondary school (middle/high school) | Specific content (word problem solving) | At risk (children with learning disorder or at risk) | Other (heuristic, authentic and semantic) |
| King et al. (2016) | Systematic review ($k = 28$) | Not specified | Not specified (different target skills) | Other (autism spectrum disorder) | Not specified |
| Mulcahy et al. (2016) | Systematic review ($k = 19$) | Secondary school | Not specified | Other (emotional and behavioral disorders) | Not specified |
| Burns et al. (2016) | MA ($k = 37$) | Not specified | Not specified (next to math, also reading) | Not specified | Not specified |
| Shin and Bryant (2015) | Systematic review ($k = 17$) | Elementary school/ Secondary school (3rd to 12th grade (9-18j) | Specific content (fractions) | At-risk (students struggling with math, only US) | Not specified (different instructional techniques) |
| Barnett and Cleary (2015) | Systematic Review ($k = 11$) | Elementary school/ Secondary school (6–21) | Not specified (different aspects of math) | Other (autism spectrum disorder) | Other (visual presentations vs. cognitive strategies) |
| Chodura et al. (2015) | MA ($k = 35$) | Elementary school | Not specified | At-risk (children with dyscalculia, high-risk or low math abilities) | Not specified |
| Mononen et al. (2014) | Systematic review ($k = 19$) | Preschool/ Elementary school | Early numeracy | At-risk | Not specified |
| Cheung and Slavin (2013) | MA ($k = 74$) | Not specified | Not specified | Not-specified | Technology-based |

*Continued*

**Table A.1** Characteristics of the included studies.—cont'd

| Studies | Methodology | Age | Content | Sample characteristics | Instructional method |
|---|---|---|---|---|---|
| De Smedt et al. (2013) | Other (review, not systematic) | Not specified | Early numeracy (symbolic and nonsymbolic training) | Not specified | Not specified |
| Fischer et al. (2013) | MA ($k = 39$) | Not specified | Not specified | Not specified | Not specified |
| Hinton, Flores, and Shippen (2013) | Systematic review ($k = 17$) | Preschool/ Elementary school | Not specified (problem solving, calculation, number sense) | At-risk (children with Specific learning disorder, at risk) | Not specified |
| Hord and Xin (2013) | Other (narrative review) | Elementary school | Specific content (word problems) | At-risk (children with difficulties with math) | Other (cognitive and metacognitive instruction) |
| Cohen Kadosh et al. (2013) | Other (review, not systematic) | Not specified | Not specified | Not specified | Not specified (focus on innovative methods such as games, neuroimaging) |
| Methe et al. (2012) | MA ($k = 11$) | Elementary school (single subject studies) | Specific content (computational skills) | At-risk (children with achievement problems or high incidence disabilities) | Not specified |
| Kroeger et al. (2012) | Systematic review | Preschool/ Elementary school (prekindergarten through 3rd-grade students) | Not specified | Not specified | Not specified |

| | | | | | |
|---|---|---|---|---|---|
| Codding et al. (2011) | MA ($k = 17$) | Elementary school | Specific content (computational skills) | Not specified | Not specified |
| Gersten et al. (2009) | MA ($k = 42$) | Not specified | Not specified | At-risk (students with learning disabilities) | Not specified |
| Robinson, Schofield, and Steers-Wentzell (2005) | Systematic review | Elementary school/ Secondary school | Not specified | Other (minority students) | Other (peer and cross-age tutoring) |
| Kroesbergen and Van Luit (2003) | MA ($k = 58$) | Preschool/ Elementary school | Not specified (interventions involving math instructions) | At-risk (children with math difficulties) | Not specified |
| Baker et al. (2002) | MA ($k = 15$) | Not specified | Not specified | At-risk (low-achieving student or at risk) | Not specified |
| Xin and Jitendra (1999) | MA ($k = 25$) | Not specified | Specific content (word problem solving) | At-risk (students with learning problems) | Not specified |
| Jitendra and Xin (1997) | Systematic review | Not specified | Specific content (word problem solving) | At-risk (students at risk for math failure and with mild disabilities) | Not specified |

## Acknowledgments

BR and SV contributed equally to all stages of this book chapter. FD and DS provided valuable feedback on the method and reporting of the metareview. BR is supported by a grant from the KU Leuven Research Fund and the Fund for Scientific Research-Flanders. DS is a postdoctoral researcher of the Fund for Scientific Research-Flanders.

## References

Baker, S., Gersten, R., & Lee, D. S. (2002). A synthesis of empirical research on teaching mathematics to low-achieving students. *The Elementary School Journal, 103*(1), 51–73.

Barnett, J. E. H., & Cleary, S. (2015). Review of evidence-based mathematics interventions for students with autism spectrum disorders. *Education and Training in Autism and Developmental Disabilities, 50*(2), 172.

Bettany-Saltikov, J. (2010a). Learning how to undertake a systematic review: Part 1. *Nursing Standard, 24*(50), 47–55.

Bettany-Saltikov, J. (2010b). Learning how to undertake a systematic review: Part 2. *Nursing Standard, 24*(51), 47–58.

Blackwood, D. (2016). Taking it to the next level...reviews of systematic reviews. *HLA News, Winter*, 13–15.

Burns, M. K., Petersen-Brown, S., Haegele, K., Rodriguez, M., Schmitt, B., Cooper, M., … Van Der Heyden, A. M. (2016). Meta-analysis of academic interventions derived from neuropsychological data. *School Psychology Quarterly, 31*(1), 28.

Byun, J., & Joung, E. (2018). Digital game-based learning for K–12 mathematics education: A meta-analysis. *School Science and Mathematics, 118*(3–4), 113–126.

Chauhan, S. (2017). A meta-analysis of the impact of technology on learning effectiveness of elementary students. *Computers & Education, 105*, 14–30.

Cheung, A. C., & Slavin, R. E. (2013). The effectiveness of educational technology applications for enhancing mathematics achievement in K-12 classrooms: A meta-analysis. *Educational Research Review, 9*, 88–113.

Chodura, S., Kuhn, J. T., & Holling, H. (2015). Interventions for children with mathematical difficulties. *Zeitschrift für Psychologie., 223*(2), 129–144.

Codding, R. S., Burns, M. K., & Lukito, G. (2011). Meta-analysis of mathematic basic-fact fluency interventions: A component analysis. *Learning Disabilities Research & Practice, 26*(1), 36–47.

Cohen Kadosh, R., Dowker, A., Heine, A., Kaufmann, L., & Kucian, K. (2013). Interventions for improving numerical abilities: Present and future. *Trends in Neuroscience and Education, 2*(2), 85–93.

Day-Hess, C., & Clements, D. H. (2017). The DREME network: Research and interventions in early childhood mathematics. In *Vol. 53. Advances in child development and behavior* (pp. 1–41). JAI.

De Smedt, B., Noël, M. P., Gilmore, C., & Ansari, D. (2013). How do symbolic and non-symbolic numerical magnitude processing skills relate to individual differences in children's mathematical skills? A review of evidence from brain and behavior. *Trends in Neuroscience and Education, 2*(2), 48–55.

Dennis, M. S., Sharp, E., Chovanes, J., Thomas, A., Burns, R. M., Custer, B., & Park, J. (2016). A meta-analysis of empirical research on teaching students with mathematics learning difficulties. *Learning Disabilities Research & Practice, 31*(3), 156–168.

Dowker, A. (2017). Interventions for primary school children with difficulties in mathematics. In *Vol. 53. Advances in child development and behavior* (pp. 255–287). JAI.

Field, A. P., & Gillett, R. (2010). How to do a meta-analysis. *British Journal of Mathematical and Statistical Psychology, 63*(3), 665–694.

Fischer, U., Moeller, K., Cress, U., & Nuerk, H. C. (2013). Interventions supporting children's mathematics school success: A meta-analytic review. *European Psychologist, 18*(2), 89.
Fuchs, L. S., Fuchs, D., & Prentice, K. (2004). Responsiveness to mathematical problem-solving instruction: Comparing students at risk of mathematics disability with and without risk of reading disability. *Journal of Learning Disabilities, 37*, 293–306.
Gersten, R., Chard, D. J., Jayanthi, M., Baker, S. K., Morphy, P., & Flojo, J. (2009). Mathematics instruction for students with learning disabilities: A meta-analysis of instructional components. *Review of Educational Research, 79*(3), 1202–1242.
Gevarter, C., Bryant, D. P., Bryant, B., Watkins, L., Zamora, C., & Sammarco, N. (2016). Mathematics interventions for individuals with autism spectrum disorder: A systematic review. *Review Journal of Autism and Developmental Disorders, 3*(3), 224–238.
Grant, M. J., & Booth, A. (2009). A typology of reviews: An analysis of 14 review types and associated methodologies. *Health Information & Libraries Journal, 26*(2), 91–108.
Haberstroh, S., & Schulte-Körne, G. (2019). The diagnosis and treatment of dyscalculia. *Deutsches Ärzteblatt International, 116*(7), 107.
Higgins, K., Huscroft-D'Angelo, J., & Crawford, L. (2019). Effects of technology in mathematics on achievement, motivation, and attitude: A meta-analysis. *Journal of Educational Computing Research, 57*(2), 283–319.
Hinton, V., Flores, M. M., & Shippen, M. (2013). Response to intervention and math instruction. *International Journal of Education in Mathematics Science and Technology, 1*(3), 190–201.
Hord, C., & Xin, Y. P. (2013). Intervention research for helping elementary school students with math learning difficulties understand and solve word problems: 1996–2010. *Learning Disabilities: A Multidisciplinary Journal, 19*(1), 3–17.
Hughes, E. M., & Yakubova, G. (2019). Addressing the mathematics gap for students with ASD: An evidence-based systematic review of video-based mathematics interventions. *Review Journal of Autism and Developmental Disorders, 6*(2), 147–158.
Hwang, J., & Riccomini, P. J. (2016). Enhancing mathematical problem solving for secondary students with or at risk of learning disabilities: A literature review. *Learning Disabilities Research & Practice, 31*(3), 169–181.
Jitendra, A., & Xin, Y. P. (1997). Mathematical word-problem-solving instruction for students with mild disabilities and students at risk for math failure: A research synthesis. *The Journal of Special Education, 30*(4), 412–438.
King, S. A., Lemons, C. J., & Davidson, K. A. (2016). Math interventions for students with autism spectrum disorder: A best-evidence synthesis. *Exceptional Children, 82*(4), 443–462.
Kiru, E. W., Doabler, C. T., Sorrells, A. M., & Cooc, N. A. (2018). A synthesis of technology-mediated mathematics interventions for students with or at risk for mathematics learning disabilities. *Journal of Special Education Technology, 33*(2), 111–123.
Kroeger, L. A., Brown, R. D., & O'Brien, B. A. (2012). Connecting neuroscience, cognitive, and educational theories and research to practice: A review of mathematics intervention programs. *Early Education & Development, 23*(1), 37–58.
Kroesbergen, E. H., & Van Luit, J. E. (2003). Mathematics interventions for children with special educational needs: A meta-analysis. *Remedial and Special Education, 24*(2), 97–114.
Lämsä, J., Hämäläinen, R., Aro, M., Koskimaa, R., & Äyrämö, S. M. (2018). Games for enhancing basic reading and maths skills: A systematic review of educational game design in supporting learning by people with learning disabilities. *British Journal of Educational Technology, 49*(4), 596–607.
Losinski, M. L., Ennis, R. P., Sanders, S. A., & Nelson, J. A. (2019). A meta-analysis examining the evidence-base of mathematical interventions for students with emotional disturbances. *The Journal of Special Education, 52*(4), 228–241.
Methe, S. A., Kilgus, S. P., Neiman, C., & Riley-Tillman, T. C. (2012). Meta-analysis of interventions for basic mathematics computation in single-case research. *Journal of Behavioral Education, 21*(3), 230–253.

Mononen, R., Aunio, P., Koponen, T., & Aro, M. (2014). A review of early numeracy interventions for children at risk in mathematics. *International Journal of Early Childhood Special Education, 6*(1), 25–54.

Mulcahy, C. A., Krezmien, M. P., & Travers, J. (2016). Improving mathematics performance among secondary students with EBD: A methodological review. *Remedial and Special Education, 37*(2), 113–128.

National Academies of Sciences Engineering and Medicine. (2018). *How people learn II: Learners, contexts, and cultures*. National Academies Press.

Nelson, G., & McMaster, K. L. (2018). The effects of early numeracy interventions for students in preschool and early elementary: A meta-analysis. *Journal of Educational Psychology, 22*, 1001–1022.

Prodia. (2016). *Protocol wiskundeproblemen en dyscalculie. Bijlage 7: RTI-model bij wiskunde*. Retrieved from http://www.prodiagnostiek.be/materiaal/W_Bijlage%207%20RTI-model%20bij%20wiskunde.pdf. (Accessed 28 January 2020).

Raghubar, K. P., & Barnes, M. A. (2017). Early numeracy skills in preschool-aged children: A review of neurocognitive findings and implications for assessment and intervention. *The Clinical Neuropsychologist, 31*(2), 329–351.

Robinson, D. R., Schofield, J. W., & Steers-Wentzell, K. L. (2005). Peer and cross-age tutoring in math: Outcomes and their design implications. *Educational Psychology Review, 17*(4), 327–362.

Roesslein, R. I., & Codding, R. S. (2019). Fraction interventions for struggling elementary math learners: A review of the literature. *Psychology in the Schools, 56*(3), 413–432.

Schölvinck, M., & Jansen, L. (2014). *Basisboek RTI. Groei meten voor passend onderwijs*. Huizen: Pica.

Shin, M., & Bryant, D. P. (2015). Fraction interventions for students struggling to learn mathematics: A research synthesis. *Remedial and Special Education, 36*(6), 374–387.

Stevens, E. A., Rodgers, M. A., & Powell, S. R. (2018). Mathematics interventions for upper elementary and secondary students: A meta-analysis of research. *Remedial and Special Education, 39*(6), 327–340.

Stocker, J. D., Jr., & Kubina, R. M., Jr. (2017). Impact of cover, copy, and compare on fluency outcomes for students with disabilities and math deficits: A review of the literature. *Preventing School Failure: Alternative Education for Children and Youth, 61*(1), 56–68.

Szűcs, D., & Myers, T. (2017). A critical analysis of design, facts, bias and inference in the approximate number system training literature: A systematic review. *Trends in Neuroscience and Education, 6*, 187–203.

Xin, Y. P., & Jitendra, A. K. (1999). The effects of instruction in solving mathematical word problems for students with learning problems: A meta-analysis. *The Journal of Special Education, 32*(4), 207–225.

SECTION IV

# Math deficiencies and difficulties

CHAPTER 13

# The complex pathways toward the development of math anxiety and links with achievements

**Lital Daches Cohen[a,b] and Orly Rubinsten[a,b]**
[a]Edmond J. Safra Brain Research Center for the Study of Learning Disabilities, University of Haifa, Haifa, Israel
[b]Department of Learning Disabilities, University of Haifa, Haifa, Israel

## Introduction

Math anxiety (Richardson & Suinn, 1972) is characterized by heterogeneous symptoms (Rubinsten, Marciano, Eidlin Levy, & Daches Cohen, 2018), including negative attitudes toward math (Dowker, Bennett, & Smith, 2012; Kucian, McCaskey, Tuura, & von Aster, 2018), an excessive and sometimes unreasonable negative emotional reaction to situations involving numerical information (Beilock & Maloney, 2015; Kucian et al., 2018; Pizzie & Kraemer, 2017; Richardson & Suinn, 1972), and avoiding math-related activities (Ashcraft & Ridley, 2005; Pizzie & Kraemer, 2017). Moreover, math anxiety is associated with poor numerical cognition skills (Ashcraft & Krause, 2007; Ashcraft & Moore, 2009; Kucian et al., 2018; Namkung, Peng, & Lin, 2019; Passolunghi, Cargnelutti, & Pellizzoni, 2019; Rubinsten & Tannock, 2010; Zhang, Zhao, & Kong, 2019), regardless of math learning disabilities (Ashcraft, 2002) and reduced working memory capacity (Ashcraft & Kirk, 2001; Passolunghi et al., 2019) and processing speed (Passolunghi et al., 2019). Over time, math-anxious individuals are less likely to integrate in math-related careers (Ahmed, 2018; Ashcraft & Ridley, 2005; Hembree, 1990; Kucian et al., 2018; Ma & Xu, 2004; Maloney & Beilock, 2012), become associated with lower socioeconomic status (Ritchie & Bates, 2013), and tend to display reduced financial literacy (Skagerlund, Lind, Strömbäck, Tinghög, & Västfjäll, 2018).

Although the understanding of risk factors for the development of math anxiety has progressed (Dowker, Sarkar, & Looi, 2016; Maloney & Beilock, 2012; Rubinsten et al., 2018; Rubinsten, Eidlin Levy, & Daches Cohen, 2019; Suárez-Pellicioni, Núñez-Peña, & Colomé, 2016), the

*Heterogeneous Contributions to Numerical Cognition*
https://doi.org/10.1016/B978-0-12-817414-2.00003-8

question of why math anxiety develops remains unresolved. Such an understanding will facilitate intervention programs aimed at reducing math anxiety and providing tools for coping with it rather than focusing solely on numerical skills (e.g., Supekar, Iuculano, Chen, & Menon, 2015). Furthermore, awareness of the broad spectrum of risk factors for the development of math anxiety may assist in early identification of at-risk students.

Here, we aim to organize the broad spectrum of risk factors for the development and maintenance of math anxiety (see Fig. 1). The current theoretical framework suggests that low math achievements may increase one's vulnerability to intrinsic and environmental causes of math anxiety and thus accelerate its development (see top of Fig. 1). On the other hand, math anxiety can lead to decreased math achievements, which in turn exacerbate emotional problems (see bottom of Fig. 1).

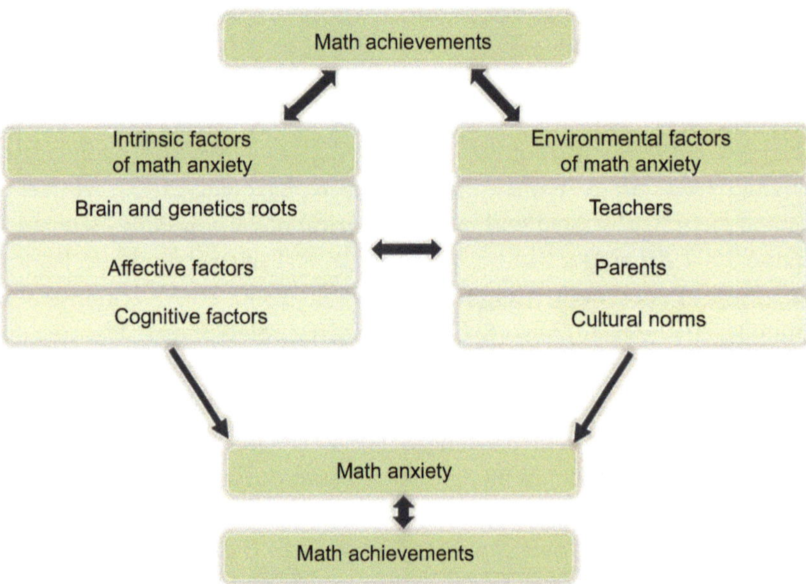


**Fig. 1** The theoretical framework of risk factors for the development and maintenance of math anxiety. *Note.* The theoretical framework highlights the dynamic interplay between intrinsic and environmental factors in the development of math anxiety. The major assumptions of the current model are that, on one hand, low math achievements can be a cause of math anxiety by increasing the individual's vulnerability to the effects of intrinsic and environmental risk factors. On the other hand, low math achievements can be the result of math anxiety and exacerbate emotional problems.

## The antecedents of math anxiety

Math anxiety is a specific anxiety (Dowker et al., 2016; Foley et al., 2017; Haase, Guimarães, & Wood, 2019; Suárez-Pellicioni et al., 2016) that is related to a distinct pattern of neural activity in brain regions associated with numerical computations (Klados, Simos, Micheloyannis, Margulies, & Bamidis, 2015; Supekar et al., 2015; Young, Wu, & Menon, 2012) and the processing of negative emotions (Lyons & Beilock, 2012; Sarkar, Dowker, & Kadosh, 2014; Supekar et al., 2015; Young et al., 2012). Accordingly, it has been shown that math anxiety is not related to general intelligence (Ashcraft & Moore, 2009; Rubinsten & Tannock, 2010) and is distinct from general anxiety (Ashcraft & Moore, 2009; Hembree, 1990) and test anxiety (Hembree, 1990; Ma, 1999), even in childhood (Carey, Hill, Devine, & Szűcs, 2017). It had been suggested that the causes of math anxiety may be both intrinsic and environmental (Luttenberger, Wimmer, & Paechter, 2018; Rubinsten et al., 2019, 2018).

## Intrinsic factors (see top left-hand side of Fig. 1)

Heightened negative emotional experiences in anxiety disorders are perceived as the result of an attentional bias toward threatening information relative to neutral stimuli (e.g., Dudeney, Sharpe, & Hunt, 2015; MacLeod, Mathews, & Tata, 1986; Okon-Singer, Alyagon, Kofman, Tzelgov, & Henik, 2011). Attentional bias is a core feature of anxiety disorders (Mathews & MacLeod, 2002; Shackman et al., 2016). Indeed, math-anxious individuals display maladaptive threat-related attentional bias toward basic numerical information (Rubinsten, Eidlin, Wohl, & Akibli, 2015; Suárez-Pellicioni, Núñez-Peña, & Colomé, 2015). Yet, there are individual differences in emotional reactivity to similar situations and stimuli, including the threshold of stimuli needed to create an emotional response and the intensity of this response (Davidson, 1998; Okon-Singer, Lichtenstein-Vidne, & Cohen, 2013). During math problem-solving, for example, high math-anxious children from ages seven to nine demonstrated greater activity in the amygdala, a brain region associated with the processing of negative emotions, compared to low math-anxious children (Young et al., 2012). Even when high math-anxious undergraduate students merely looked at math expressions without being requested to solve them, the same pattern of results was recorded (Pizzie & Kraemer, 2017).

The suggested intrinsic factors for this heightened negative emotional experience in math anxiety include the individual's predispositions related to brain malfunctions (Young et al., 2012), genetic roots (Malanchini et al., 2017, 2020; Wang et al., 2014), and a tendency toward anxiety in general (Ashcraft & Moore, 2009; Hembree, 1990; O'Leary, Fitzpatrick, & Hallett, 2017). For example, math-anxious individuals demonstrated reduced activity in brain regions associated with mathematical reasoning and cognitive control (Young et al., 2012), increased activity and abnormal connectivity between brain regions associated with the processing and regulating of negative emotions (Artemenko, Daroczy, & Nuerk, 2015; Lyons & Beilock, 2012; Suárez-Pellicioni, Núñez-Peña, & Colomé, 2013; Young et al., 2012), and general-anxiety-related neurophysiological responses (Cargnelutti, Tomasetto, & Passolunghi, 2017).

Genetic factors accounted for about 40% of the variance in math anxiety, and genes related to general anxiety explained 9% of the total variance in math anxiety (Malanchini et al., 2017; Wang et al., 2014). Recently, Malanchini et al. (2020) demonstrated through multivariate quantitative genetic analyses among adolescent twin pairs that general- and math anxiety shared about 20% of their genetic variance, but general anxiety did not account for the substantial genetic overlap (~75%) between math anxiety, attitudes toward math, math ability, and achievements (see also Lukowski et al., 2019). However, math anxiety was also influenced by nonfamilial environmental risk factors involved in general anxiety (Malanchini et al., 2017; Wang et al., 2014).

Deficient basic numerical skills (Lindskog, Winman, & Poom, 2017; Ma & Xu, 2004), spatial processing (Sokolowski, Hawes, & Lyons, 2019), and executive functions (Orbach, Herzog, & Fritz, 2020) have also been proposed as significantly related to math anxiety. Indeed, math anxiety has been associated with basic numerical skills, such as simple counting (Maloney, Risko, Ansari, & Fugelsang, 2010; Núñez-Peña & Suárez-Pellicioni, 2014; Rubinsten, Bialik, & Solar, 2012; Rubinsten & Tannock, 2010), and children with diagnosed mathematical disabilities exhibited more math anxiety compared to a control group (Passolunghi, 2011; Rubinsten & Tannock, 2010). However, recent findings showed that despite the significant cooccurrence, there was a dissociation between cognitive and emotional math problems in a large sample of children (Devine, Hill, Carey, & Szűcs, 2018).

Other intrinsic factors that have been suggested as contributors to or protectors against the development of math anxiety include affective factors, such as math motivation (Chang & Beilock, 2016; Lyons & Beilock, 2011; Wang et al., 2015), attitudes toward math and math learning (Hembree, 1990), math self-efficacy (Hoffman, 2010; Lee, 2009; Mammarella, Donolato,

Caviola, & Giofrè, 2018), and self-concept (Justicia-Galiano, Martín-Puga, Linares, & Pelegrina, 2017). For instance, a negative linear correlation was found between math anxiety and math performance in adults and adolescents with low math motivation, whereas an inverted-U curvilinear relation was observed in more motivated students (Wang et al., 2015). Accordingly, high activation in motivation-related brain regions was associated with reduced negative effects of high math anxiety on performance (Lyons & Beilock, 2011). Likewise, Mammarella et al. (2018) demonstrated that low self-concept increases the risk of developing anxiety. However, others claimed that low self-concept is the result of math anxiety (Ahmed, Minnaert, Kuyper, & van der Werf, 2012).

Taken together, the reviewed literature suggests that intrinsic factors provide a partial explanation for the development of math anxiety (Rubinsten et al., 2018). Similarly, evidence supporting the diathesis-stress models demonstrated that life stressors moderate associations between intrinsic factors, such as temperamental and neural vulnerabilities, and subsequent internalizing psychopathology (Klein & Finsaas, 2017; Kushner, 2015; Swartz, Knodt, Radtke, & Hariri, 2015).

## Environmental factors (see top right-hand side of Fig. 1)

The proposed environmental causes of math anxiety include adverse pedagogical and social math-related events with teachers (Beilock, Gunderson, Ramirez, & Levine, 2010; Gunderson, Ramirez, Levine, & Beilock, 2012; Novak & Tassell, 2017; Ramirez, Hooper, Kersting, Ferguson, & Yeager, 2018) and parents (Daches Cohen & Rubinsten, 2017; Gunderson et al., 2012; Maloney, Ramirez, Gunderson, Levine, & Beilock, 2015). In the school setting, inappropriate math instruction strategies (Baroody & Hume, 1991; Jackson & Leffingwell, 1999) and poor modeling by teachers may contribute to the development of math anxiety (Beilock et al., 2010). Particularly, females are more sensitive to the effects of their math-anxious female teachers by acquiring negative stereotypes about girls and math, which are known to be related to math anxiety (Hembree, 1990). Accordingly, math achievements of females, but not of their male peers, have been found to be influenced by the level of math anxiety displayed by their elementary school female teachers (Beilock et al., 2010).

In the home setting, children's math anxiety was found to be affected by parent pressure to maintain high achievements (Daches Cohen & Rubinsten, 2017) and involvement of parents in their children's math-learning processes (Roberts & Vukovic, 2011). Especially if parents are math-anxious, they can trigger their children's math anxiety through involvement in their children's math-learning processes (Daches Cohen & Rubinsten, 2017; Maloney et al., 2015). However, the transfer of math anxiety from parents to children is mostly affected by mothers (Casad, Hale, & Wachs, 2015; Daches Cohen & Rubinsten, 2017; Maloney et al., 2015), particularly in mother-daughter dyads (Daches Cohen & Rubinsten, 2017).

From a broader social perspective, cultural norms too have a significant effect on math anxiety (Foley et al., 2017; Stoet, Bailey, Moore, & Geary, 2016) and its expression (Stoet et al., 2016). For instance, while the prevalence of math anxiety in Asian countries (e.g., Korea, Japan, and Thailand) tends to be high, Western European countries (e.g., Austria, Germany, and Switzerland; Lee, 2009) and economically developed and gender-equal countries showed a low prevalence of math anxiety compared to the 41 countries who participated in the Program for International Student Assessment (PISA) in 2003 (Stoet et al., 2016).

Recently, it has been suggested that math anxiety results from a dynamic interplay between intrinsic and environmental factors (Rubinsten et al., 2019, 2018). The dynamic relationship between intrinsic and environmental factors affects the negative emotional experience of math-anxious individuals, and hence the symptoms are heterogeneous and the intensity of the emotional response may vary (Rubinsten et al., 2018).

## Assessment of math anxiety: Methods and difficulties

The primary method for evaluating math anxiety is through **explicit questionnaires** that typically assess accessible self-representations (for review, see Luttenberger et al., 2018). The two most widely used questionnaires are the Mathematics Anxiety Rating Scale (MARS; Richardson & Suinn, 1972) and its short version, the Revised Mathematics Anxiety Rating Scale (R-MARS; Plake & Parker, 1982). Although the reliability of these explicit tools has generally been found to be good (Dowker et al., 2016), gender differences in math anxiety that are not due to anxiety per se could arise. For example, women have consistently been found to score higher than men on explicit self-report questionnaires of trait anxiety (e.g., Costa Jr, Terracciano, & McCrae, 2001; Egloff & Schmukle, 2004).

**Implicit measures**, on the other hand, may eliminate gender differences (Egloff & Schmukle, 2004). These tools usually assess inaccessible cognitive structures that are processed automatically after seeing a salient affective stimulus (for review, see Rubinsten, 2015). In the case of math anxiety, for instance, high math-anxious individuals demonstrated an attentional bias toward math-related stimuli relative to neutral stimuli (Rubinsten et al., 2015; Suárez-Pellicioni et al., 2015). The differences between explicit and implicit tools were further highlighted in our previous research (Daches Cohen & Rubinsten, 2017). This research indicated that while arithmetic skills predicted sixth-grade children's explicit, conscious, and accessible self-representations of math anxiety, mothers' behaviors and attitudes toward math were more substantial in the prediction of children's implicit and unconscious cognitive structures of math anxiety. However, differences between explicit and implicit measures of math anxiety should be examined in further research.

Since implicit tools may be more effective than explicit measures, more implicit tools which aim to measure specific intrinsic cognitive aspects are needed (Rubinsten et al., 2019). Hence, the current comprehensive review of the broad spectrum of risk factors for the development and maintenance of math anxiety can be beneficial.

## The role of low math achievements

A negative relationship between math anxiety and achievements (Hembree, 1990; Ma, 1999; Ma & Xu, 2004; Namkung et al., 2019; Rubinsten & Tannock, 2010; Skaalvik, 2018) has been found within and across countries in PISA studies (Lee, 2009; OECD, 2013). This negative link leads math-anxious individual to avoid different types of mathematical problems (Choe, Jenifer, Rozek, Berman, & Beilock, 2019). Several explanations of the negative association between math anxiety and math achievements have been proposed.

According to a leading theory, math-anxious individuals do not necessarily exhibit math learning disabilities (Ashcraft & Kirk, 2001; Ashcraft & Krause, 2007). Rather, they suffer from math anxiety-induced ruminations (repetitive thinking about negative personal concerns and worries; Nolen-Hoeksema, Blair, & Sonja, 2008). These ruminations deplete one's cognitive resources, such as attentional resources of the working memory (Ashcraft & Kirk, 2001; Luttenberger et al., 2018; Maloney & Beilock, 2012), and disrupt thinking processes (Chang & Beilock, 2016). As a result, math-anxious individuals tend to demonstrate low math achievements (Beilock, 2010; Maloney & Beilock, 2012; Ramirez, Chang, Maloney, Levine, & Beilock, 2016). Specifically, math-anxious individuals with higher working memory may be more vulnerable to anxiety-related impairments in math performance because of their tendency to use and rely on working memory-intensive strategies (Ramirez et al., 2016; Ramirez, Gunderson, Levine, & Beilock, 2013). Indeed, math-anxious children with the highest working memory capacity demonstrated the lowest math achievements compared to their peers with the highest levels of math anxiety and a decreased working memory capacity (Ramirez et al., 2016).

In contrast, it has recently been shown that the relationship between math anxiety and working memory is mediated by math performance, except for a specific math-related working memory subtest which has been shown to be negatively related to math anxiety (Korem & Rubinsten, 2020). In line with these findings, researchers have argued that there is a bidirectional link between math anxiety and math achievements (Carey, Hill, Devine, & Szücs, 2016; see bottom of Fig. 1). Math anxiety can also be the result of decreased math achievements (Ma & Xu, 2004; Wang, Rimfeld, Shakeshaft, Schofield, & Malanchini, 2020). A reasonable explanation is that children create their interpretive constructions of emotional situations based on their experiences of success or failure in achieving their personal goals (e.g., Graham & Weiner, 1986; Stein & Levine, 1989; Thompson, 1990; see top of Fig. 1). The novel suggestion in the current theoretical framework is that low math achievements can be both a cause and a result of math anxiety (see Fig. 1).

## The current framework

Although the association between elevated math anxiety and low math achievements is well documented (Hembree, 1990; Ma, 1999; Ma & Xu, 2004; Namkung et al., 2019; Rubinsten & Tannock, 2010; Skaalvik, 2018), the directionality of this correlation is still under debate. The current framework suggests that math anxiety can be the cause of low math achievements (see bottom of Fig. 1) through the depletion of one's cognitive resources (Beilock, 2010; Maloney & Beilock, 2012; Ramirez et al., 2016) or directly by the high arousal and negative affect, which are detrimental to academic functioning (Bohn-Gettler & Rapp, 2011; Mason, Scrimin, Tornatora, & Zaccoletti, 2017; Scrimin, Altoè, Moscardino, Pastore, & Mason, 2016; Scrimin & Mason, 2015; Scrimin, Mason, Moscardino, & Altoè, 2015; Wunsch, Meier, Ueberholz, Strahler, & Kasten, 2019). For example, recently Skagerlund and colleagues (Skagerlund, Östergren, Västfjäll, & Träff, 2019) have suggested that math anxiety leads to low math achievements either directly or through low working memory abilities or symbolic numerical processing.

Longitudinal studies found that math anxiety in the second grade predicted low math achievements in the third grade among Chinese (Ching, 2017) and US children (Vukovic, Kieffer, Bailey, & Harari, 2013). Similarly, Cargnelutti et al. (2017) showed the same pattern of results among Italian second and third graders and demonstrated that the effects of math anxiety on performance emerged only in the third grade.

On the other hand, math anxiety can also be the result of low math achievements (see top of Fig. 1). Low math achievements can lead to adverse pedagogical and social math-related events, such as higher levels of parental stress and involvement in children's learning processes as well as stereotypical statements from teachers, particularly for female low achievers. Parental pressure to maintain high achievements (Daches Cohen & Rubinsten, 2017) and involvement in math-learning processes (Roberts & Vukovic, 2011) and negative stereotypical beliefs of teachers about the performance of females in math (Hembree, 1990) have been found to be related to math anxiety. Indeed, in a yearlong longitudinal study with primary school students, it was found that arithmetic achievement predicted later anxiety about failure in math, but math anxiety did not predict later achievement (Sorvo et al., 2019). In older children, math achievements predicted subsequent development in math anxiety in a large nationally representative sample of middle and high school students in the US (Ma & Xu, 2004) and among Italian high school students (Wang et al., 2020). The cumulative evidence suggests that the mechanisms underlying the relationship between math anxiety and achievements are age-dependent (Thomas & Dowker, 2000; Wang et al., 2020).

To sum, the present theoretical framework incorporates risk factors for the development and maintenance of math anxiety and suggests that low math achievements may result from math anxiety (see bottom of Fig. 1) but also cause its development by intensifying one's vulnerability to its environmental and intrinsic causes (see top of Fig. 1).

## Conclusions

The current theoretical framework focuses on the complex and dynamic interplay between environmental factors and intrinsic elements, in increasing the risk of developing math anxiety (see top of Fig. 1). Namely, math anxiety results from multiple causal interacting influences between biological, cognitive, and social factors. Specifically, the current theoretical framework demonstrates the unique role of math achievements, which can be both a cause and a result of math anxiety (see Fig. 1).

Therefore, for instance, intrinsically biologically vulnerable individuals who are exposed to adverse math learning experiences and low achievements are at-risk for developing math anxiety (see top of Fig. 1). On the other hand, math anxiety, which develops through dynamic interactions between intrinsic and environmental factors, may lead to low math achievements (see bottom of Fig. 1), which in turn will reinforce emotional problems. For example, avoidance behavior exacerbates feelings of anxiety in math-related situations and increases learning gaps. As a result, math achievements are impaired, with a consequent rise in pressure applied by environmental factors, such as parents, as well as in intrinsic biological vulnerability. The overall effect is that math anxiety is intensified and low achievements are maintained.

Future studies should investigate the developmental course of the relationship between math anxiety and achievements in order to identify different groups of individuals with distinct trajectories. Of particular interest would be to examine the relationship between math anxiety and basic numerical skills that were found to predict math achievements, such as spontaneous focusing on numerosity (SFON; Edens & Potter, 2013; Hannula-Sormunen, 2015; Hannula-Sormunen, Lehtinen, & Räsänen, 2015; Lepola & Hannula-Sormunen, 2019; Nanu, McMullen, Munck, & Hannula-Sormunen, 2018). Previous studies showed that SFON did not significantly correlate with math-related emotional constructs, such as motivational orientations (Edens & Potter, 2013; Lepola & Hannula-Sormunen, 2019; Nanu et al., 2018) and math self-efficacy (Nanu et al., 2018). However, to the best of our knowledge, the direct relation between math anxiety and SFON has not yet been examined. Educators, psychologists, and policy makers should also take into account the complex pathways toward the development of math anxiety. Namely, awareness of the heterogeneity in both causes and results of math anxiety will enable early identification of at-risk students. Moreover, intervention programs need to give greater attention to the development of emotional skills for reducing emotional problems and promoting academic achievements.

## References

Ahmed, W. (2018). Developmental trajectories of math anxiety during adolescence: Associations with STEM career choice. *Journal of Adolescence*, *67*, 158–166.

Ahmed, W., Minnaert, A., Kuyper, H., & van der Werf, G. (2012). Reciprocal relationships between math self-concept and math anxiety. *Learning and Individual Differences*, *22*(3), 385–389.

Artemenko, C., Daroczy, G., & Nuerk, H. C. (2015). Neural correlates of math anxiety – An overview and implications. *Frontiers in Psychology*, *6*, 1333. https://doi.org/10.3389/fpsyg.2015.01333.
Ashcraft, M. H. (2002). Math anxiety: Personal, educational, and cognitive consequences. *Current Directions in Psychological Science*, *11*(5), 181–185.
Ashcraft, M. H., & Kirk, E. P. (2001). The relationships among working memory, math anxiety, and performance. *Journal of Experimental Psychology: General*, *130*(2), 224–237.
Ashcraft, M. H., & Krause, J. A. (2007). Working memory, math performance, and math anxiety. *Psychonomic Bulletin & Review*, *14*(2), 243–248.
Ashcraft, M. H., & Moore, A. M. (2009). Mathematics anxiety and the affective drop in performance. *Journal of Psychoeducational Assessment*, *27*(3), 197–205.
Ashcraft, M. H., & Ridley, K. S. (2005). Math anxiety and its cognitive consequences: A tutorial review. In J. I. D. Campbell (Ed.), *Handbook of mathematical cognition* (pp. 315–327). New York, NY: Psychology Press.
Baroody, A. J., & Hume, J. (1991). Meaningful mathematics instruction: The case of fractions. *Remedial and Special Education*, *12*(3), 54–68.
Beilock, S. (2010). *Choke: What the secrets of the brain reveal about getting it right when you have to*. New York, NY: Simon and Schuster.
Beilock, S. L., Gunderson, E. A., Ramirez, G., & Levine, S. C. (2010). Female teachers' math anxiety affects girls' math achievement. *Proceedings of the National Academy of Sciences*, *107*(5), 1860–1863.
Beilock, S. L., & Maloney, E. A. (2015). Math anxiety a factor in math achievement not to be ignored. *Policy Insights From the Behavioral and Brain Sciences*, *2*(1), 4–12.
Bohn-Gettler, C. M., & Rapp, D. N. (2011). Depending on my mood: Mood-driven influences on text comprehension. *Journal of Educational Psychology*, *103*(3), 562–577.
Carey, E., Hill, F., Devine, A., & Szücs, D. (2016). The chicken or the egg? The direction of the relationship between mathematics anxiety and mathematics performance. *Frontiers in Psychology*, *6*, 1987. https://doi.org/10.3389/fpsyg.2015.01987.
Carey, E., Hill, F., Devine, A., & Szűcs, D. (2017). The modified abbreviated math anxiety scale: A valid and reliable instrument for use with children. *Frontiers in Psychology*, *8*, 11. https://doi.org/10.3389/fpsyg.2017.00011.
Cargnelutti, E., Tomasetto, C., & Passolunghi, M. C. (2017). How is anxiety related to math performance in young students? A longitudinal study of Grade 2 to Grade 3 children. *Cognition Emotion*, *31*(4), 755–764.
Casad, B. J., Hale, P., & Wachs, F. L. (2015). Parent-child math anxiety and math-gender stereotypes predict adolescents' math education outcomes. *Frontiers in Psychology*, *6*, 1597. https://doi.org/10.3389/fpsyg.2015.01597.
Chang, H., & Beilock, S. L. (2016). The math anxiety-math performance link and its relation to individual and environmental factors: A review of current behavioral and psychophysiological research. *Current Opinion in Behavioral Sciences*, *10*, 33–38.
Ching, B. H. H. (2017). Mathematics anxiety and working memory: Longitudinal associations with mathematical performance in Chinese children. *Contemporary Educational Psychology*, *51*, 99–113.
Choe, K. W., Jenifer, J. B., Rozek, C. S., Berman, M. G., & Beilock, S. L. (2019). Calculated avoidance: Math anxiety predicts math avoidance in effort-based decision-making. *Science Advances*, *5*(11), eaay1062.
Costa, P., Jr., Terracciano, A., & McCrae, R. R. (2001). Gender differences in personality traits across cultures: Robust and surprising findings. *Journal of Personality and Social Psychology*, *81*(2), 322–331.
Daches Cohen, L., & Rubinsten, O. (2017). Mothers, intrinsic math motivation, arithmetic skills, and math anxiety in elementary school. *Frontiers in Psychology*, *8*, 1939. https://doi.org/10.3389/fpsyg.2017.01939.

Davidson, R. J. (1998). Affective style and affective disorders: Perspectives from affective neuroscience. *Cognition & Emotion, 12*(3), 307–330.
Devine, A., Hill, F., Carey, E., & Szűcs, D. (2018). Cognitive and emotional math problems largely dissociate: Prevalence of developmental dyscalculia and mathematics anxiety. *Journal of Educational Psychology, 110*(3), 431–444.
Dowker, A., Bennett, K., & Smith, L. (2012). Attitudes to mathematics in primary school children. *Child Development Research, 2012*, 1–8.
Dowker, A., Sarkar, A., & Looi, C. Y. (2016). Mathematics anxiety: What have we learned in 60 years? *Frontiers in Psychology*, 7, 508. https://doi.org/10.3389/fpsyg.2016.00508.
Dudeney, J., Sharpe, L., & Hunt, C. (2015). Attentional bias towards threatening stimuli in children with anxiety: A meta-analysis. *Clinical Psychology Review, 40*, 66–75.
Edens, K. M., & Potter, E. F. (2013). An exploratory look at the relationships among math skills, motivational factors and activity choice. *Early Childhood Education Journal, 41*(3), 235–243.
Egloff, B., & Schmukle, S. C. (2004). Gender differences in implicit and explicit anxiety measures. *Personality and Individual Differences, 36*(8), 1807–1815.
Foley, A. E., Herts, J. B., Borgonovi, F., Guerriero, S., Levine, S. C., & Beilock, S. L. (2017). The math anxiety-performance link: A global phenomenon. *Current Directions in Psychological Science, 26*(1), 52–58.
Graham, S., & Weiner, B. (1986). From an attributional theory of emotion to developmental psychology: A round-trip ticket? *Social Cognition, 4*(2), 152–179.
Gunderson, E. A., Ramirez, G., Levine, S. C., & Beilock, S. L. (2012). The role of parents and teachers in the development of gender-related math attitudes. *Sex Roles, 66*(3–4), 153–166.
Haase, V. G., Guimarães, A. P. L., & Wood, G. (2019). Mathematics and emotions: The case of math anxiety. In *International handbook of mathematical learning difficulties* (pp. 469–503). Cham: Springer.
Hannula-Sormunen, M. M. (2015). Spontaneous focusing on numerosity and its relation to counting and arithmetic. In *The Oxford handbook of numerical cognition* (pp. 275–290).
Hannula-Sormunen, M. M., Lehtinen, E., & Räsänen, P. (2015). Preschool children's spontaneous focusing on numerosity, subitizing, and counting skills as predictors of their mathematical performance seven years later at school. *Mathematical Thinking and Learning, 17*(2–3), 155–177.
Hembree, R. (1990). The nature, effects, and relief of mathematics anxiety. *Journal for Research in Mathematics Education, 21*(1), 33–46.
Hoffman, B. (2010). "I think I can, but I'm afraid to try": The role of self-efficacy beliefs and mathematics anxiety in mathematics problem-solving efficiency. *Learning and Individual Differences, 20*, 276–283.
Jackson, C. D., & Leffingwell, R. J. (1999). The role of instructors in creating math anxiety in students from kindergarten through college. *The Mathematics Teacher, 92*(7), 583–586.
Justicia-Galiano, M. J., Martín-Puga, M. E., Linares, R., & Pelegrina, S. (2017). Math anxiety and math performance in children: The mediating roles of working memory and math self-concept. *British Journal of Educational Psychology, 87*(4), 573–589.
Klados, M. A., Simos, P., Micheloyannis, S., Margulies, D., & Bamidis, P. D. (2015). ERP measures of math anxiety: How math anxiety affects working memory and mental calculation tasks? *Frontiers in Behavioral Neuroscience, 9*, 282. https://doi.org/10.3389/fnbeh.2015.00282.
Klein, D. N., & Finsaas, M. C. (2017). The stony brook temperament study: Early antecedents and pathways to emotional disorders. *Child Development Perspectives, 11*(4), 257–263.
Korem, N., & Rubinsten, O. (2020). *How do working memory, general anxiety and math anxiety affect female students' math performance?*. February 28. https://doi.org/10.31234/osf.io/dbp7t.

Kucian, K., McCaskey, U., Tuura, R. O. G., & von Aster, M. (2018). Neurostructural correlate of math anxiety in the brain of children. *Translational Psychiatry*, *8*(1), 1–11.
Kushner, S. C. (2015). A review of the direct and interactive effects of life stressors and dispositional traits on youth psychopathology. *Child Psychiatry & Human Development*, *46*(5), 810–819.
Lee, J. (2009). Universals and specifics of math self-concept, math self-efficacy, and math anxiety across 41 PISA 2003 participating countries. *Learning and Individual Differences*, *19*(3), 355–365.
Lepola, J., & Hannula-Sormunen, M. (2019). Spontaneous focusing on numerosity and motivational orientations as predictors of arithmetical skills from kindergarten to grade 2. *Educational Studies in Mathematics*, *100*(3), 251–269.
Lindskog, M., Winman, A., & Poom, L. (2017). Individual differences in nonverbal number skills predict math anxiety. *Cognition*, *159*, 156–162.
Lukowski, S. L., DiTrapani, J., Rockwood, N. J., Jeon, M., Thompson, L. A., & Petrill, S. A. (2019). Etiological distinction across dimensions of math anxiety. *Behavior Genetics*, *49*(3), 310–316.
Luttenberger, S., Wimmer, S., & Paechter, M. (2018). Spotlight on math anxiety. *Psychology Research and Behavior Management*, *11*, 311–322.
Lyons, I. M., & Beilock, S. L. (2011). Mathematics anxiety: Separating the math from the anxiety. *Cerebral Cortex*, *22*, 2102–2110.
Lyons, I. M., & Beilock, S. L. (2012). When math hurts: Math anxiety predicts pain network activation in anticipation of doing math. *PLoS One*, 7(10), e48076. 1–6.
Ma, X. (1999). A meta-analysis of the relationship between anxiety toward mathematics and achievement in mathematics. *Journal for Research in Mathematics Education*, *30*(5), 520–540.
Ma, X., & Xu, J. (2004). The causal ordering of mathematics anxiety and mathematics achievement: A longitudinal panel analysis. *Journal of Adolescence*, *27*(2), 165–179.
MacLeod, C., Mathews, A., & Tata, P. (1986). Attentional bias in emotional disorders. *Journal of Abnormal Psychology*, *95*(1), 15–20.
Malanchini, M., Rimfeld, K., Shakeshaft, N. G., Rodic, M., Schofield, K., Selzam, S., et al. (2017). The genetic and environmental aetiology of spatial, mathematics and general anxiety. *Scientific Reports*, 7, 42218. https://doi.org/10.1038/srep42218.
Malanchini, M., Rimfeld, K., Wang, Z., Petrill, S. A., Tucker-Drob, E. M., Plomin, R., et al. (2020). Genetic factors underlie the association between anxiety, attitudes and performance in mathematics. *Translational Psychiatry*, *10*(1), 1–11.
Maloney, E. A., & Beilock, S. L. (2012). Math anxiety: Who has it, why it develops, and how to guard against it. *Trends in Cognitive Sciences*, *16*(8), 404–406.
Maloney, E. A., Ramirez, G., Gunderson, E. A., Levine, S. C., & Beilock, S. L. (2015). Intergenerational effects of parents' math anxiety on children's math achievement and anxiety. *Psychological Science*, *26*(9), 1480–1488.
Maloney, E. A., Risko, E. F., Ansari, D., & Fugelsang, J. (2010). Mathematics anxiety affects counting but not subitizing during visual enumeration. *Cognition*, *114*(2), 293–297.
Mammarella, I. C., Donolato, E., Caviola, S., & Giofrè, D. (2018). Anxiety profiles and protective factors: A latent profile analysis in children. *Personality and Individual Differences*, *124*, 201–208.
Mason, L., Scrimin, S., Tornatora, M. C., & Zaccoletti, S. (2017). Emotional reactivity and comprehension of multiple online texts. *Learning and Individual Differences*, *58*, 10–21.
Mathews, A., & MacLeod, C. (2002). Induced processing biases have causal effects on anxiety. *Cognition & Emotion*, *16*(3), 331–354.
Namkung, J. M., Peng, P., & Lin, X. (2019). The relation between mathematics anxiety and mathematics performance among school-aged students: A meta-analysis. *Review of Educational Research*, *89*(3), 459–496.

Nanu, C. E., McMullen, J., Munck, P., & Hannula-Sormunen, M. M. (2018). Spontaneous focusing on numerosity in preschool as a predictor of mathematical skills and knowledge in the fifth grade. *Journal of Experimental Child Psychology*, *169*, 42–58.

Nolen-Hoeksema, S., Blair, E. W., & Sonja, L. (2008). Rethinking rumination. *Perspectives of Psychological Science*, *3*, 400–424.

Novak, E., & Tassell, J. L. (2017). Studying preservice teacher math anxiety and mathematics performance in geometry, word, and non-word problem solving. *Learning and Individual Differences*, *54*, 20–29.

Núñez-Peña, M. I., & Suárez-Pellicioni, M. (2014). Less precise representation of numerical magnitude in high math-anxious individuals: An ERP study of the size and distance effects. *Biological Psychology*, *103*, 176–183.

Okon-Singer, H., Alyagon, U., Kofman, O., Tzelgov, J., & Henik, A. (2011). Fear-related pictures deteriorate the performance of university students with high fear of snakes or spiders. *Stress*, *14*(2), 185–193.

Okon-Singer, H., Lichtenstein-Vidne, L., & Cohen, N. (2013). Dynamic modulation of emotional processing. *Biological Psychology*, *92*(3), 480–491.

O'Leary, K., Fitzpatrick, C. L., & Hallett, D. (2017). Math anxiety is related to some, but not all, experiences with math. *Frontiers in Psychology*, *8*, 2067. https://doi.org/10.3389/fpsyg.2017.02067.

Orbach, L., Herzog, M., & Fritz, A. (2020). State-and trait-math anxiety and their relation to math performance in children: The role of core executive functions. *Cognition*, *200*, 104271.

Organisation for Economic Co-operation and Development (OECD). (2013). *PISA 2012 results: Ready to learn: students' engagement, drive and self-beliefs (volume III): preliminary Version*. Paris, France: OECD.

Passolunghi, M. C. (2011). Cognitive and emotional factors in children with mathematical learning disabilities. *International Journal of Disability, Development and Education*, *58*(1), 61–73.

Passolunghi, M. C., Cargnelutti, E., & Pellizzoni, S. (2019). The relation between cognitive and emotional factors and arithmetic problem-solving. *Educational Studies in Mathematics*, *100*(3), 271–290.

Pizzie, R. G., & Kraemer, D. J. (2017). Avoiding math on a rapid timescale: Emotional responsivity and anxious attention in math anxiety. *Brain and Cognition*, *118*, 100–107.

Plake, B. S., & Parker, C. S. (1982). The development and validation of a revised version of the Mathematics Anxiety Rating Scale. *Educational and Psychological Measurement*, *42*(2), 551–557.

Ramirez, G., Chang, H., Maloney, E. A., Levine, S. C., & Beilock, S. L. (2016). On the relationship between math anxiety and math achievement in early elementary school: The role of problem solving strategies. *Journal of Experimental Child Psychology*, *141*, 83–100.

Ramirez, G., Gunderson, E. A., Levine, S. C., & Beilock, S. L. (2013). Math anxiety, working memory, and math achievement in early elementary school. *Journal of Cognition and Development*, *14*, 187–202.

Ramirez, G., Hooper, S. Y., Kersting, N. B., Ferguson, R., & Yeager, D. (2018). Teacher math anxiety relates to adolescent students' math achievement. *AERA Open*, *4*(1), 1–13.

Richardson, F. C., & Suinn, R. M. (1972). The mathematics anxiety rating scale: Psychometric data. *Journal of Counseling Psychology*, *19*(6), 551–554.

Ritchie, S. J., & Bates, T. C. (2013). Enduring links from childhood mathematics and reading achievement to adult socioeconomic status. *Psychological Science*, *24*(7), 1301–1308.

Roberts, S. O., & Vukovic, R. K. (2011). *The relation between parental involvement and math anxiety: Implications for mathematics achievement*. Society for Research on educational effectiveness. Retrieved from ERIC database. (ERIC document reproduction service no. ED 528 879).

Rubinsten, O. (2015). Link between cognitive neuroscience and education: The case of clinical assessment of developmental dyscalculia. *Frontiers in Human Neuroscience, 9*, 304.
Rubinsten, O., Bialik, N., & Solar, Y. (2012). Exploring the relationship between math anxiety and gender through implicit measurement. *Frontiers in Human Neuroscience, 6*, 279. https://doi.org/10.3389/fnhum.2012.00279.
Rubinsten, O., Eidlin, H., Wohl, H., & Akibli, O. (2015). Attentional bias in math anxiety. *Frontiers in Psychology, 6*, 1539. https://doi.org/10.3389/fpsyg.2015.01539.
Rubinsten, O., Eidlin Levy, H., & Daches Cohen, L. (2019). Probing the nature of deficits in math anxiety: Drawing connections between attention and numerical cognition. In *Mathematics Anxiety* (pp. 156–177). Routledge.
Rubinsten, O., Marciano, H., Eidlin Levy, H., & Daches Cohen, L. (2018). A framework for studying the heterogeneity of risk factors in math anxiety. *Frontiers in Behavioral Neuroscience, 12*, 291. https://doi.org/10.3389/fnbeh.2018.00291.
Rubinsten, O., & Tannock, R. (2010). Mathematics anxiety in children with developmental dyscalculia. *Behavioral and Brain Functions, 6*(1), 1–13.
Sarkar, A., Dowker, A., & Kadosh, R. C. (2014). Cognitive enhancement or cognitive cost: Trait-specific outcomes of brain stimulation in the case of mathematics anxiety. *Journal of Neuroscience, 34*(50), 16605–16610.
Scrimin, S., Altoè, G., Moscardino, U., Pastore, M., & Mason, L. (2016). Individual differences in emotional reactivity and academic achievement: A psychophysiological study. *Mind, Brain, and Education, 10*(1), 34–46.
Scrimin, S., & Mason, L. (2015). Does mood influence text processing and comprehension? Evidence from an eye-movement study. *British Journal of Educational Psychology, 85*(3), 387–406.
Scrimin, S., Mason, L., Moscardino, U., & Altoè, G. (2015). Externalizing behaviors and learning from text in primary school students: The moderating role of mood. *Learning and Individual Differences, 43*, 106–110.
Shackman, A. J., Stockbridge, M. D., Tillman, R. M., Kaplan, C. M., Tromp, D. P., Fox, A. S., et al. (2016). The neurobiology of dispositional negativity and attentional biases to threat: Implications for understanding anxiety disorders in adults and youth. *Journal of Experimental Psychopathology*, 7(3), 311–342.
Skaalvik, E. M. (2018). Mathematics anxiety and coping strategies among middle school students: Relations with students' achievement goal orientations and level of performance. *Social Psychology of Education, 21*(3), 709–723.
Skagerlund, K., Lind, T., Strömbäck, C., Tinghög, G., & Västfjäll, D. (2018). Financial literacy and the role of numeracy – How individuals' attitude and affinity with numbers influence financial literacy. *Journal of Behavioral and Experimental Economics, 74*, 18–25.
Skagerlund, K., Östergren, R., Västfjäll, D., & Träff, U. (2019). How does mathematics anxiety impair mathematical abilities? Investigating the link between math anxiety, working memory, and number processing. *PLoS One, 14*(1).
Sokolowski, H. M., Hawes, Z., & Lyons, I. M. (2019). What explains sex differences in math anxiety? A closer look at the role of spatial processing. *Cognition, 182*, 193–212.
Sorvo, R., Koponen, T., Viholainen, H., Aro, T., Räikkönen, E., Peura, P., et al. (2019). Development of math anxiety and its longitudinal relationships with arithmetic achievement among primary school children. *Learning and Individual Differences, 69*, 173–181.
Stein, N. L., & Levine, L. J. (1989). The causal organisation of emotional knowledge: A developmental study. *Cognition & Emotion, 3*(4), 343–378.
Stoet, G., Bailey, D. H., Moore, A. M., & Geary, D. C. (2016). Countries with higher levels of gender equality show larger national sex differences in mathematics anxiety and relatively lower parental mathematics valuation for girls. *PloS One, 11*(4), e0153857. 1–24.
Suárez-Pellicioni, M., Núñez-Peña, M. I., & Colomé, A. (2013). Mathematical anxiety effects on simple arithmetic processing efficiency: An event-related potential study. *Biological Psychology, 94*(3), 517–526.

Suárez-Pellicioni, M., Núñez-Peña, M. I., & Colomé, À. (2015). Attentional bias in high math-anxious individuals: Evidence from an emotional Stroop task. *Frontiers in Psychology*, *6*, 1577. https://doi.org/10.3389/fpsyg.2015.01577.

Suárez-Pellicioni, M., Núñez-Peña, M. I., & Colomé, À. (2016). Math anxiety: A review of its cognitive consequences, psychophysiological correlates, and brain bases. *Cognitive, Affective, & Behavioral Neuroscience*, *16*(1), 3–22.

Supekar, K., Iuculano, T., Chen, L., & Menon, V. (2015). Remediation of childhood math anxiety and associated neural circuits through cognitive tutoring. *Journal of Neuroscience*, *35*(36), 12574–12583.

Swartz, J. R., Knodt, A. R., Radtke, S. R., & Hariri, A. R. (2015). A neural biomarker of psychological vulnerability to future life stress. *Neuron*, *85*(3), 505–511.

Thomas, G., & Dowker, A. (2000). Mathematics anxiety and related factors in young children. In *British psychological society developmental section conference, September.*

Thompson, R. A. (1990, January). Emotion and self-regulation. In *Vol. 36. Nebraska symposium on motivation* (pp. 367–467). Lincoln, Nebraska: University of Nebraska Press.

Vukovic, R. K., Kieffer, M. J., Bailey, S. P., & Harari, R. R. (2013). Mathematics anxiety in young children: Concurrent and longitudinal associations with mathematical performance. *Contemporary Educational Psychology*, *38*(1), 1–10.

Wang, Z., Hart, S. A., Kovas, Y., Lukowski, S., Soden, B., Thompson, L. A., et al. (2014). Who is afraid of math? Two sources of genetic variance for mathematical anxiety. *Journal of Child Psychology and Psychiatry*, *55*(9), 1056–1064.

Wang, Z., Lukowski, S. L., Hart, S. A., Lyons, I. M., Thompson, L. A., Kovas, Y., et al. (2015). Is math anxiety always bad for math learning? The role of math motivation. *Psychological Science*, *26*(12), 1863–1876.

Wang, Z., Rimfeld, K., Shakeshaft, N., Schofield, K., & Malanchini, M. (2020). The longitudinal role of mathematics anxiety in mathematics development: Issues of gender differences and domain-specificity. *Journal of Adolescence*, *80*, 220–232.

Wunsch, K., Meier, M., Ueberholz, L., Strahler, J., & Kasten, N. (2019). Acute psychosocial stress and working memory performance: The potential of physical activity to modulate cognitive functions in children. *BMC Pediatrics*, *19*(1), 271. https://doi.org/10.1186/s12887-019-1637-x.

Young, C. B., Wu, S. S., & Menon, V. (2012). The neurodevelopmental basis of math anxiety. *Psychological Science*, *23*(5), 492–501.

Zhang, J., Zhao, N., & Kong, Q. P. (2019). The relationship between math anxiety and math performance: A meta-analytic investigation. *Frontiers in Psychology*, *10*, 1–17.

CHAPTER 14

# What predicts the development of fact retrieval speed and calculation accuracy in children with (and without) arithmetic disabilities (AD). What can we learn from longitudinal studies?

**Annemie Desoete**
Ghent University, Faculty of Psychology and Educational Sciences, Department of Experimental-Clinical and Health Psychology, Ghent, Belgium

## Introduction

Arithmetic is essential to Google on the internet, to use a mobile phone, to travel, to sort out household bills, etc. (Budd, 2015). Already at the very start of the first year of formal education, individual differences exist amongst children in terms of arithmetic proficiency development (Claessens, Duncan, & Engel, 2009; Claessens & Engel, 2013). As teaching continues, differences amongst children live on (Dowker, 2015).

## Arithmetic disability

For some children, the differences in arithmetic proficiency are the result of a specific learning disorder (American Psychiatric Association, 2013; Desoete & Baten, 2017), causing arithmetic skills substantially below those expected for the individual's chronological age and persisting arithmetic problems despite the provision of interventions that target those difficulties (Fuchs & Vaughn, 2012; Vaughn & Fuchs, 2003). Arithmetic disabilities (AD) has neurobiological underpinnings (Ashkenazi, Black, Abrams, Hoeft, & Menon, 2013; Docherty et al., 2010; Grigorenko et al., 2020), causing abnormalities at the cognitive level, resulting in arithmetic underachievement and persistent difficulties that become fully manifest in elementary school when the requirements for numerical skills exceed the individual's limited

*Heterogeneous Contributions to Numerical Cognition*
https://doi.org/10.1016/B978-0-12-817414-2.00007-5

capacities (e.g., as in timed tests or long calculation tasks). The prevalence of AD varies from 4% to 8% (Barbaresi, Katusic, Colligan, Weaver, & Jacobsen, 2005; Desoete, Roeyers, & De Clercq, 2004; Grigorenko et al., 2020; Moll, Kunze, Neuhoff, Bruder, & Schulte-Körne, 2014). The prevalence for siblings ranges from 40% to 64% (Desoete, Praet, Titeca, & Ceulemans, 2013; Shalev et al., 2001).

## Fact retrieval speed and calculation accuracy

Although some researchers have doubts about the relevance of subtyping children with AD, there is a general consensus about some heterogeneity of arithmetic tasks (such as fact retrieval speed and calculation accuracy), difficulties, and deficiencies (Dowker, 2015; Dowker, De Smedt, & Desoete, 2019; Henik, Rubinstein, & Ashkenazi, 2015).

To study this heterogeneity, a data-driven model-based clustering was used to identify subgroups of individuals with relatively homogeneous profiles on arithmetic skills (fact retrieval and calculation) in 391 Belgian children (Pieters, Roeyers, Rosseel, Van Waelvelde, & Desoete, 2015). Such a clustering approach is based on the assumption that the observed data come from a population consisting of several subpopulations. Each subpopulation is modeled by a (multivariate) normal distribution. Several different types of normal distributions (spherical, ellipsoidal, or diagonal shape, same or different shape for each component) were fitted, each one for a number of cluster solutions, ranging from 1 up to 10 clusters. The Bayesian information criterion (BIC) was used to compare the different models and the different cluster solutions, whereby a higher BIC value indicated a better fit of the model. The BIC plot (see Fig. 1) helped in deciding on the best model, looking for a peak in the plot and a clear separation between the clusters on the scatter plot matrix. The best cluster solution in this study was an unconstrained ellipsoidal model with three clusters.

In Fig. 1, the BIC plots of the 10 different models are outlined. In this figure, an obvious peak can be seen. The scatter plot matrix (see Fig. 2) confirmed the three clusters across the two variables (fact retrieval speed and procedural calculation accuracy) indicating a clear separation of clusters (see Fig. 2).

Three emergent clusters (see Fig. 2) displayed profiles that conformed to a group without arithmetic problems, and two groups with arithmetic problems characterized by different profiles regarding their scores for fact retrieval speed and procedural calculation accuracy, respectively.

Cluster 1 included average achieving (AA) children without learning disabilities ($n = 209$ of whom 84 girls) and with high scores on both fact

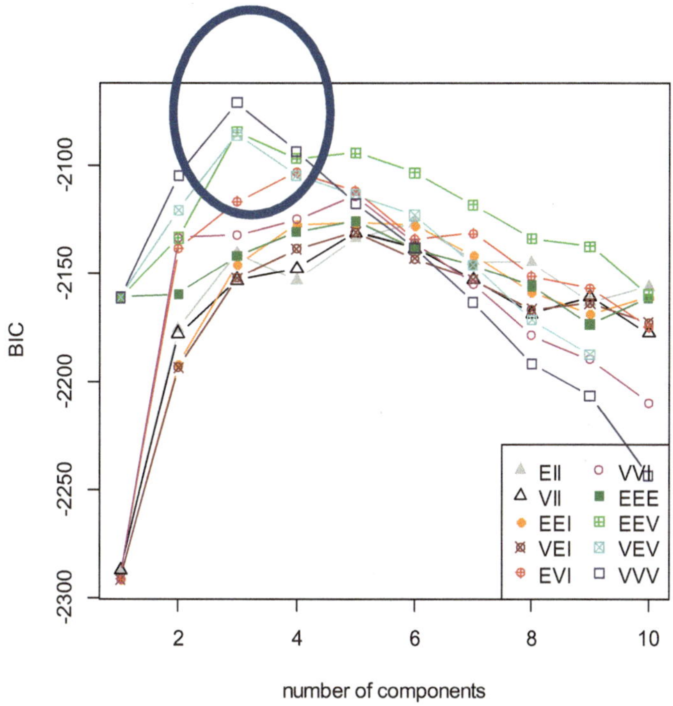


**Fig. 1** Cluster analysis. *Note. BIC*, Bayesian information criterion of different cluster solutions; *EEE*, ellipsoidal, equal volume, equal shape, equal orientation; *EEI*, diagonal, equal volume, equal shape, coordinate axes orientation; *EEV*, ellipsoidal, equal volume, equal shape, variable orientation; *EII*, spherical, equal volume, equal shape; *VEI*, diagonal, variable volume, equal shape, coordinate axes orientation; *VEV*, ellipsoidal, variable volume, equal shape, variable orientation; *VII*, spherical, variable volume, equal shape; *VVI*, diagonal, variable volume, variable shape, coordinate axes orientation; *VVV*, ellipsoidal, variable volume, variable shape, variable orientation.

retrieval speed and calculation accuracy. Cluster 2 ($n$=112 of whom 51 girls) manifested a pattern of the lowest scores for both speed and accuracy in children with AD. Cluster 3 ($n$=70 of whom 36 girls) exhibited also children with AD. However, these children had low scores for accuracy, but not for fact retrieval speed. There were no significant differences in socioeconomic status (SES) between these children allocated to the three clusters. Linear regressions revealed that the higher the value for the posterior probability of belonging to cluster 1 (AA children), the higher the score for decoding words and nonwords, spelling, motor skills, handwriting quality, and handwriting speed. The higher the probability of belonging to cluster 2 (speed and accuracy problems in AD), the lower the score for reading of

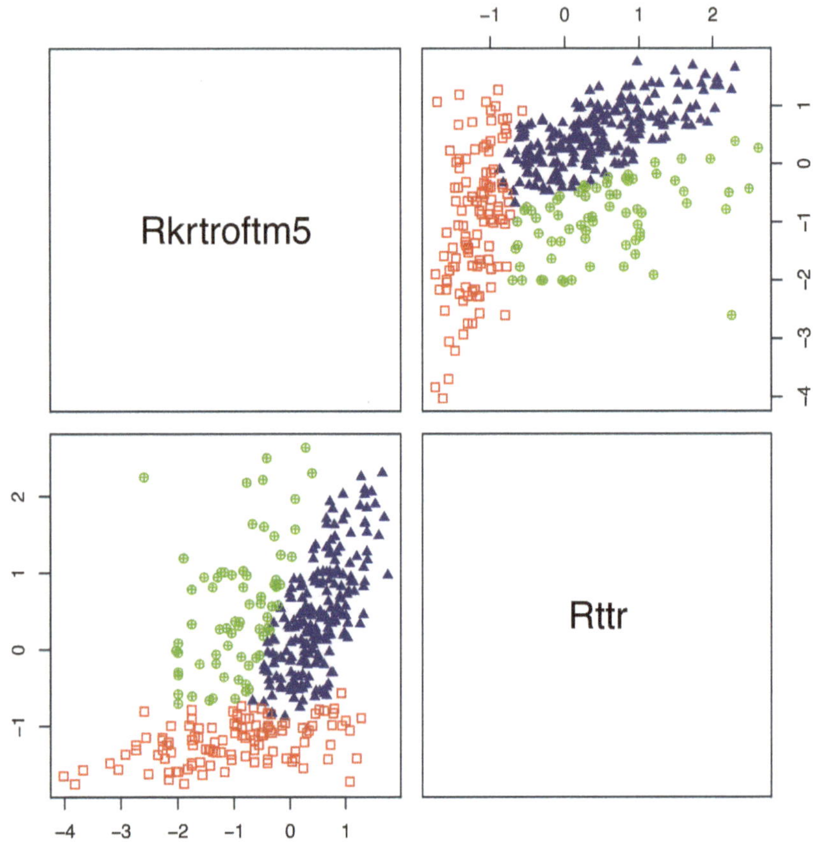


**Fig. 2** Scatter plot matrix showing three clusters.

existing and nonwords, motor skills, handwriting quality, and handwriting speed. A higher probability of belonging to cluster 3 (especially accuracy problems in AD) was significantly related to lower scores for reading of existing words, spelling, and motor skills (Pieters et al., 2015).

## Predictors of arithmetic abilities

Several skills play a role in the development of arithmetic ability.

### *Logical abilities*

Seriation and classification have been found to be involved in arithmetic development (Grégoire, 2005). Classification seems relevant to know the cardinal of a set (e.g. "How many balls can you see in this picture?"), whereas seriation is needed when dealing with ordinal numbers (e.g."Circle the third ball from the beginning."). Nunes et al. (2006) revealed that these abilities and working memory of 6-year-old children predicted arithmetic abilities 16 months later. Even after checking working memory, logical abilities remained a strong predictor of arithmetic abilities. Thus it might be interesting to look whether children with AD may be detected through their problems with seriation and classification as preschoolers.

### Counting abilities

Children's counting skills also play an important role in the development of arithmetic abilities. Aunola, Leskinen, Lerkkanen, and Nurmi (2004) followed 194 children from preschool through second grade and concluded that the initial level of counting knowledge in preschool was the best predictor of subsequent arithmetic abilities and for the amount of growth in arithmetic abilities over years. Although much of the research viewed counting as a unitary ability, Dowker (2005) suggested that counting consists of procedural and conceptual aspects. Procedural counting knowledge is defined as children's ability to perform an arithmetic task, for example, when a child can successfully determine that there are five objects in an array (Lefevre et al., 2009). Conceptual knowledge reflects a child's understanding of how a procedure works or whether a procedure is legitimate (Lefevre et al., 2009). It might be interesting to look whether children with AD can be detected by their procedural and conceptual counting skills as preschoolers.

### Comparison, subitizing

There are some arguments for the predictive value of "number *comparison*" (De Smedt, Verschaffel, & Ghesquière, 2009) tasks as key precursors of arithmetical development, with deficits leading to AD (e.g., Moeller, Pixner, Kaufmann, & Nuerk, 2009; Piazza et al., 2010). In addition, according to the triple code model (Dehaene & Cohen, 1995; Schmithorst & Brown, 2004; Siemann & Petermann, 2018), there are three types of *representations* for numbers. Two of them are symbolic and format-dependent: a visual Arabic number form (e.g., "5") and a verbal word frame with number-words (e.g., "five"), and one is nonsymbolic and format-independent: the analog magnitude representation (e.g., five dots). *Subitizing* is the rapid apprehension of small numbers (Hannula, Räsänen, & Lehtinen, 2007). It might be interesting to look whether children with AD can be detected by their comparison and subitizing skills before entering elementary school and formal arithmetic instruction.

### *Language*

Although children process numbers way before the acquisition of language (Dehaene, 2001), the value of including language has been stressed for the prediction of numeracy development. Some studies (Barner, Chow, & Yang, 2009; Negen & Sarnecka, 2012) showed that general language development measures likewise predicted number-word knowledge, although other studies (e.g., Ansari et al., 2003) did not find such a link. Additional longitudinal studies seem indicated.

### *Noncognitive predictors: Motivation and anxiety*

Finally, there might also exist some noncognitive predictors for arithmetic abilities such as motivation (Froiland, Davison, & Worrell, 2016; Ryan & Deci, 2000, 2017) and anxiety (Eunsook, 1999). In a meta-analysis, Taylor et al. (2014) highlighted a positive relationship between autonomous motivation (motivation where the force to fulfill a task is internal, e.g., passion) and general school achievement, in addition to a negative relationship between controlled motivation (motivation where the force to fulfill a task is external, e.g., reward-related) and academic achievement or learning. In addition, difficult exercises seem related to self-reported test anxiety (Baten & Desoete, 2018; Eunsook, 1999). Although the importance of autonomous has been documented (Kriegbaum, Jansen, & Spinath, 2015; Ryan & Deci, 2017) even for arithmetic (Taylor et al., 2014), the question of motivational differences and the relationship with math anxiety in children with AD has received limited attention in longitudinal studies.

## Purpose and research questions

Although the number of studies in the area of "numerical cognition" is growing rapidly, especially studies with a longitudinal design are needed to make predictions on individual differences in learning. In the present chapter, such longitudinal studies are described to address the following three research questions (RQ).

RQ 1. What are potential predictors for numerical competences for infants and toddlers?

RQ 2. What are kindergarten predictors for numerical competences and AD in elementary school? Do logical abilities (seriation, classification), counting, and comparison matter? Is there a difference between comparison using Arabic numbers, number words, and dots? Is general number-line estimation, intelligence, and language in kindergarten a predictor for mathematical proficiency in elementary school? Are there noncognitive predictors for AD?

RQ 3. Can children outgrow AD and what are the best predictors for persistent AD?

## Results

### Findings in infants and toddlers

So as to detect predictors for infants and toddlers, and answering the first research question, 31 children were tested at the age of 8, 24, and 48 months (Ceulemans et al., 2015; Desoete, Ceulemans, Rousseau, & Roelants, 2021). Participants were part of a birth cohort of babies born between May 2008 and April 2009, living in different Flemish districts in Belgium. They were recruited within the scope of a longitudinal study for the Belgian government (http://www.steunpuntwvg.be).

Infants (Time 1; 8 months) completed a subitizing task using a habituation paradigm with one and three-element arrays of red dots on a white background (see Fig. 3). The task consisted of a habituation and a test phase. Six different displays were presented in (repeating) random order until the infant met the habituation criterion (a 50% reduction in looking time over three consecutive trials, relative to the first three trials) or until 14 trials were completed. All infants reached the habituation criterion. Afterward, infants were presented with six test displays containing the habituated (old) or the new number of dots in alternation (counterbalanced for order).

Furthermore, a manual search task within the subtilizing range was administered to the toddlers (Time 2; 24 months). A wooden box (25 cm × 12.5 cm × 31.5 cm) had a slit at the front oriented to the toddlers and an opening at the backside oriented to the researcher who faced the child at an—besides the box—empty table. Parents were told that some balls would be hidden by the researcher (through the slit in front of the box) to find out how children reacted, parents were also told that no wrong reaction existed. Children could search through the slit for 10 s after each type of trial. It was expected that children would search longer after the *more remaining* than after the *box empty* trials (see Fig. 4). This would indicate successful discrimination. Cumulative searching time was coded manually afterward using *The Observer XT* software (http://www.noldus.com).

In kindergarten (Time 3; 48 months), the early numerical skills of children were tested through the TEDI-MATH (Grégoire, Noel, & Nieuwenhoven, 2004).

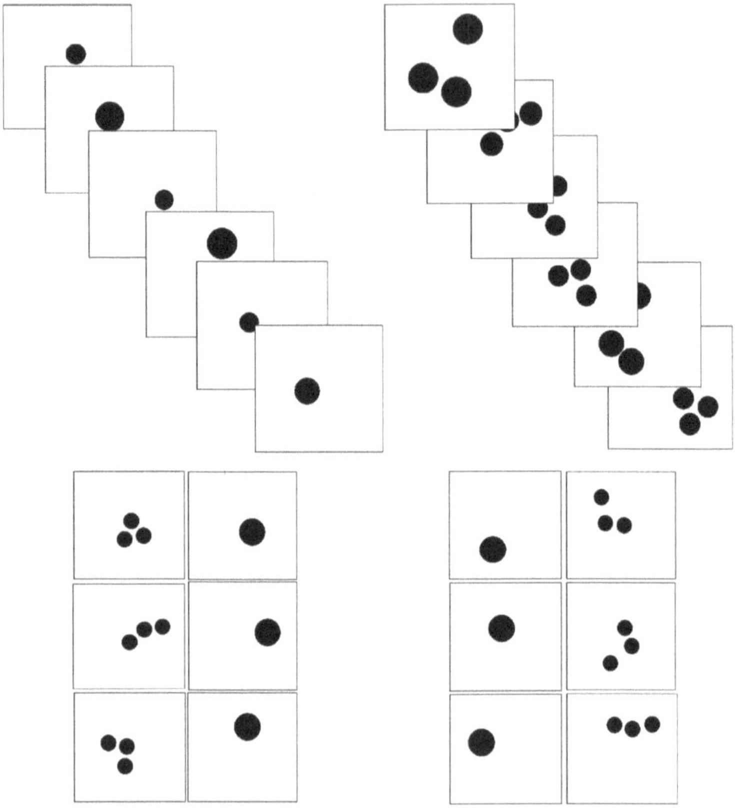

**Fig. 3** Example of stimuli used in the habituation and test phase (for each test trial pair).

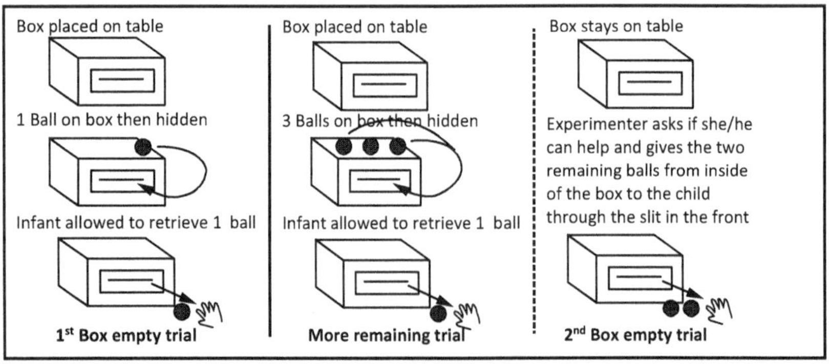


**Fig. 4** Different trial types of the manual search task. *Adapted from Feigenson, L., & Carey, S. (2005). On the limits of infants' quantification of small object arrays. Cognition, 97, 301. Copyright 2004 by Elsevier B.V.*

On top of that, during a structured play situation at 24 months (T2) and 48 months (T3), the frequency of numerical stimulation by the parent was recorded on video, coded, and counted. Parent and child sat on a soft pink plastic carpet and were instructed to build a house with a set of Duplo blocks according to a model. The model contained ordinary blocks and two blocks with numerical information. The set of blocks for mother and child was almost identical as the set for the model. A notable difference was two additional blocks, one with a number of rabbits and one with the number two. The purpose was to give all participants the opportunity to focus on these numerical cues, inconsistency during the play. Mother and child were blind for the true intention of the structured play observation. They were asked to play in a similar way as at home. After the full instruction was given, parent and child were left alone in the room, and after 5 min, the play was impeded. The structured play was recorded on tape and all numerical actions of mothers and/or children were coded.

The results of the present study revealed that subitizing on the manual search task of toddlers (T2) was significantly related to their conceptual counting ($r=0.31$) and calculation ($r=0.38$) proficiency in kindergarten (T3). Even on top of intelligence, a marginally significant effect remained for conceptual counting, $B=0.57$, SE $(B)=0.29$, $\beta=0.30$, $t$ $(27)=1.97$, $P=0.060$.

The linear regression analysis of numerical mother–child interaction through parental talk in toddlerhood (T2) revealed a marginally significant relationship with conceptual counting knowledge (T3), $F(1,27)=3.60$, $P=0.068$, $R^2=0.118$ with an effect size of $r=0.34$. A significant quadratic (positive) relationship was found between numerical mother–child interaction in toddlerhood (T2) and arithmetic operations in kindergarten (T3), $F(2,26)=3.68$, $P=0.039$, $R^2=0.22$ with an effect size of $r=0.47$. There were significant relationships between arithmetic operations (T3) and numerical mother–child interaction in toddlerhood (T3) and kindergarten (T3), respectively (Desoete et al., 2021).

## Findings in kindergarten

To study the second research question on kindergarten predictors, seven separate longitudinal studies were set up so as to look for kindergarten predictors. In the first study, the predictive value of logical abilities (seriation and classification), counting, and dot comparison were investigated. Moreover in study 2, in line with the Triple Code model, the comparison skills of children with and without AD using Arabic numbers, number words, and magnitudes were compared in kindergarten and in grade 2. In study 3 and 4, the predictive value of number-line estimation, intelligence, and general language abilities in kindergarten were looked at. Finally in study 5, 6, and 7, we looked for noncognitive predictors for AD.

### Study 1: Logical abilities, counting, and dot comparison

Children were tracked from kindergarten till grade 2 (Stock, Desoete, & Roeyers, 2010). Logical abilities such as seriation and classification, procedural and conceptual counting abilities, and magnitude comparison in kindergarten tested through the TEDI-MATH (Grégoire et al., 2004). Procedural knowledge of counting was assessed using accuracy in counting numbers, counting forward to an upper bound (e.g., "Count up to 6"), counting forward from a lower bound (e.g.,"Count from 3"), and counting forward with an upper and lower bound (e.g.,"Count from 5 up to 9"; see the appendix). Conceptual knowledge of counting was assessed with judgments about the validity of counting procedures. Children had to judge the counting of linear and random patterns of drawings and counters. Logical abilities were assessed using seriation and classification tasks. Children had to seriate numbers (e.g.,"Sort the [six] cards from the one with the fewest trees to the one with the most trees"). Children had to make groups of nine cards in order to assess the classification of numbers (e.g., "Make [three] groups with the cards that go together"). Magnitude comparison was assessed by comparison of dot sets (e.g., 4 dots vs 6 dots or 7 dots vs. 2 dots). All row item scores were summed and converted to $z$ scores and percentile scores were computed.

Results were compared for children with AD, low achievers, and for typical achievers in the first grades of elementary school. Children were retrospectively classified as having mathematical learning disabilities (AD) if they scored below the 10th percentile on at least one of the arithmetic achievement tests, both in first and second grade ($n=16$). Children who scored between the 10th and 25th percentiles on at least one of the arithmetic achievement tests, both in first and second grade, were classified as low achieving (LA; $n=27$). This group also included children who scored below the 10th percentile, in first or second grade, and who scored between the 10th and 25th percentiles the other year. The third group consisted of children who scored above the 25th percentile on all arithmetic achievement tests in both grades, these children were classified as average achievers (AA, $n=319$).

A discriminant analysis on kindergarten tests investigated the overall accurateness of the predicted classifications in the AD, LA, or AA groups (at age 7–8). Fisher's linear discriminant function was used to investigate the accurateness of the predicted classifications in the AD, LA, and AA groups. The overall Wilks' Lambda was significant, $L=0.831$, $c^2$ (10, $N=357)=65.263$, $P<0.001$, indicating that overall the kindergarten predictors differentiated amongst the AD, LA, and AA group.

In Table 1, the standardized weights of the predictors are presented. Based on these coefficients, magnitude comparison and seriation demonstrated the strongest relationships with the arithmetic achievement of the children. The means on the discriminant function were consistent with this interpretation (see Table 1). The AA children did better on the preparatory arithmetic abilities. Based on the scores for these kindergarten predictors, 88.0% were classified correctly into the AD, LA, or AA group in grades 1 and 2, whereas 88.0% of the cross-validated grouped cases were classified correctly. However, only 18.8% of the children with AD, none of the LA group, and 99.0% of the AA children were classified correctly. Especially dot comparison classified AD children correctly ($L=0.919$, $c^2$ (2, $N=357)=30.119$, $P<0.001$, with 99% of the AA children, none of the LA children, and 18.8% of the AD children classified correctly). The other predictors only classified 100% of the AA children but none of the other groups correctly, with an overall Wilks' Lambda for seriation ($L=0.929$, $c^2$ (2, $N=358)=26.256$,

**Table 1** Means (of *z*-scores) and standard deviations (in parentheses) (*F* values and standardized canonical discriminant function coefficients of the kindergarten predictors for the three groups).

| | AD | LA | AA | *F*(2, 340) |
|---|---|---|---|---|
| | M (SD) | M (SD) | M (SD) | |
| Procedural knowledge | −0.47 (0.92) | 0.16 (0.96) | 0.28 (0.92) | 3.73* |
| Conceptual knowledge | −0.38 (0.82) | 0.24 (0.74) | 0.24 (0.91) | 2.78* |
| Seriation | $-0.58_b$ (1.10) | $-0.52_b$ (1.08) | $0.24_a$ (0.88) | 11.92** |
| Classification | −0.43 (0.98) | 0.29 (1.05) | 0.19 (0.96) | 3.34* |
| Dot comparison | $-0.76_b$ (1.77) | $-0.05_a$ (1.11) | $0.19_a$ (0.52) | 15.39** |

*Note*: *AA*, average achievers; *AD*, Arithmetic disabilities group; *LA*, low achieving group. *$P\leq 0.05$; **$P\leq 0.008$ after Bonferroni adjustment. All item scores were converted to *z* scores. Means with different subscripts (a, b, or c) are significantly different by Bonferroni post hoc tests. Children in TA group (subscript *a*) were better than those in the AD group (subscript *b*) on seriation and magnitude comparison tasks. Children in the TA group were better than those in the LA group on seriation tasks (subscript *ab*) but not on magnitude comparison tasks (subscript *aa*). Children in the LA group were better than those in the AD group on magnitude comparison tasks (subscript *ab*) but not on seriation tasks (subscript *bb*).

$P<0.001$), classification ($L=0.982$, $c^2$ $(2, N=358)=6.577$, $P=0.037$), procedural counting ($L=0.971$, $c^2$ $(2, N=357)=10.365$, $P=0.006$), and conceptual counting ($L=0.981$, $c^2$ $(2, N=357)=6.901$, $P=0.032$) that were significant (Stock et al., 2010). In addition, it was investigated if the use of clinical cutoff scores (using percentile scores) in kindergarten (at age 5–6) was consistent with the previous analyses on $z$ scores. The percentage of true-positive children with AD (at age 7–8) correctly detected in kindergarten (at age 5–6) by the combination of the seriation, classification, procedural counting, conceptual counting, and magnitude comparison predictors was 87.50%. There were 18.25% false positives, or children with AD not failing at age 5–6 on the kindergarten measures but developing AD (at age 7–8) in elementary school. In our sample, none of the children with AD in elementary school failed on the seriation or classification tasks as preschoolers. There were 12.50% of the children with AD in elementary school with clinical scores on procedural counting tasks assessed in kindergarten. Conceptual counting assessed in kindergarten detected 31.25% of the children with AD in elementary school. Finally, 43.75% of the children with AD in elementary school already had clinical scores on magnitude comparison tasks in kindergarten. The clinical specificity was 89.40% in the AA group. In our data set, 0.15% of the AA children scored below the clinical cutoff on procedural counting tasks in preschool, and 5.95% scored below this cutoff on conceptual counting tasks. No false negatives were found on seriation, classification, and magnitude comparison tasks in kindergarten (Stock et al., 2010).

### *Study 2: Comparison of dots, number words, and Arabic numbers*

To compare the three types of number representations, the mean accuracy on symbolic (e.g., "5" and "five") and nonsymbolic (e.g., five dots) tasks were analyzed with tasks of the TEDI-MATH in 16 AA children, 64 LA, and 315 AA children in kindergarten and in grade 2 (Desoete, Ceulemans, De Weerdt, & Pieters, 2012). In this study, children completed subtests of an existing test, namely, the TEDI-MATH (Grégoire et al., 2004). Nonsymbolic magnitude comparison was assessed by comparison a collection of dots (as in study 1). Symbolic verbal number-word comparison was assessed by three kinds of tasks. In the first tasks, children have to judge if a spoken verbal numeral is a number word. In the second tasks, children have to judge if a number word is syntaxically correct. In a third task, children have to judge which of two spoken verbal numbers is the larger one. Symbolic Arabic number comparison was assessed by two kinds of tasks.

In the first tasks, children have to judge if a written Arabic symbol is a number. In the second tasks, children have to judge which of two written Arabic numbers the larger one is. The present study compared the mean accuracy of dots, number-words Arabic numbers. To compare between subtests, the scores were converted into *z*-scores.

The multivariate analysis of variance (MANOVA) revealed kindergarten differences between the children with AD, LA, and AA (see Table 2) on the multivariate level, $F(6,778)=7.434$, $P \leq 0.0005$, $\eta^2=0.05$. AA and LA performers scored better than AD performers on the comparison of dots and Arabic numbers, but not on the comparison of number words in kindergarten. AA children also had better scores than LA children for dot comparison tasks in kindergarten.

Children also differed in grade 2 (see Table 2) on the multivariate level, $F(6,750)=2.808$, $P \leq 0.01$, $\eta^2=0.02$. AA performers scored better than AD performers for the comparison of Arabic numbers and number words. No significant differences were revealed between LA children and the AA or AD children for the comparison of dots in grade 2 (Desoete et al., 2012).

### *Study 3 and 4: Number-line estimation, language, and intelligence*

In study 3, the prediction of number-line estimation, general kindergarten language, and intelligence for arithmetics in grade 1 and 2 was looked into (Praet & Desoete, 2014).

**Table 2** Means (of *z*-scores) and standard deviations (in parentheses) (*F* values of the predictors for the three groups in kindergarten and in grade 2).

| | AD | LA | AA | |
|---|---|---|---|---|
| | M (SD) | M (SD) | M (SD) | *F*(2,391) |
| **Kindergarten skills** | | | | |
| Arabic numbers | $-1.02_b$ (1.03) | $-0.16_a$ (1.06) | $0.08_a$ (0.93) | 10.2465** |
| Number words | −0.32 (0.98) | −0.12 (0.90) | 0.05 (1.01) | 1.644 |
| Dots | $-0.97_c$ (1.61) | $-0.22_b$ (1.29) | $0.09_a$ (0.85) | 10.558** |
| | | | | *F*(2,377) |
| **Grade 2 skills** | | | | |
| Arabic numbers | $-0.45_b$ (0.72) | −0.34 (0.89) | $0.09_a$ (1.01) | 6.721* |
| Number words | $-0.30_b$ (1.03) | −0.29 (0.89) | $0.08_a$ (1.00) | 4.345* |
| Dots | −0.42 (1.04) | −0.07 (1.01) | 0.04 (0.99) | 1.735 |

*Note*: *AA*, average achieving group; *AD*, arithmetic disabilities group; *LA*, low achieving group; $*P<0.01$, $**P \leq 0.001$, $_{(abc)}$ posthoc indexes $P<0.05$. The item scores were converted to *z* scores.

Number estimation was evaluated through the Number Line Estimation (NLE) test. Hundred and thirty-two children were required to put a single mark on 0–100 number line, using symbolic stimuli (Arabic numbers, number words) or nonsymbolic (dots) stimuli, so as to indicate the location of a number. The Percentage Absolute Error (PAE)—the amount by which their estimated deviated from the correct values—was calculated per child as a measure of children's estimation accuracy. To get a picture of their language skills, all children were tested in kindergarten with the Clinical Evaluation of Language Fundamentals, or the CELF-$_4$Nl (Kort, Schittekatte, & Compaan, 2008; Semel, Wiig, & Secord, 2008). The CELF-$_4$Nl resulted in a core language score, a receptive language index, an expressive language index, a language content index, and a language structure index. Intelligence was assessed with the Wechsler Preschool and Primary Scale of Intelligence or the WPPSI-III-NL (Hendriksen & Hurks, 2009; Wechsler et al., 2002). All children were followed up from kindergarten till grade 2.

The study revealed that most children in kindergarten had logarithmic or no valid representations on the number-to-position task of the 0–100 number line using symbolic stimuli (Arabic numbers, number words) or nonsymbolic (dots) stimuli. At the start of grade 1, when children received instruction on numbers 0–10, almost all children had valid but logarithmic representations for Arabic numbers and number words. At the end of grade 1, with children becoming familiar with numbers up to 20, about one-third had linear representations, whereas two-thirds still had more logarithmic representations of numbers 0–100 on all formats. At the start of grade 2, more than half of the children had linear representations of Arabic numbers, whereas this was only the case for 41.7% and 46.7% of the representation with number words or dots as stimuli. In the middle of grade 2, the representation became linear for nearly 60% of the children on all formats (Praet & Desoete, 2014). When analyzing if the linearity of the distribution ($R^2_{lin}$ values of linear fits) could predict arithmetic learning, this was marginally the case in kindergarten ($F(1,129) = 3.766$, $P = 0.055$, $R^2 = 0.029$) but not at the beginning of grade 1 ($F(1,125) = 2.347$, $P = 0.128$, $R^2 = 0.019$). At the end of grade 1, the regression analysis was significant for procedural calculation ($F(1,124) = 20.758$, $P < 0.001$, $R^2 = 0.144$), but not for fact retrieval speed ($F(1,124) = 3.551$, $P = 0.062$, $R^2 = 0.028$). Moreover, procedural calculation learning could be predicted by the number line linearity in the middle of grade 2 ($F(1,59) = 18.832$, $P < 0.001$, $R^2 = 0.245$), but this was not the case for fact retrieval speed at that moment ($F(1,54) = 1.821$, $P = 0.183$, $R^2 = 0.033$; Praet & Desoete, 2014).

So as to investigate whether intelligence and language could explain some of the inter individual variation in the "growth curves," chi-square test were used to compare the models with and without the interaction in all children. Our analysis revealed that intelligence (assessed in kindergarten) explained part of the variability in intercepts and slopes, whereas language (also assessed in kindergarten) explained variation when children enter the school system (in kindergarten) but not in the evolution of growth curves (Praet & Desoete, 2014).

In addition, half of the children of study 3 (Praet & Desoete, 2014) were participated in a follow-up study at the end of grade 2 to investigate language as a possible predictor for procedural calculation and fact retrieval at that moment (Praet & Desoete, 2016). Kindergarten language predicted procedural calculation at the end of grade 2 ($F(4,59)=6.209$, $P<0.001$, $R^2=0.31$), with significant results for expressive language ($P=0.012$). The kindergarten language components also predicted fact retrieval skills at the end of grade 2 ($F(4,54)=3.098$, $P=0.024$, $R^2=0.19$), with marginally significant results for receptive language ($P=0.057$) and significant results for the language structure index ($P=0.036$), pointing to the value of assessing general language in kindergarten (Praet & Desoete, 2016).

### *Study 5, 6, and 7: Noncognitive predictors*

In this section, the outcome of two longitudinal studies on noncognitive predictors, namely, on motivation and math anxiety, are summarized.

In study 5 (Vanrooy, 2019), a group of 38 children with AD and 31 children age-matched peers without AD and similar results on intelligence ($\chi^2$ $(1, N=67)$ $=34.31$, $P=0.310$), aged between 9 and 11 years were compared on arithmetic and motivation during a follow-up study.

Procedural calculation and fact retrieval as well as children's motivation for mathematics were measured during two consecutive years. To assess procedural calculation, all children completed the Kortrijk Arithmetic Test Revision (Baudonck et al., 2006), requiring them to solve calculations in a number-problem (e.g., $16-12=\ldots$) or in a word-problem (e.g., 1 less than 8 is ...) format. To assess fact retrieval they all solved the Arithmetic Number Fact Test (de Vos, 2002), whereas many simple additions, subtractions, divisions, multiplications, and mixed exercises have to be solved as possible in 5 min. Motivation was testing with an adapted measure of the Self-Regulation Questionnaire for Reading Motivation (De Naeghel, Van Keer, Vansteenkiste, & Rosseel, 2012) and items tapping into autonomous motivation (e.g., "I count because arithmetic is fun") and controlled motivation (e.g., "I count because I do not want to disappoint others").

The results revealed that arithmetic proficiency predicted motivation 1 year later. Procedural calculation accuracy predicted 15% of the variance in autonomous motivation ($R^2=0.15$, $F(1,59)=10.68$, $P=0.002$) and 16% of the variance in controlled motivation ($R^2=0.16$, $F(1,59)=10.99$, $P=0.002$) 1 year later. In addition, fact retrieval proficiency predicted 12% of the variation in autonomous motivation ($R^2=0.12$, $F(1, 59)=8.10$, $P=0.006$) and also 12% of the variation in controlled motivation ($R^2=0.12$, $F(1,59)=7.50$, $P=0.002$) of children 1 year later. Better arithmetic abilities correlated with more autonomous and less controlled motivation. However, the prediction from motivation to arithmetic was more complicated. More autonomous motivation predicted significantly 8% of the variance in fact retrieval speed ($R^2=0.08$, $F(1,61)=5.05$, $P=0.028$) and less controlled motivation predicted significantly 15% of the variance in procedural calculation accuracy ($R^2=0.15$, $F(1,61)=10.80$, $P=0.002$) 1 year later. However, autonomous motivation was not significantly predicting the variance in calculation accuracy ($R^2=0.04$, $F(1,61)=2.48$, $P=0.120$) 1 year later, and controlled motivation was only marginally significant in the prediction of the fact retrieval speed ($R^2=0.05$, $F(1,61)=3.47$, $P=0.067$) of children 1 year later.

Children with AD differed from peers without AD (see Table 3) for autonomous and for controlled motivation at both measuring points (Time 1: $F(2,60)=9.48$, $P<0.001$; Time 2: $F(2,66)=18.25$, $P<0.001$). Children with AD were less autonomously motivated and more controlled motivated when compared to their peers without AD, see Table 3.

**Table 3** Means (of *z*-scores) and standard deviations (in parentheses).

| | AA | AD | *F*(1,61) |
|---|---|---|---|
| | M (SD) | M (SD) | |
| **Time 1 (grade 3, 4, or 5)** | | | |
| Autonomous motivation | 3.69 (0.87) | 2.88 (0.96) | 12.00* |
| Controlled motivation | 2.72 (0.78) | 3.19 (0.80) | 5.51* |
| | | | ***F*(1,67)** |
| **Time 2 (grade 4, 5, or 6)** | | | |
| Autonomous motivation | 3.71 (0.81) | 2.71 (0.88) | 23.71** |
| Controlled motivation | 2.14 (0.73) | 2.58 (0.75) | 6.10* |

*F* values of the motivational predictors for the two Groups in elementary school.
*Note:* **$P<0.01$, *$P<0.05$, *AA*, average achievers; *AD*, children with arithmetic disabilities.

In study 6 (Landuyt, 2020), the math anxiety was compared for 54 elementary school children. Procedural calculation and fact retrieval as well as children's math anxiety were measured during two consecutive years. All children completed the same arithmetic test as in study 5. In addition, the Mathematics Anxiety Scale for Children (MASC), the child version of the Mathematics Anxiety Rating Scale (MARS; Richardson & Suinn, 1972), was completed by all participants rating items from not at all (0) to very much (3) and higher scores pointing to more anxiety.

The results revealed that math anxiety predicted arithmetic abilities 1 year later. There was a significant negative relationship between more anxiety and less fact retrieval proficiency ($\beta = -1.35$; $t(50) = -3.90$; $P = 0.001$) and between more anxiety and less procedural calculation accuracy ($\beta = -0.43$; $t(50) = -2.78$; $P = 0.008$) 1 year later. Intelligence and math anxiety predicted 42% of the variance in procedural calculation accuracy ($R^2 = 0.42$; $F(2,49) = 17.63$; $P \leq 0.001$) and 32% of the variance in fact retrieval speed ($R^2 = 0.32$; $F(2,49) = 11.45$; $P \leq 0.001$) 1 year later. However, the prediction from arithmetic proficiency to math anxiety was more complicated. There was a significant negative relationship between poor fact retrieval skills and more math anxiety ($\beta = -0.26$; $t(47) = -4.08$; $P < 0.001$) 1 year later, but the relationship between procedural calculation and math anxiety 1 year later was not significant ($\beta = -0.03$; $t(38) = -0.22$; $P = 0.827$).

In study 7 (Van Ryckeghem, 2019), the math anxiety was assessed and compared for 31 AA and 38 CE children with the same instruments as in study 6. The results confirmed the findings of study 6, revealing that intelligence predicted calculation accuracy 1 year later in the AD group ($\beta = 0.02$; $t(35) = 2.25$; $P = 0.031$). When math anxiety was added, the model remained significantly predicting procedural calculation tested 1 year later ($\beta = 0.02$; $t(34) = 2.42$; $P = 0.021$). Moreover, also in the AA group, math anxiety was predictively associated with calculation accuracy 1 year later even when controlling for intelligence ($\beta = -0.02$; $t(27) = -2.36$; $P = 0.026$). Intelligence and math anxiety explained 28.9% of the variance in procedural calculation 1 year later ($R^2 = 0.29$; $F(2, 27) = 5.48$; $P = 0.010$). The data of this study also revealed that girls with AD were more significantly more anxious than girls without AD ($U = 112.0$, $z = -2.77$, $P = 0.006$) and that boys with AD were significantly less anxious than girls with AD ($U = 66.0$, $z = -2.11$, $P = 0.035$), pointing to the fact to also keep math anxiety in mind when looking at variance in arithmetic abilities especially in girls with AD.

## Findings on the stability of AD diagnoses

To answer research question 3, we looked at differences between children outgrowing and not outgrowing an AD diagnosis in elementary school. Children (13 children with a diagnosis of AD and 13 AA peers) were tested at the age of 5, 6, 7, and 10 years. In line with previous research (De Weerdt, Desoete, & Roeyers, 2013; Passolunghi & Cornoldi, 2008; Passolunghi & Lanfranchi, 2012; Passolunghi & Mammarella, 2010; Passolunghi, Vercelloni, & Schadee, 2007), the role of the working memory was looked into. Working memory was tested in elementary school, at the age of 10. On top of that, (backward) digit-, word list-, listening-, and block recall, backward word list recall and backward block recall were used. In addition, all children were subjected to the spatial span test (see Fig. 5).

The study (Desoete, De Weerdt, Vanderswalmen, & De Bondt, 2014) showed that 54% (7 out of 13 children) of the children diagnosed as "children with AD" at the age of 7 years still experienced difficulties with procedural calculation or fact retrieval compared to the AA children at the age of 10 years (Desoete et al., 2014). The differences between the three groups of children are presented in Table 4.

Logistic regression analyses were performed, producing odds ratios, as indication of the fact that a condition was likely to occur in a group (see Table 5).

The father's SES turned out to be a significant predictor, $\chi^2$ $(2, n=26)=6.80$, $P=0.033$, Nagelkerke $R^2=0.26$. The model could correctly categorize 53.8% of the children, 57.1% of the children not outgrowing AD, 0% of the children outgrowing AD, and 76.9% of the control group.

Besides, seriation appeared to be the only significant kindergarten predictor, $\chi^2$ $(2, n=26)=6.37$, $P=0.041$, Nagelkerke $R^2=0.25$. Globally, 57.7% of the children were correctly categorized: 57.1% of the children not outgrowing AD, 0% of the children outgrowing AD, and 84.6% of the control group. The fact that none of the other kindergarten abilities (such as classification, dot comparison, or counting) was of importance might be explained by the dynamic aspect of arithmetic abilities. One may expect a decreasing influence of kindergarten abilities on the persistence of AD over time (see Table 5). Indeed, the study of Toll, Van der Ven, Kroesbergen, and Van Luit (2011) could correctly categorize 76.9% of the children if persistence was not taken into account, against 57.1% when we controlled for chronicity.

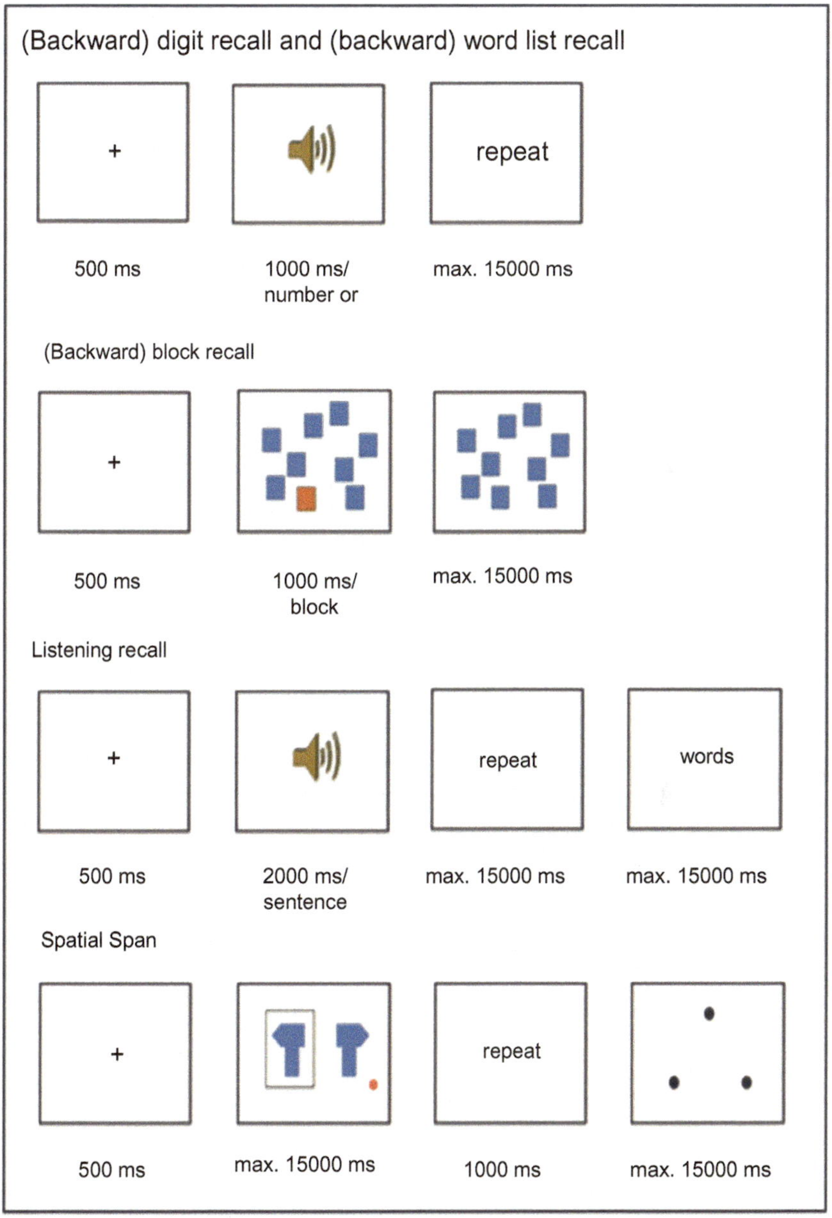


**Fig. 5** Visualization of the working memory tasks. In backward digit recall, backward word-list recall, and backward block recall, children are required to recall sequences of digits, words, or squares in the reverse order as measure of the central executive component of working memory. In the listening recall dual task, children had to verify sentences by stating "true" or "false" and memorize the final word for each sentence. In the spatial span dual task, children had to identify whether the shape on the right side was the same or opposite of the shape on the left. In addition, they had to recall the location of a *red dot* (see De Weerdt et al., 2013).

**Table 4** Means and standard deviations on predictors of stability of performance.

| | AA control children (*n* = 13) | Not outgrowing AD (*n* = 7) | Outgrowing AD (*n* = 6) |
|---|---|---|---|
| | M (SD) | M (SD) | M (SD) |
| Age | 123.08 (2.41) | 122.57 (2.97) | 119.50 (4.01) |
| SES mother | 15.46 (0.45) | 14.43 (0.48) | 16.33 (1.69) |
| SES father | 16.15 (0.62) | 13.57 (0.57) | 16.67 (1.78) |
| Intelligence | 104.54 (2.95) | 100.43 (3.32) | 109.33 (4.60) |
| *Kindergarten abilities* | 0.80 (0.23) | 0.33 (0.32) | 0.84 (0.19) |
| *(z-scores)* | 0.53 (0.10) | −0.20 (0.42) | 0.22 (0.19) |
| Conceptual counting | 0.64 (0.38)[a] | −0.16 (1.08)b | −0.01 (1.00)[ab]* |
| Procedural counting | −0.06 (0.70) | −0.62 (0.70) | −0.10 (1.01) |
| Seriation | 0.04 (0.23) | −0.19 (0.41) | −0.31 (0.47) |
| Classification | 1.00 (1.86)[a] | −0.26 (0.15)[b] | 0.18 (0.22)[b]*** |
| Dot comparison | 1.05 (0.15)[a] | −0.33 (0.18)[b] | 0.36 (0.26)[b]*** |
| *Reading and spelling* | 1.05 (0.10)[a] | −0.03 (0.20)[b] | 0.25 (0.26)[b]*** |
| *(z-scores)* | 4.89 (0.14)[a] | 3.95 (0.14)[b] | 4.39 (0.25)[ab]** |
| Reading existing words | 3.92 (0.11) | 3.67 (0.15) | 4.06 (0.14) |
| Reading pseudo words | 4.82 (0.15) | 4.41 (0.08) | 4.36 (0.08) |
| Spelling | 3.59 (0.17) | 3.12 (0.18) | 3.22 (0.07) |
| *Working memory* | 3.03 (0.12) | 2.79 (0.14) | 2.92 (0.10) |
| *Phonological loop* | 2.23 (0.13) | 1.93 (0.13) | 2.11 (0.13) |
| Digit recall | 4.40 (0.10) | 4.16 (0.06) | 4.11 (0.16) |
| Word-list recall | 3.60 (0.27)[a] | 2.60 (0.38)[ab] | 2.11 (0.24)[b]** |
| *Visuospatial sketchpad* | | | |
| Block recall | | | |
| *Central executive* | | | |
| Backward digit recall | | | |
| Backward word recall | | | |
| Listening recall dual task | | | |
| Backward block recall | | | |
| Spatial span dual task | | | |

*Note: AA*, average achieving participant, control group without arithmetic disability; *AD*, arithmetic disability; *SES*, socioeconomic status, as measured by years of parental education.
*$P < 0.05$; **$P \leq 0.01$; ***$P \leq 0.001$.
[a]Post hoc indices at $p < 0.05$.
[b]Post hoc indices at $p < 0.05$.

Moreover, the best model of predicting chronicity of AD consisted of nonword reading and spelling, $\chi^2$ (4, $n$ = 26) = 27.21, $P$ < 0.001, Nagelkerke $R^2$ = 0.74. Log-likelihood tests showed significant results for both nonword reading ($\chi^2$ (2, $n$ = 26) = 7.32, $P$ = 0.026) and spelling ($\chi^2$ (2, $n$ = 26) = 7.57, $P$ = 0.023). About 71.4% of the children not outgrowing AD, 50% of the children outgrowing AD, and 92.3% of the AA control group were correctly

**Table 5** Multinomial logistic regression models for predicting children outgrowing or not arithmetic disability (AD).

| Group comparison | Model | OR | 95% CI for OR Lower | Upper | Wald (*df*) |
|---|---|---|---|---|---|
| **SES** | | | | | |
| N-AD vs AA[a] | SES father | 0.57 | 0.33 | 0.97 | 4.23 (1)* |
| O-AD vs AA | SES father | 1.07 | 0.76 | 1.50 | 0.14 (1) |
| N-AD vs O_AD[b] | SES father | 0.53 | 0.30 | 0.96 | 4.42 (1)* |
| **Intelligence** | | | | | |
| N-AD vs O-AD | Intelligence | 1.11 | 0.96 | 1.28 | 2.02 (1) |
| **Kindergarten abilities** | | | | | |
| N-AD vs AA | Seriation | 0.18 | 0.03 | 1.06 | 3.59 (1) |
| O-AD vs AA | Seriation | 0.21 | 0.03 | 1.28 | 2.87 (1) |
| N-AD vs O-AD | Seriation | 0.85 | 0.27 | 2.70 | 0.08 (1) |
| **Reading and spelling** | | | | | |
| N-AD vs AA | Reading | 0.01 | 0.00 | 1.04 | 3.79 (1)* |
| | Spelling | 0.05 | 0.00 | 11.39 | 1.17 (1) |
| O-AD vs AA | Reading | 0.24 | 0.01 | 4.34 | 0.94 (1) |
| | Spelling | 0.01 | 0.00 | 0.74 | 4.36 (1)* |
| N-AD vs O-AD | Reading | 0.03 | 0.00 | 2.41 | 2.50 (1) |
| | Spelling | 6.36 | 0.14 | 300.33 | 0.88 (1) |
| **Working memory** | | | | | |
| N-AD vs AA | Digit recall | 0.02 | 0.00 | 0.72 | 4.59 (1)* |
| | Spatial span | 0.52 | 0.13 | 2.08 | 0.86 (1) |
| O-AD vs AA | Digit recall | 0.29 | 0.02 | 4.02 | 0.84 (1) |
| | Spatial span | 0.17 | 0.03 | 0.92 | 4.24 (1)* |
| N-AD vs O-AD | Digit recall | 0.07 | 0.00 | 2.11 | 2.34 (1) |
| | Spatial span | 3.03 | 0.57 | 16.10 | 1.69 (1) |

*Note*: *AA*, average achieving children; *CI*, confidence interval; *OR*, odds ratio; *N-AD*, not outgrowing AD; *O-AD*, outgrowing AD; *SES*, socioeconomic status, as measured by years of parental education; reading, reading nonwords.

*$P \leq 0.05$.

[a]AA group as reference category.

[b]O-AD group as reference category.

categorized. Globally, spelling and nonword reading, as measured at the age of 10, could categorize 76.9% of the children not outgrowing AD, children outgrowing AD, and AA control children. In line with previous studies on the importance of spelling (Pieters et al., 2015), spelling appeared to be the best predictor for outgrowing or not outgrowing AD.

Finally, in line with De Weerdt et al. (2013), the working memory was the strongest predictor for outgrowing AD or not, even stronger than the reading and spelling scores of the children (Desoete et al., 2014). Results revealed, in line with Geary, Hoard, Byrd-Craven, Nugent, and Numtee (2007) and Vukovic and Siegel (2010), that children not outgrowing AD performed poorer on digit recall than children without AD. In addition, the digit recall task (numerical component) and the spatial span task (central executive component) appeared to be the best working memory predictors. Whereas spatial span best predicted children outgrowing AD, digit recall turned out to be the best predictor of not outgrowing AD. The higher the numerical (phonological) and visuospatial central executive working memory scores, the higher the probability to belong to the control group. The model could correctly categorize 76.9% of the children, 66.7% were correctly categorized in the group of children outgrowing AD (Desoete et al., 2014).

## Discussion

Longitudinal studies were used to look for potential predictors for fact retrieval speed and calculation accuracy in children with and without AD.

### What are potential predictors for numerical competences for infants and toddlers?

Our findings in infants and toddlers (Ceulemans et al., 2015) confirmed that subitizing on the manual search task of toddlers was significantly related to their conceptual counting and calculation proficiency in kindergarten. In addition, there was a relationship between the numerical focus of mothers during the interactions in toddlerhood and the arithmetic abilities of their child in kindergarten. However, while the relationship in kindergarten age (48 months) was linear, the relationship at toddler age (24 months) was quadratic in nature. In kindergarten, this implied that more numerical focus through parental talk was associated with higher arithmetic learning. In toddlerhood, however, it seemed that more numerical focus of mothers only predicted higher arithmetic learning in kindergarten to some extent, needing the appropriate engagement at the right time. At higher rates, later learning declined again (Desoete et al., 2021). Future research needs to clarify this finding and the clinical relevance more in detail.

## What are kindergarten predictors for numerical competences and AD in elementary school?

Our findings in kindergarten (Stock et al., 2010) revealed, in line with Geary (2015), that it was possible to correctly categorize three out of four children in the AD, LA, or AA group in elementary school, based on their *counting knowledge*, *seriation*, *classification*, and dot *comparison scores* in kindergarten (Stock et al., 2010). However, it was easier to predict typical achievement (AA) than to detect children that were going to develop AD in elementary school, based on these kindergarten measurers. In addition, children with AD were also found to already have deficits for nonsymbolic and symbolic Arabic number comparison in kindergarten, whereas in grade 2 their deficits for processing symbolic information remained (Desoete et al., 2012).

In addition, general *language* explained some variation when children entered the school system (in kindergarten), but not the variation in the evolution of number-line estimation growth curves (Praet & Desoete, 2014). Kindergarten language predicted about one-third of the variation in procedural calculation and about one-fifth of the variation in fact retrieval skills at the end of grade 2 (Praet & Desoete, 2016).

Moreover, also studies on noncognitive predictors resulted in important findings. Our longitudinal data revealed that arithmetic abilities predicted *motivation* 1 year later (Vanrooy, 2019). Better fact retrieval or procedural calculation abilities resulted in more autonomous or internal motivation (where the force to fulfill a math task is internal, e.g., passion) and in less controlled (or reward-dependent) motivation. However, the prediction from motivation to arithmetic 1 year later was more complicated. More autonomous (passion driven) motivation only predicted significantly better fact retrieval skills. Less controlled motivation (reward dependent) only predicted significantly more procedural calculation accuracy in children 1 year later, pointing to the componential nature of arithmetic abilities. Elementary school children with AD were less autonomously motivated and more controlled (reward dependent) motivated compared to their peers without AD.

In addition, also *math anxiety* seems involved in the development of arithmetic abilities (Landuyt, 2020; Van Ryckeghem, 2019). Intelligence and math anxiety explained more than one-fourth of the variance in procedural calculation 1 year later (Van Ryckeghem, 2019). Anxious children were less accurate but also slower in less fact retrieval 1 year later (Landuyt, 2020).

However, the prospective prediction from arithmetical abilities to math anxiety was more complicated. Children that were slow in fact retrieval had more math anxiety the next year. The prediction from procedural calculation to math anxiety the next year was not significant, again pointing to the complexity and componential nature of arithmetic abilities (Landuyt, 2020). In addition, math anxiety was especially present in girls with AD (Van Ryckeghem, 2019).

## Can children outgrow AD?

Our findings (Desoete et al., 2014) revealed that less than half of the children with AD at the age of 6 and 7 outgrow this AD diagnosis at the age of 10. The less educated the father was, the higher the child's probability not outgrowing AD. The accuracy to spell and read nonwords contributed to the prediction of children outgrowing AD or not. Working memory correctly categorized more than three out of four children, and digit recall turned out to be the best predictor for not outgrowing (or persistent) AD. Results emphasize the dynamic aspect of AD and the importance of assessing the working memory, as well as reading and spelling by children with AD.

## Limitations

These results should be interpreted with care, since there are some limitations to the present studies. In some of the studies, the sample size was rather small and the number of independent variables was rather large, which might cause repercussions on the results. We should take into account that some significant associations or differences on population level might not be detected within this sample due to power issues. Future research should collect primary data from larger sample sizes. In addition, the present study differs from previous research exploring reaction time or Weber fraction and manipulating ratios in comparison tasks. In the current studies in kindergarten, children completed an existing and validated standardized arithmetic test, namely, the TEDI-MATH (Grégoire et al., 2004). This had some benefits in that the effects of various factors on mean arithmetic accuracy could be indexed against population norms. However, using this existing test had the disadvantage that the ratios or numerical distances could be less carefully controlled than we would have wished. Additionally, we only tested a limited number of predictors, and, in doing so, the importance and unique explained variance of these predictors might be overestimated. Byrnes and colleagues developed the Opportunity-Propensity (O–P) framework, aiming to combine predictors and differentiate between opportunity and propensity factors to explain variance and individual differences in development (Byrnes & Miller, 2016; Byrnes & Wasik, 2009). Thus additional data and analyses using this comprehensive O–P model seem indicated.

## Conclusion

Despite the limitations of these studies, our longitudinal data allowed us to confirm that many processes (counting, seriation, classification, comparison, language, motivation, math anxiety, working memory) and comorbid reading and spelling problems can differ between average achieving children (AA) and children with arithmetic disabilities (AD). However, with the current data it is difficult to arrive at conclusions about the unique contributions of these factors. Additional research is needed to combine the predictors and add some nuance to the literature on predictors for AA and AD. One of the comprehensive models to combine predictors is the Opportunity-Propensity (O–P) model (Byrnes, 2020; Wang, Shen, & Byrnes, 2013).

### Propensity predictors

In the O–P model, propensity predictors (P) refer to child variables that make people able (e.g., intelligence, math anxiety, working memory, or co-morbidity with reading and spelling problems) and/or willing (e.g., motivation) to learn.

Based on the findings of the current longitudinal studies, we can conclude that *counting, seriation, and classification* (Stock et al., 2010), number *comparison* (Desoete et al., 2012; Stock et al., 2010), and *language* (Praet & Desoete, 2014, 2016) might be candidates to be included as "propensity" factors in the prediction of arithmetic abilities and AD in elementary school. In addition, *motivation* might also be included in the comprehensive model as P-factor, since autonomous motivation predicted arithmetic speed and controlled motivation involved in calculation accuracy 1 year later, and children with AD were less autonomously and more controlled (or reward dependently) motivated compared to their peers without AD. However, the relationship between motivation and arithmetic seems bidirectional, since better arithmetic abilities were also related to higher motivation 1 year later (Vanrooy, 2019). Moreover, also *math anxiety* might also be included as P-factor in the development of arithmetic abilities, especially in girls with AD (Van Ryckeghem, 2019). Anxious children were less accurate in calculation and slower in fact retrieval 1 year later (Landuyt, 2020). However, also this relationship was bidirectional since children slow in fact retrieval experienced more math anxiety 1 year later (Landuyt, 2020). Finally, a longitudinal study (Desoete et al., 2014) also revealed the importance of including the accuracy to *spell and read nonwords* and *working memory* as propensity predictors of arithmetic abilities (AA) and disability (AD).

## Opportunity predictors

Opportunity factors (O) refer in the O–P model to contexts and variables that expose children to learning content (e.g., classroom instruction or home numeracy experiences and numerical focus of mothers during interactions with children).

Preliminary data about children with and without AD within this comprehensive model revealed the importance of the school environment, with teachers' experience contributing as O factor, and explaining about 6% of the variance in arithmetic abilities (Baten & Desoete, 2018). The longitudinal findings included in this chapter (Ceulemans et al., 2015) revealed in line with Levine, Suriyakham, Rowe, Huttenlocher, and Gunderson (2011) and Kleemans, Peeters, Segers, and Verhoeven (2012) the relationship between the numerical focus of mothers and the arithmetic learning as outcome. However, this learning might not only depend on the amount of numerical parental talk as "opportunity" factor, but also on the fact that the empowerment of "opportunities" should be within children's zone of proximal development (Desoete et al., 2021).

## Outgrowing AD

Additional longitudinal data and analyses combining these Propensity (P) and Opportunity (O) predictors and studying if these factors either interact or cancel each other out, using this comprehensive O–P model, seem indicated, especially since our longitudinal data also revealed that about 46% of the children with AD at the age of 6 and 7 outgrow AD at the age of 10 years (Desoete et al., 2014), stressing the dynamic aspect of AD.

## References

American Psychiatric Association. (2013). Neurodevelopmental disorders. In *Diagnostic and Statistical Manual of Mental Disorders* (5th ed.). APA. DSM-5.

Ansari, D., Donlan, C., Thomas, M. S. C., Ewing, S.A., Peen, T., & Karmiloff-Smith, A. (2003). What makes counting count? Verbal and visuo-spatial contributions to typical and atypical counting development. *Journal of Experimental Child Psychology*, *85*, 50–62. https://doi.org/10.1016/S0022-0965(03)00026-2.

Ashkenazi, S., Black, J. M., Abrams, D. A., Hoeft, F., & Menon, V. (2013). Neurobiological underpinnings of math and reading learning disabilities. *Journal of Learning Disabilities*, *46*, 549–569. https://doi.org/10.1177/0022219413483174.

Aunola, K., Leskinen, E., Lerkkanen, M.-K., & Nurmi, J.-E. (2004). Developmental dynamics of mathematical performance from preschool to grade 2. *Journal of Educational Psychology*, *96*, 699–713. https://doi.org/10.1037/0022-0663.96.4.699.

Barbaresi, W. J., Katusic, S. K., Colligan, R. C., Weaver, A. L., & Jacobsen, S. J. (2005). Learning disorder: Incidence in a population-based birth cohort, 1976-82, Rochester, Minn. *Ambulatory Pediatrics*, *5*, 281–289. https://doi.org/10.1367/A04-209R.1.

Barner, D., Chow, K., & Yang, S. (2009). Finding one's meaning: A test of the relation between quantifiers and integers in language development. *Cognitive Psychology*, *58*, 195–219. https://doi.org/10.1016/j.cogpsych.2008.07.001.

Baten, E., & Desoete, A. (2018). Mathematical (dis)abilities within the opportunity- propensity model: The choice of math test matters. *Frontiers in Psychology*, *9*. https://doi.org/10.3389/fpsyg.2018.00667. Art. 667 Developmental Psychology.

Baudonck, M., Debusschere, A., Dewulf, B., Samyn, F., Vercaemst, V., & Desoete, A. (2006). *De Kortrijkse Rekentest revision KRT-R. [the Kortrijk arithmetic test revision KRT-R]*. Kortrijk: CAR Overleie.

Budd, C. J. (2015). Promoting maths to the general public. In R. C. Kadosh, & A. Dowker (Eds.), *The oxford handbook of numerical cognition* (pp. 3–16). Oxford UK: Oxford University Press.

Byrnes, J. P. (2020). The potential utility of an opportunity-propensity framework for understanding individual and group differences in developmental outcomes: A retrospective progress report. *Developmental Review*, *56*. https://doi.org/10.1016/j.dr.2020.100911.

Byrnes, J. P., & Miller, D. (2016). The growth of mathematics and reading skills in segregated and diverse schools: An opportunity-propensity analysis of a national database. *Contemporary Educational Psychology*, *46*, 34–51. https://doi.org/10.1016/j.cedpsych.2016.04.002.

Byrnes, J. P., & Wasik, B. A. (2009). Factors predictive of mathematics achievement in kindergarten, first and third grades: An opportunity—Propensity analysis. *Contemporary Educational Psychology*, *34*, 167–183. https://doi.org/10.1016/j.cedpsych.2009.01.002.

Ceulemans, A., Titeca, D., Loeys, T., Hoppenbrouwers, K., Rousseau, S., & Desoete, A. (2015). The sense of small number discrimination: The predictive value in infancy and toddlerhood for numerical competencies in kindergarten. *Learning and Individual Differences*, *39*, 150–157. https://doi.org/10.1016/j.lindif.2015.03.009.

Claessens, A., Duncan, G., & Engel, M. (2009). Kindergarten skills and fifth-grade achievement: Evidence from the ECLS-K. *Economics of Education Review*, *28*, 415–427. https://doi.org/10.1016/j.econedurev.2008.09.003.

Claessens, A., & Engel, M. (2013). How important is where you start? Early mathematics knowledge and later school success. *Teachers College Record*, *115*(6), 060306.

De Naeghel, J., Van Keer, H., Vansteenkiste, M., & Rosseel, Y. (2012). The relation between elementary students' recreational and academic reading motivation, reading frequency, engagement, and comprehension: A self-determination theory perspective. *Journal of Educational Psychology*, *104*, 1006–1021. https://doi.org/10.1037/a0027800.

De Smedt, B., Verschaffel, L., & Ghesquière, P. (2009). The predictive value of numerical magnitude comparison for individual differences in mathematics achievement. *Journal of Experimental Child Psychology*, *103*, 469–479. https://doi.org/10.1016/j.jecp.2009.01.010.

de Vos, T. (2002). *Tempo test Rekenen (TTR). [The arithmetic number fact test]*. Amsterdam: Harcourt.

De Weerdt, F., Desoete, A., & Roeyers, H. (2013). Working memory in children with reading and/or mathematical disabilities. *Journal of Learning Disabilities*, *46*, 461–472. https://doi.org/10.1177/0022219412455238.

Dehaene, S. (2001). Précis of "the number sense". *Mind and Language*, *16*, 16–32. https://doi.org/10.1111/1468-0017.00154.

Dehaene, S., & Cohen, L. (1995). Towards an anatomical and functional model of number processing. *Mathematical Cognition*, *1*, 83–120.

Desoete, A., & Baten, E. (2017). Indicators for a specific learning disorder in mathematics or dyscalculia in toddlers and in kindergarten children. *Belgian Journal of Paedicatrics, 19*(2), 122–124.

Desoete, A., Ceulemans, A., De Weerdt, F., & Pieters, S. (2012). Can we predict mathematical learning disabilities from symbolic and non-symbolic comparison tasks in kindergarten? *British Journal of Educational Psychology, 82*, 64–81. https://doi.org/10.1348/2044-8279.002002.

Desoete, A., Ceulemans, A., Rousseau, S., & Roelants, M. (2021). The relative importance of "parental talk" as a predictor of the diversity in mathematics learning in young children. In A. Fritz, E. Guersoy, & M. Herzog (Eds.), *Diversity dimensions in mathematics and language learning*. Berlin: DeGruyter.

Desoete, A., De Weerdt, F., Vanderswalmen, R., & De Bondt, A. (2014). How persistent is a diagnosis of mathematical disorder at an early age? A longitudinal study. *International Journal of Research on Learning Disabilities, 2*(1), 42–71.

Desoete, A., Praet, M., Titeca, D., & Ceulemans, A. (2013). Cognitive phenotype of mathematical learning disabilities: What can we learn from siblings? *Research on Developmental Disabilities, 34*, 404–412. https://doi.org/10.1016/j.ridd.2012.08.022.

Desoete, A., Roeyers, H., & De Clercq, A. (2004). Children with mathematics learning disabilities in Belgium. *Journal of Learning Disabilities, 37*, 50–61. https://doi.org/10.1177/00222194010340050 5.

Docherty, S. J., Davis, O. S. P., Kovas, Y., Meaburn, E. L., Dale, P. S., Petrill, S. A., et al. (2010). A genome-wide association study identifies multiple loci associated with mathematics ability and disability. *Genes, Brain, and Behavior, 9*, 234–247. https://doi.org/10.1111/j.1601-183X.2009.00553.x.

Dowker, A. (Ed.). (2005). Individual differences in arithmetic. In *Implications for psychology, neuroscience and education*. Hove, UK: Psychology Press.

Dowker, A. (2015). Individual differences in arithmetical abilities. The componential nature of arithmetic. In *The oxford handbook of mathematical cognition* (pp. 862–878). Oxford: Medicine UK.

Dowker, A., De Smedt, B., & Desoete, A. (2019). Editorial. Individual differences in arithmetical development. *Frontiers in Psychology, 10*. https://doi.org/10.3389/fpsyg.2019.02672. article 2672. Developmental Psychology.

Eunsook, H. (1999). Test anxiety, perceived test difficulty, and test performance: Temporal patterns of their effects. *Learning and Individual Differences, 11*(4), 431–447. https://doi.org/10.1016/S1041-6080(99)80012-0.

Froiland, M. J., Davison, M. L., & Worrell, F. C. (2016). Aloha teachers: Teacher autonomy support promotes native Hawaiian and Pacific islander students' motivation, school belonging, course-taking and math achievement. *Social Psychology of Education, 19*, 879–894. https://doi.org/10.1007/s11218-016-9355-9.

Fuchs, L. S., & Vaughn, S. (2012). Responsiveness-to-intervention: A decade later. *Journal of Learning Disabilities, 45*, 195–203. https://doi.org/10.1177/0022219412442150.

Geary, D. C. (2015). Preschool children's quantitative knowledge and long-term risk for functional innumeracy. In S. Chinn (Ed.), *The international handbook for mathematical difficulties and dyscalculia* (pp. 235–242). London/New York: Routledge.

Geary, D. C., Hoard, M. K., Byrd-Craven, J., Nugent, L., & Numtee, C. (2007). Cognitive mechanisms underlying achievement deficits in children with mathematical learning disability. *Child Development, 78*, 1343–1359. https://doi.org/10.1111/j.1467-8624.2007.01069.x.

Grégoire, J. (2005). Développement logique et compétences arithmétiques. Le modèle piagétien est-il toujours actuel? In M. Crahay, L. Verschaffel, E. De Corte, & J. Grégoire (Eds.), *Enseignement et apprentissage des mathématiques* (pp. 57–77). Brussels: De Boeck.

Grégoire, J., Noel, M., & Nieuwenhoven, V. (2004). *TEDI-MATH*. TEMA: Brussel/Harcourt: Antwerpen.

Grigorenko, E. L., Compton, D., Fuchs, L. S., Wagner, R. K., Willcutt, E. G., & Fletcher, J. M. (2020). Understanding, educating, and supporting children with specific learning disabilities: 50 years of science and practice. *American Psychologist, 75*, 37–51. https://doi.org/10.1037/amp0000452.

Hannula, M. M., Räsänen, P., & Lehtinen, E. (2007). Development of counting skills: Role on numerosity and subitizing-based enumeration. *Mathematical Thinking, 9*, 51–57.

Hendriksen, J., & Hurks, P. (2009). *WPPSI-III-NL | Wechsler preschool and primary scale of intelligence.*

Henik, A., Rubinstein, O., & Ashkenazi, S. (2015). Developmental dyscalculia as a heterogenous disability. In R. C. Kadosh, & A. Dowker (Eds.), *The Oxford handbook of numerical cognition* (pp. 662–667). Oxford, UK: Oxford University Press.

Kleemans, T., Peeters, M., Segers, E., & Verhoeven, L. (2012). Child and home predictors of early numaneracy skills in kindergarten. *Early Childhood Research Quarterly, 27*, 471–477. https://doi.org/10.1016/j.ecresq.2011.12.004.

Kort, W., Schittekatte, M., & Compaan, E. (2008). *CELF-4-NL Test voor diagnose en evaluatie van taalproblemen. Handleiding [Test for the evaluation of language problems. Manual].* Amsterdam: Pearson.

Kriegbaum, K., Jansen, M., & Spinath, B. (2015). Motivation: A predictor of PISA's mathematical competence beyond intelligence and prior test achievement. *Learning and Individual Differences, 43*, 140–148. https://doi.org/10.1016/j.lindif.2015.08.026.

Landuyt, J. (2020). *Bang om te rekenen. Een verkennend onderzoek naar rekenangst en rekenvaardigheden bij lagereschoolkinderen. [Afraid of mathematics. An exploratory study on math anxiety and arithmetics in elementary school children].* Unpublished Master Thesis Gent: Ghent University.

Lefevre, J.-A., Skwarchuk, S.-L., Smith-Chant, B. L., Fase, L., Kamawar, D., Bisanz, J., et al. (2009). Home numeracy experiences and Children's math performance in the early school years. *Canadian Journal of Behavioural Science, 41*, 55–66. https://doi.org/10.1037/a0014532.

Levine, S. C., Suriyakham, L. W., Rowe, M. L., Huttenlocher, J., & Gunderson, E. A. (2011). What counts in the development of young children's number knowledge? *Developmental Psychology, 47*, 302. https://doi.org/10.1037/a0019671.

Moeller, K., Pixner, S., Kaufmann, L., & Nuerk, H.-C. (2009). Children's early mental numberline: Logarithmic of decomposed linear? *Journal of Experimental Child Psychology, 103*, 503–515. https://doi.org/10.1016/j.jecp.2009.02.006.

Moll, K., Kunze, S., Neuhoff, N., Bruder, J., & Schulte-Körne, G. (2014). Specific learning disorder: Prevalence and gender differences. *PLoS One, 9*. https://doi.org/10.1371/journal.pone.0103537, e103537.

Negen, J., & Sarnecka, B. W. (2012). Number-concept acquisition and general vocabulary development. *Child Development, 83*, 2019–2027. https://doi.org/10.1111/j.1467-8624.2012.01815.x.

Nunes, T., Bryant, P., Evans, D., Bell, D., Gardner, A., Gardner, A., et al. (2006). The contribution of logical reasoning to the learning of mathematics in primary school. *British Journal of Developmental Psychology, 00*, 1–21.

Passolunghi, M. C., & Cornoldi, C. (2008). Working memory failures in children with arithmetical difficulties. *Child Neuropsychology, 14*, 387–400. https://doi.org/10.1080/09297040701566662.

Passolunghi, M. C., & Lanfranchi, S. (2012). Domain-specific and domain-general precursors of mathematical achievement: A longitudinal study from kindergarten to first grade. *British Journal of Educational Psychology, 82*, 42–63. https://doi.org/10.1111/j.2044-8279.2011.02039.x.

Passolunghi, M. C., & Mammarella, I. C. (2010). Spatial and visual working memory ability in children with difficulties in arithmetic word problem solving. *European Journal of Cognitive Psychology, 22*, 944–963. https://doi.org/10.1080/09541440903091127.

Passolunghi, M. C., Vercelloni, B., & Schadee, H. (2007). The precursors of mathematics learning: Working memory, phonological ability and numerical competence. *Cognitive Development, 22*, 165–184. https://doi.org/10.1016/j.cogdev.2006.09.001.

Piazza, M., Facoetti, A., Trussardi, A. N., Berteletti, I., Conte, S., Lucangeli, D., et al. (2010). Developmental trajectory of number acuity reveals a severe impairment in developmental dyscalculia. *Cognition*, *116*, 33–41. https://doi.org/10.1016/j.cognition.2010.03.012.

Pieters, S., Roeyers, H., Rosseel, Y., Van Waelvelde, H., & Desoete, A. (2015). Identifying subtypes among children with developmental coordination disorder and mathematical learning disabilities, using model-based clustering. *Journal of Learning Disabilities*, *48*(1), 83–95. https://doi.org/10.1177/0022219413491288.

Praet, M., & Desoete, A. (2014). Number line estimation from kindergarten to grade 2: A longitudinal study. *Learning and Instruction*, *33*, 19–28. https://doi.org/10.1016/j.learninstruc.2014.02.003.

Praet, M., & Desoete, A. (2016). Language in the prediction of mathematical (dis)abilities? Evidence from a longitudinal study following-up children from kindergarten till grade 2. *SCIREA Journal of Education*, *1*(1), 37–64.

Richardson, F. C., & Suinn, R. M. (1972). The mathematics anxiety rating scale: Psychometric data. *Journal of Counseling Psychology*, *19*, 551–554. https://doi.org/10.1037/h0033456.

Ryan, R., & Deci, E. (2000). Self-determination theory and the facilitation of intrinsic motivation, social development, and well-being. *The American Psychologist*, *55*, 68–78. https://doi.org/10.1037/0003-066X.55.1.68.

Ryan, R. M., & Deci, E. L. (2017). *Self-determination theory. Basic psychological needs in motivation, development and wellness*. New York, NY: Guilford Press.

Schmithorst, V. J., & Brown, R. D. (2004). Empirical validation of the triple-code model of numerical processing for complex math operations using functional MRI and group Independent Component Analysis of the mental addition and subtraction of fractions. *NeuroImage*, *22*, 1414–1420. https://doi.org/10.1016/j.neuroimage.2004.03.021.

Semel, E., Wiig, E. H., & Secord, W. A. (2008). *CELF 4 Nl clinical evaluation of language fundamentals*. Amsterdam: Pearson.

Shalev, R., Manor, O., Kerem, B., Ayali, M., Badichi, N., Friedlander, Y., et al. (2001). Developmental dyscalculia is a familial learning disability. *Journal of Learning Disabilities*, *34*, 59–65. https://doi.org/10.1177/002221940103400105.

Siemann, J., & Petermann, F. (2018). Evaluation of the triple code model of numerical processing—Reviewing past neuroimaging and clinical findings. *Research in Developmental Disabilities*, *72*, 106–117. https://doi.org/10.1016/j.ridd.2017.11.001.

Stock, P., Desoete, A., & Roeyers, H. (2010). Detecting children with arithmetic disabilities from kindergarten: Evidence form a three year longitudinal study on the role of preparatory arithmetic abilities. *Journal of Learning Disabilities*, *43*, 250–268. https://doi.org/10.1177/0022219409345011.

Taylor, G., Jungert, T., Mageau, G. A., Schattke, K., Dedic, H., Rosenfield, S., et al. (2014). A self-determination theory approach to predicting school achievement over time: The unique role of Intrinsic Motivation. *Contemporary Educational Psychology*, *39*, 342–358. https://doi.org/10.1016/j.cedpsych.2014.08.002.

Toll, S. W. M., Van der Ven, S. H. G., Kroesbergen, E. H., & Van Luit, J. E. H. (2011). Executive functions as predictors of math learning disabilities. *Journal of Learning Disabilities*, *44*, 521–532. https://doi.org/10.1177/0022219410387302.

Van Ryckeghem, L. (2019). *Hoe ik me voel doet ertoe… Een onderzoek naar rekenangst bij kinderen met en zonder dyscalculie in de lagere school. [How I feel, matters… a study on math anxiety in children with and without arithmetic disabilities in elementary school]*. Unpublished Master Thesis Gent: Ghent University.

Vanrooy, J. (2019). *Doet goesting beter rekenen of krijg je goesting door beter te rekenen? Een verkennende studie naar de rol van motivatie bij kinderen met en zonder dyscalculie. [Does motivation makes you better in mathematics or does mathematics makes you more motivated? An exploratory study on the role of motivation of children with and without arithmetic disabilities]*. Unpublished Master Thesis Ghent: Ghent University.

Vaughn, S., & Fuchs, L. S. (2003). Redefining learning disabilities as inadequate response to instruction: The promise and potential problems. *Learning Disabilities Research & Practice, 18*, 137–146.

Vukovic, R. K., & Siegel, L. S. (2010). Academic and cognitive characteristics of persistent mathematics difficulty from first through fourth grade. *Learning Disabilities Research & Practice, 25*, 25–38. https://doi.org/10.1111/j.1540-5826.2009.00298.x.

Wang, A. H., Shen, F., & Byrnes, J. P. (2013). Does the opportunity–propensity framework predict the early mathematics skills of low-income pre-kindergarten children? *Contemporary Educational Psychology, 38*, 259–270. https://doi.org/10.1016/j.cedpsych.2013.04.004.

Wechsler, D., Kort, W., Schittekatte, M., Bosmans, M., Compaan, E. L., Dekker, P. H., et al. (2002). *Wechsler intelligence scale for children-III-Nl*. Amsterdam, The Netherlands: Harcourt.

# CHAPTER 15

# Early neurocognitive development of dyscalculia

**Karin Landerl, Stephan E. Vogel, and Roland H. Grabner**
Institute of Psychology, University of Graz, Graz, Austria

## Introduction

According to the latest version of the International Classification of Diseases (ICD-11) of the World Health Organization (2020), dyscalculia, a "developmental learning disorder with impairment in mathematics, is characterized by significant and persistent difficulties in learning academic skills related to mathematics or arithmetic, such as number sense, memorization of number facts, accurate calculation, fluent calculation, and accurate mathematic reasoning." This definition covers a very wide range of maths-related skills: (a) numerical abilities, i.e., understanding and processing of nonsymbolic numerosities and symbolic numbers; (b) arithmetic abilities, i.e., mental calculations, retrieval of arithmetic facts, and calculation procedures; and (c) maths reasoning. The latter component is a very broad concept in itself and will not be in the focus of the current chapter.

While prevalence rates of dyscalculia are similarly high as for dyslexia/developmental reading disorder (see Landerl, Göbel, & Moll, 2013, for an overview), the paramount importance of numerical and mathematical competencies is not yet reflected in the accumulated amount of research. In particular, we know little about early developmental trajectories of children who later on develop impairments in mathematics. This is understandable as dyscalculia can only be diagnosed once children fall behind their peers in school. There is, however, a lot of theoretical as well as empirical knowledge on early predictors of and potential risk factors for later deficits in math learning. This chapter will give an overview of the current state of research on cognitive predictors and their potential neurobiological foundations as well as an outlook toward future research.

*Heterogeneous Contributions to Numerical Cognition*
https://doi.org/10.1016/B978-0-12-817414-2.00011-7

## Interplay of domain-specific and domain-general factors in math development

Mathematical abilities constitute a multidimensional construct. Two important types of cognitive predictors have been discussed in association with the development of mathematical abilities: Domain-specific predictors are those that are specifically relevant to the domain of mathematics. They mostly comprise different aspects of numerical abilities (e.g., representation of symbolic numbers—such as Arabic digits and number words—and nonsymbolic numerosities—representation of the number of elements in a set), as symbolic numbers and nonsymbolic numerosities are an integral part of mathematics and essential in other mathematical domains such as arithmetic. The second type is domain-general predictors, referring to cognitive and personality factors that are not specific to mathematics (e.g., working memory, executive functions, spatial reasoning, etc.), but are relevant for learning more generally (e.g., Szucs, Devine, Soltesz, Nobes, & Gabriel, 2014; see also Fig. 1). During development, various domain-specific and domain-general predictors can be assumed to interact with each other in a dynamic and complex way. However, our knowledge of this interplay is so far quite limited. It is largely unclear which factors contribute to math development at what time (from early infancy to adulthood), in which populations (from poor to high levels of math abilities), and for which math demand (from basic arithmetic to higher order mathematics). In addition, since most of the existing studies are cross-sectional, the factors that are predictive of math development and, therefore, potential markers for atypical developmental trajectories are largely unknown.

## Domain-specific predictors

Efficient processing of nonsymbolic numerosities and symbolic numbers is seen as a core mechanism of mathematical cognition. Current neurocognitive theories consider dyscalculia as a consequence of a possibly innate core deficit to represent nonsymbolic numerosities and/or symbolic numbers, even though the exact nature of such difficulties is as yet unclear. The *defective number module* hypothesis (Butterworth, 1999, 2010; Landerl, Bevan, & Butterworth, 2004) postulates that dyscalculia is induced by a highly selective impairment in the capacity to understand and represent cardinal numbers, while proponents of the *number sense deficit* hypothesis (Dehaene, Piazza, Pinel, & Cohen, 2003; Wilson & Dehaene, 2007) postulate an impairment in the ability to quickly approximate and manipulate nonsymbolic numerosities.

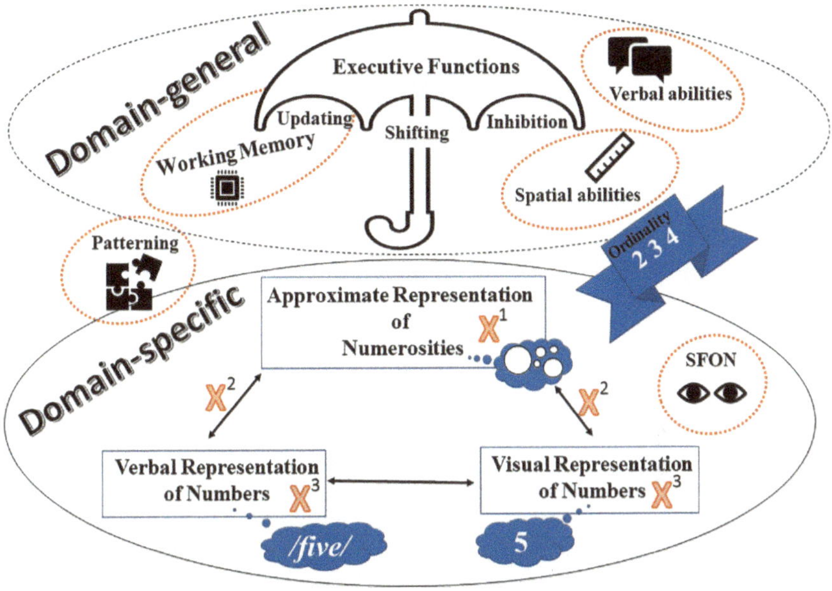


**Fig. 1** Schematic illustration of domain-general and domain-specific factors in the development of mathematical abilities. Both factors interact and individual differences can be attributed to differences in these interactions. Ordinality and patterning are located at the edges of domain-specific and domain-general abilities as it is not yet clear whether these processes are uniquely related to domain-specific knowledge or whether domain general mechanisms play a role. The red X's relate to common theories about the cognitive impairment in children with dyscalculia. 1 = "Core deficit model" predicts a deficit in the representation of numerosities; 2 = "access deficit model" predicts a deficit in the access of numerosities via symbolic representations; 3 = "deficit number model" predicts a deficit in the exact representation of symbolic numbers.

The *access deficit hypothesis* (Rousselle & Noël, 2007) assumes that numerical impairments in individuals with dyscalculia arise from a deficit in establishing an efficient association between nonsymbolic numerosities and symbolic numbers, rather than a deficit to internally represent nonsymbolic numerosities per se. All three hypotheses acknowledge that identifying, comparing, and ordering nonsymbolic numerosities and symbolic numbers are central aspects of numerical processing. Current research addresses a number of important issues within the domain of numerical processing.

## Nonsymbolic representation of numerosities

A measure that is often used to investigate nonsymbolic numerosity processing is the dot comparison task. In this task, participants are asked to decide as fast and as accurately as possible, which of two presented dot arrays contains more dots. The accuracy and reaction times in selecting the larger set of dots has been found to correlate with measures of school arithmetic (e.g., Mazzocco, Feigenson, & Halberda, 2011). While some studies also reported that children with dyscalculia are significantly less accurate and slower than age-matched controls in this task (e.g., Landerl, Fussenegger, Moll, & Willburger, 2009; Piazza et al., 2010), other studies did not confirm such a deficit (e.g., Rousselle & Noël, 2007).

Even infants have been shown to be attentive toward the numerosities of objects in the world; research has demonstrated that 6-month old infants are able to discriminate between nonsymbolic numerosities in specific habituation designs (Xu & Spelke, 2000). However, it is as yet under debate (i.e., longitudinal investigations are missing) whether this early attentiveness toward nonsymbolic numerosities is causally related to the later development of arithmetic. Current evidence rather suggests that it is the ability to understand and efficiently access symbolic representations of numbers that are crucial for arithmetic development (De Smedt, Noël, Gilmore, & Ansari, 2013; Schneider et al., 2017).

## Symbolic representation of numerosities

Long before children start doing school math, they acquire the understanding that number words and Arabic numbers represent nonsymbolic numerosities. Children's ability to understand symbolic representations of numerosities (i.e., number) is regarded as a crucial foundation of the development of math competencies (Ansari, 2008). There are positive correlations between performance in symbolic number comparison tasks (e.g., tasks in which participants are asked to indicate which of two presented numbers is numerically larger) and math competencies, in particular arithmetic (Schneider et al., 2017), and an impairment in understanding symbolic numbers is considered as one of the hallmark deficits in children with dyscalculia (Butterworth, 2011).

For instance, Rousselle and Noël (2007) found that children with dyscalculia showed lower performance when asked to select the numerically larger of two Arabic digits (i.e., symbolic numbers), but not on a dot comparison task (i.e., nonsymbolic numerosities). Based on a systematic review, De Smedt et al. (2013) concluded that symbolic number processing is a more robust predictor of arithmetic than nonsymbolic numerosity processing, as many studies failed to find a significant correlation between nonsymbolic comparison tasks and arithmetic. This assumption was recently confirmed by two metaanalyses (Fazio, Bailey, Thompson, & Siegler, 2014;

Schneider et al., 2017) reporting a significantly stronger association with mathematics for symbolic than for nonsymbolic processing. In studies that assessed both, symbolic and nonsymbolic processing, the latter typically does not contribute additional variance to the prediction of arithmetic above and beyond symbolic number processing (e.g., Lyons & Ansari, 2015), leading some researchers to the conclusion that nonsymbolic numerosity processing skills are "not particularly critical for children's development of school-relevant mathematical competencies" (De Smedt et al., 2013, p. 54).

However, it is still possible that adequate processing of nonsymbolic numerosities provides a foundation for the early mapping of symbolic number representations on specific numerosities. But evidence on early developmental periods is as yet scarce and mixed. A large study ($N = 539$) following children through their last year in kindergarten indicated that understanding of symbolic number representation drives development in nonsymbolic numerosity processing and not the other way round (Lyons, Bugden, Zheng, De Jesus, & Ansari, 2018). But at the same time, there is some evidence that symbolic number processing may be a mediator between early nonsymbolic processing and arithmetic development (Finke, Freudenthaler, & Landerl, 2020; van Marle, Chu, Li, & Geary, 2014). It is also possible that nonsymbolic numerosity processing is only impaired in a particular subtype of dyscalculia (Wilson & Dehaene, 2007). Such an association for a relatively small group of children may be concealed in large-scale studies of the general population and may only be uncovered in studies that focus on early developmental trajectories of children who later on develop deficits in arithmetic performance. In the future, it will be important to investigate whether an impairment in symbolic number understanding is already evident during early developmental periods, before problems in arithmetic become manifest.

## Ordinality

Another distinct conceptual dimension of symbolic number representation was recently discovered as being relevant for the development of arithmetic skills, i.e., knowledge about the relative rank or position of a number within a particular sequence (Lyons, Vogel, & Ansari, 2016). Understanding of order (or ordinality) is usually assessed by tasks asking participants whether or not sequences of Arabic numerals are consistently ascending or descending (e.g., 3–4–5 or 5–4–3 vs. 3–5–4). This task has been found to be a unique and reliable predictor of arithmetic skills in children (Lyons, Price, Vaessen, Blomert, & Ansari, 2014) and adults (Vogel et al., 2017, 2019). Recent evidence also suggests that the relative importance of ordinal processing to explain arithmetic competencies increases during the first years of formal education (Lyons et al., 2014) and fully mediates the association between nonsymbolic numerosity processing and arithmetic in children (Sasanguie & Vos, 2018). However, little is currently known about the early development of ordinal processing in typically developing children or children with dyscalculia and it is as yet unclear whether ordinality should be regarded as a domain-specific ability, since it is also relevant in other domains (e.g., the order of the alphabet).

## Cross-modal integration of number words and Arabic digits

When individuals with competent arithmetic skills talk about numbers, they usually do not specify whether they are referring to Arabic numbers or number words, indicating that those two symbolic notations are closely linked in their cognitive system. During early development, children typically learn number words first and later on associate Arabic digits with their corresponding number words (see Landerl, 2019 for an overview). However, we currently know hardly anything about the development of cross-modal integration of Arabic digits and number words. This is surprising given that research on letter-sound integration has demonstrated that even 5th graders who have ample reading experience do not yet show an adult-like EEG pattern of automatic letter-sound integration (Froyen, Van Atteveldt, Bonte, & Blomert, 2008) and that 11-year-old dyslexic individuals' neurocognitive pattern of letter-sound integration corresponds to typically developing 2nd graders (Froyen, Willems, & Blomert, 2011). While we know that individuals with dyscalculia have difficulties to access numerosities when presented with symbolic representations in terms of Arabic numbers or number words (Rousselle & Noël, 2007), it is as yet unclear whether they are also impaired in the process of cross-modal integration of verbal and visual-Arabic symbolic number forms (Dehaene, 1992, see Fig. 1).

## Understanding and transcoding of multidigit numbers

Tasks assessing the understanding of number and ordinality are mostly limited to single-digit numerals. However, when children start counting and doing arithmetic, they soon have to deal with the complexities of multidigit numbers. The parsimony of the Arabic number system most likely explains why it is used in most languages today (Chrisomalis, 2010). Its ingenious place-value system is based on a limited set of 10 single digits. The value of each digit is defined by its position within the sequence. Thus children must work out that the same symbol has very different meanings depending on its position in the number string. This is complicated by the fact that the structure of multidigit number words is language-dependent and not always consistent with Arabic multidigits. An important and quite frequent discrepancy between Arabic and number word systems is the inversion rule (e.g., German, Dutch, Slovenian), which requires to invert the order of units and decades (92 is literally pronounced as *two-and-ninety*).

There is evidence that the language-specific characteristics of number word structure can delay or accelerate the development of number transcoding. For example, German-speaking children have more difficulties with digit order in two-digit numbers than English-speaking children (e.g., Clayton et al., 2020; Steiner et al., 2021) or Japanese-speaking children who acquire a highly transparent number word system (Moeller, Zuber, Olsen, Nuerk, & Willmes, 2015). Early skills in identifying and writing of multidigit numbers were found to be related to arithmetic development in both, languages with inversion property (Imbo, Vanden Bulcke, De Brauwer, & Fias, 2014; Moeller, Pixner, Zuber, Kaufmann, & Nuerk, 2011; Van der Ven, Klaiber, & Van der Maas, 2017) and without inversion property (Göbel, Watson, Lervåg, & Hulme, 2014; Simmons, Willis, & Adams, 2012). Still, there is hardly any evidence on the specific difficulties in multidigit number processing experienced by individuals with dyscalculia.

In a longitudinal study investigating development of numerical processing in children with typical and atypically poor development of arithmetic skills (Landerl, 2013), children performed a number comparison task with two-digit numbers. The most impressive finding was that at the beginning of Grade 2, the children with low arithmetic skills performed at chance when the unit of the larger number was smaller than the unit of the smaller number (e.g., 41–26), showing that they still had marked difficulties to integrate decade and unit digits into one numerical magnitude. While it is notable that problems in numerical processing among children with poor math performance are even evident for very simple-single digit tasks, these problems are likely to markedly increase when it comes to processing of and doing arithmetic with multidigit numbers.

## Spontaneous math focusing tendencies

Children strongly differ in their tendency to spontaneously focus their attention on math information in the surroundings. This tendency has been found to be associated with their math development (McMullen, Chan, Mazzocco, & Hannula-Sormunen, 2019). Children's spontaneous focusing on (nonsymbolic) numerosities (SFON; Hannula & Lehtinen, 2005) has been found to predict numerical abilities (e.g., counting) and arithmetic abilities, and children with dyscalculia were observed to show lower SFON than typically developing children (Kucian et al., 2012). Spontaneous focusing on (quantitative) relations (SFOR; McMullen, Hannula-Sormunen, & Lehtinen, 2018) was found to be associated with (later) conceptual knowledge of fractions. Most recently, kindergarteners' spontaneous focusing on (Arabic) number symbols (SFONS; Rathé, Torbeyns, De Smedt, & Verschaffel, 2019) was found to be related with their numerical abilities. The interrelations of the different tendencies and the relationships with other cognitive factors across development, however, are not well understood yet. We also do not know whether children who later on develop dyscalculia show lower levels of spontaneously focusing attention on numerical information during early development.

## Patterning

Patterning refers to the extraction of regularities and structures and has been regarded as a basic skill in early mathematics. It comprises a number of different activities (e.g., copying, creating, extending) of regular object arrangements (e.g., repeating patterns like ABAB or AABAAB, growing patterns like 2–4–6, or spatial structure patterns like ::::, for current critical reviews, see Burgoyne, Witteveen, Tolan, Malone, & Hulme, 2017; Wijns, Torbeyns, De Smedt, & Verschaffel, 2019). There is evidence that pattern understanding correlates with concurrent measures of arithmetic (e.g., Lee, Ng, Bull, Pe, & Ho, 2011; MacKay & De Smedt, 2019) and a small number of recent studies also report unique longitudinal prediction of later arithmetic (e.g., Rittle-Johnson, Zippert, & Boyce, 2019). However, patterning skills are also related to general cognitive skills and it is as yet unclear to what extent the involved processes are domain-specific or rather reflective of domain-general abilities such as intelligence. We also do not know whether or not poor understanding of patterns can indeed induce the severe impairments of mathematical cognition associated with dyscalculia.

## Math anxiety

Math anxiety can be considered to be a widespread problem in math development. For instance, in a recent Program for International Student Assessment (PISA) report (OECD, 2013), between 30% and 60% of the students reported signs of math anxiety. It is defined as "a feeling of tension, apprehension, or fear that interferes with math performance"

(Ashcraft, 2002, p. 181). The interference during doing arithmetic is mostly related to the fact that task-irrelevant anxiety-related thoughts of rumination block the domain-general component of working-memory (see below) capacities that would be necessary to solve the math problem (Ashcraft & Kirk, 2001). Math anxiety is associated with lower math performance, a more negative math self-concept, and higher avoidance of math activities and career paths (Dowker, Sarkar, & Looi, 2016). Cohen and Rubinsten (2021) point out the dynamic interplay between intrinsic and environmental factors in the development of math anxiety: poor math achievements can cause math anxiety by increasing the individual's vulnerability to the effects of intrinsic and environmental risk factors, but at the same time, poor math skills can result from math anxiety and exacerbate emotional problems.

## Domain-general factors

Mathematics also depends strongly on a wide range of general cognitive abilities. In this section, the most relevant of these general skills and their association with mathematics performance will be reviewed.

### Working memory and executive functions

Working memory refers to the ability to temporarily hold information available for processing. A recent metaanalysis revealed a medium correlation between individual differences in different subcomponents of working memory (verbal, numerical, spatial) and math competencies, especially mental calculation and word problem solving (Peng, Namkung, Barnes, & Sun, 2016). Thus the relationship between working memory and math depends on the math demand, the individual's age as well as competence level, and the working memory component required. Peng et al. found that the relation was stronger among individuals with poor arithmetic performance who also showed other impairments in their domain-general cognitive processing skills. In a metaanalysis of dyscalculia studies, Haberstroh and Schulte-Körne (2019) found a larger effect size for visuospatial than verbal working memory. Spatial working memory, the ability to hold spatial information actively in mind, supports spatial reasoning (see below) by providing a limited capacity mental workspace to move and manipulate information.

There is also evidence that preschool working memory skills predict later math performance in school (Gathercole, Brown, & Pickering, 2003; Mazzocco & Thompson, 2005). It is thus plausible that in individuals with dyscalculia, working memory deficits appear even before the problems in arithmetic development become manifest. However, evidence on preschool markers of dyscalculia is as yet scarce.

The term "executive functions" is an umbrella term for cognitive control processes regulating cognition and action (Miyake et al., 2000). These are typically differentiated into three distinct subcomponents: updating (i.e., monitoring and updating working memory contents), shifting (i.e., switching between tasks), and inhibition (i.e., deliberate overriding of dominant or prepotent responses). Current evidence suggests that the updating function is uniquely related to math competencies (e.g., Bellon, Fias, & De Smedt, 2019; Cragg & Gilmore, 2014), whereas the results regarding the other two components are less clear (Bellon et al., 2019; Van der Ven, Kroesbergen, Boom, & Leseman, 2012; but see also Haberstroh & Schulte-Körne, 2019).

Just like mathematical cognition, working memory and executive functions are multicomponential constructs. Longitudinal studies are needed to investigate the complex and dynamic interplay of the multifaceted relations during development. A question that is of particular interest is whether intact working memory and/or executive functions can act as protective factors for children who have a core deficit in numerical processing.

## Verbal abilities

There is an inextricable link between linguistic skills and mathematical performance. Mathematics instruction depends heavily on oral explanations, and mathematics also entails a specialized vocabulary (e.g., sum, volume, product). A number of studies have focused on the relation of phonological processes with encoding and retrieving number facts (Simmons & Singleton, 2008). Findings also suggest that verbal ability plays a more general role in children's arithmetic performance, perhaps by influencing mathematical thinking that involves the symbolic number system (LeFevre et al., 2010; Vukovic & Lesaux, 2013). There is also evidence that the language used in word problems influences how children represent and solve them (Brissiaud & Sander, 2010). Neuroscientific evidence shows that in addition to a specialized quantitative circuit in the parietal lobe, a linguistic circuit in the left angular gyrus supports the manipulation of numbers in verbal form (Dehaene et al., 2003). Although mathematical thinking can exist independent of language, children need language to express, understand, and learn mathematics, and closer investigation of the contribution of verbal abilities to mathematics especially during early development is called for.

## Spatial reasoning

Spatial reasoning is broadly defined as the ability to mentally manipulate and understand the spatial relations between and within objects. Just like mathematical cognition, spatial reasoning is a multidimensional construct encompassing different and dissociable spatial skills like mental rotation or perspective taking (Hegarty & Waller, 2004). Evidence from typically developing children reveals a close connection between spatial reasoning and mathematics achievement (Mix & Battista, 2018). Spatial reasoning is a significant predictor of later mathematics outcomes even when controlling for a range of variables. For example, at age 3 spatial reasoning uniquely predicted 27% of the variance in mathematics skills at age 4 (Verdine, Irwin, Golinkoff, & Hirsh-Pasek, 2014).

The association between spatial reasoning and numerical processing is usually explained by the fact that reasoning about numerical magnitude is supported by a cognitive representation referred to as the mental number line (e.g., de Hevia & Spelke, 2010). Indeed, Gunderson, Ramirez, Beilock, and Levine (2012) found that number line estimation acuity at age 6 mediated the relation between spatial visualization (assessed by mental rotation) at age 5 and mathematics performance at age 8. This evidence suggests that spatial visualization supports representation of an accurate mental number line, which, in turn, supports math performance. Neuroimaging studies also demonstrated that number line tasks engage specific regions of the parietal cortex, which are thought to be involved in processing spatial representation of numerical relationships (Vogel, Goffin, & Ansari, 2015). Spatial visualization may thus help children to represent and compare numerosities on their mental number line.

Interestingly, while studies on dyscalculia consistently show deficient performance on number line tasks compared to typically developing control groups (e.g., Landerl, 2013; Landerl et al., 2009), empirical evidence on deficits in spatial reasoning skills is surprisingly limited. Szucs et al. (2014), for instance, reported intact performance in a mental rotation task. It should be noted, however, that standard tasks of nonverbal IQ that require spatial abilities (e.g., block design) are often used for sample matching, which may explain why observed group differences are often small (Landerl et al., 2013). It is also possible that deficits in spatial reasoning are only evident during early development, when the representation of a number line is constructed (de Hevia & Spelke, 2010), and are less obvious later on. Thus it would be important to investigate preschool development of children who develop dyscalculia later in life.

## Neurobiological evidence

Dyscalculia is recognized as a neurodevelopmental disorder (World Health Organization, 2020), meaning that it originates from impairments in brain development that arise early and are probably genetic. While our current knowledge of deviations in early brain development and their genetic basis is very limited, we do see differences in brain structure and brain function between individuals with good versus poor mathematical competencies. The current evidence is summarized in the next section.

## Brain mechanisms associated with dyscalculia

Our current understanding of the brain networks involved in mathematics is largely based on adult studies (Ansari & Vogel, 2013; Arsalidou & Taylor, 2011; Menon, 2015). Only few studies have so far investigated alterations in brain structure (e.g., white matter and gray matter volume) and function (brain activation) in atypical development—in particular children with dyscalculia (e.g., Kaufmann, Wood, Rubinsten, & Henik, 2011; Kucian, Kaufmann, & van Aster, 2015; Peters & De Smedt, 2018). The available evidence suggests significant differences in the functional and structural organization of specific brain regions that process and represent mathematical information. Although children with math difficulties seem to use the same networks as those with adequate development, the efficiency of these brain regions appears to be reduced (Peters & De Smedt, 2018). More specifically, evidence from various brain imaging studies with children and adults have indicated that multiple, and partially overlapping, functional and structural brain networks are associated with processing mathematical information (Arsalidou & Taylor, 2011; Menon, 2015; Peters & De Smedt, 2018). The brain regions of these networks can be broadly classified into regions that are related to domain-general and domain-specific functions (see also Fig. 2). Domain-general brain areas are mainly, but not exclusively, associated with the ventral and dorsal prefrontal cortex and the posterior superior parietal cortex; other areas include regions in the anterior and medial temporal lobe (i.e., hippocampus), regions of the visual cortex (e.g., fusiform gyrus), and subcortical regions (e.g., basal ganglia). Domain-specific regions are especially associated with the intraparietal sulcus and the angular gyrus of the parietal cortex.

Brain regions of the domain-general network have been shown to be modulated by the complexity of the mathematical task (Arsalidou & Taylor, 2011). For instance, stronger brain activation in these regions is typically

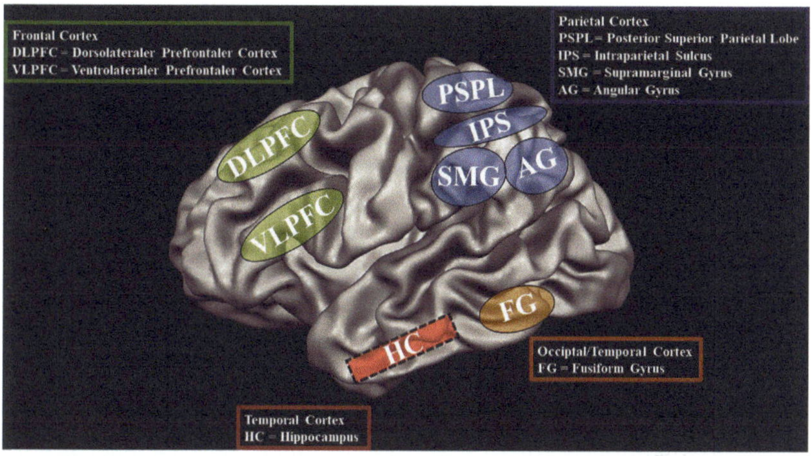


**Fig. 2** Brain regions of the mathematical network. The relative contributions of these domain-general and domain-specific regions depend on the task, the task complexity, age, and individual differences.

observed when participants use procedural strategies to solve a complex mathematical problem (e.g., using multiple steps to calculate the answer) compared to an easy problem for which the answer is retrieved via fact retrieval (e.g., knowing the answer to the problem; Grabner et al., 2009). Since calculating the answer via multiple steps affords greater engagement of executive functions and working memory, these brain regions tend to show a greater involvement compared to when the answer is retrieved.

Hyper- as well as hypoactivation of frontal brain regions has been shown in several functional imaging studies that have contrasted the brain activation of children with and without dyscalculia during mathematical tasks (Ashkenazi, Rosenberg-Lee, Tenison, & Menon, 2012; Berteletti, Prado, & Booth, 2014; Davis et al., 2009; Iuculano et al., 2015; Kucian et al., 2006; Rosenberg-Lee et al., 2015; Simos et al., 2008). While there is an ongoing debate about the nature of these seemingly conflicting results, increased brain activation is often interpreted to reflect the engagement of additional cognitive resources (e.g., compensatory mechanisms) that children with dyscalculia may need to perform a "complex" mathematical task (e.g., Kaufmann et al., 2009). In contrast, deactivations are often related to a failure to engage appropriate domain-general or domain-specific resources to solve certain mathematical operations (Berteletti et al., 2014). Although a number of studies have indicated that the engagement of these domain-general regions differs as a function of learning (difficulties), age, and task complexity, the causality of the observed activation differences is still unknown.

The parietal cortex is thought to reflect a mosaic of domain-general (e.g., spatial abilities in the PSPL) as well as domain-specific (e.g., representation of numerosities) brain regions. Nevertheless, most functional brain imaging studies investigating dyscalculia focused on the IPS as "the central hub" of domain-specific number processing (Kucian et al., 2015). There is indeed a large body of evidence that suggest that the IPS is associated with the processing of nonsymbolic numerosities (e.g., Piazza, Izard, Pinel, Le Bihan, & Dehaene, 2004; Piazza, Pinel, Le Bihan, & Dehaene, 2007; Price, Holloway, Räsänen, Vesterinen, & Ansari, 2007; Wilkey, Barone, Mazzocco, Vogel, & Price, 2017) and symbolic numbers (e.g., Goffin, Sokolowski, Slipenkyj, & Ansari, 2019; Holloway, Battista, Vogel, & Ansari, 2012; Piazza et al., 2007; Vogel et al., 2015, 2017), as well as with numerical order (Matejko, Hutchison, & Ansari, 2019; Sommerauer, Grass, Grabner, & Vogel, 2019). It is this brain region that is often engaged when nonsymbolic numerosities and symbolic numbers are manipulated or the order of numbers has to be judged (Ansari & Vogel, 2013; Lyons et al., 2016). Several neuroimaging studies found atypical functional (Kaufmann et al., 2009; Mussolin, Mejias, & Noël, 2010; Price et al., 2007) and structural (Ranpura et al., 2013; Rotzer et al., 2008; Rykhlevskaia, Uddin, Kondos, & Menon, 2009) brain activation patterns in this brain region in children with dyscalculia.

One potential explanation for this deviant brain activation is an atypical functional specialization of the IPS to represent nonsymbolic numerosities or symbolic numbers. Developmental studies have provided significant evidence that the brain activation of the IPS increases with age and becomes increasingly attuned to represent nonsymbolic numerosities and symbolic numbers (Ansari, 2007; Matejko et al., 2019; Piazza et al., 2007; Vogel et al., 2015). This developmental functional specialization (i.e., specific regions become attuned to process specific information due to the interaction with other brain regions, Johnson, 2001) appears to be impaired in children with dyscalculia. A neuroimaging study with adults also confirmed that dyscalculia is likely due to a specific processing difficulty of numerical quantities (i.e., numerosities) and not due to deficits in judging nonnumerical continuous quantities (e.g., size or length; McCaskey, von Aster, O'Gorman Tuura, & Kucian, 2017). This explanation is also in line with the behavioral evidence suggesting that numerical abilities are deficient in individuals with dyscalculia.

Only few studies have thus far investigated anatomical differences between children with typical versus atypical development. Structural measures such as voxel-based morphometry found significant alterations in both gray and whiter matter volume in children with dyscalculia. Specifically, reduced gray matter volume was found in the IPS (Rotzer et al., 2008; Rykhlevskaia et al., 2009), the posterior parietal cortex (Ranpura et al., 2013), and the middle frontal cortex (Rotzer et al., 2008); reduced white matter volume was also found in frontal regions and the temporoparietal regions (Ranpura et al., 2013; Rotzer et al., 2008; Rykhlevskaia et al., 2009). In their longitudinal study, McCaskey, von Aster, O'Gorman, and Kucian (2020) found persistently reduced gray and withe matter volumes in children with dyscalculia over a time span of 4 years in a large network of frontal, parietal, temporal, and occipital brain regions. These findings indicate the existence of structural brain differences across multiple regions.

Other studies have investigated connectivity patterns (i.e., fiber tracts that connect distant areas of the brain) between different brain regions. The first study to investigate the integrity of white matter fibers using fractional anisotropy (FA) found significantly lower FA values (i.e., less integrity) in children with dyscalculia in the inferior longitudinal fasciculus (ILF)—a tract that connects the occipital lobe with the temporal lobe—and the inferior frontal-occipital fasciculs (IFOF)—a tract that connects the occipital lobe with the frontal lobe (Rykhlevskaia et al., 2009). Another study found reduced FA values in the superior longitudinal fasciculus (SLF), especially in regions near the parietal lobe (Kucian et al., 2014). Although sparse, these findings confirm that dyscalculia is associated with the IPS but they also demonstrate that aberrations are not solely related to one particular brain region. The abovementioned connectional approach has also sparked new work that has started to investigate functional connectivity patterns in children with dyscalculia. For instance, a resting state study (Jolles et al., 2016) found that the IPS of children with dyscalculia expresses a hyperconnectivity with other, mostly frontal and parietal brain regions. Hyperconnectivity in children with dyscalculia was confirmed by Michels, O'Gorman, and Kucian (2018), who also showed that a numerical intervention can reduce patterns of hyperconnectivity. Although the reasons for this hyperconnectivity are not yet determined, it is plausible that it is associated with an undifferentiated functional specialization to efficiently process relevant numerical information.

Overall, neuroimaging findings suggest that the deficiencies in mathematical abilities associated with dyscalculia may be due to an atypical functional specialization of specific brain regions and networks to process numerical information. Nevertheless, as the current evidence comes predominantly from cross-sectional studies (for a recent exception, see

McCaskey et al., 2018) with relatively small sample sizes, it is unclear whether any peculiarities in the dyscalculic brain are cause or effect of the maths deficits. Furthermore, most functional studies focused on restricted age ranges (mostly between 7 and 14 years). This is partly attributable to difficulties to investigate young children with functional imaging techniques (Vogel, Matejko, & Ansari, 2016), but mostly due to the fact that dyscalculia can only be diagnosed once mathematical deficiencies become manifest at school age. Therefore we currently have not much information about the underlying mechanism that drives the brain activation differences in preschool children. What we observe is the outcome of a developmental trajectory that has probably started well before the onset of formal education.

## Heritability of dyscalculia

Mathematical cognition is a highly complex human ability, and identifying its polygenic basis is challenging. Maths problems clearly run in families. Based on a sample of 39 children with dyscalculia, Shalev et al. (2001) reported that 66% of mothers, 40% of fathers, 53% of siblings, and 44% of second-degree relatives also qualified for a dyscalculia diagnosis. The prevalence rate among siblings was therefore 10 times larger than in the general population. In their analysis of 198 monozygotic and 198 dizygotic twin pairs with low maths achievement, Haworth, Kovas, Petrill, and Plomin (2007) found heritability estimates between 0.69 and 0.75 for tasks related to numerical and mathematical processing. The fact that there is as yet no convincing evidence on candidate genes related to dyscalculia is probably owed to the high heterogeneity of the disorder. While in dyslexia research word recognition skills have been identified as the crucial phenotype and measures of word recognition are relatively comparable across samples (see, for example, Carrion-Castillo, Franke, & Fisher, 2013, for a current review of findings), there is as yet no crucial phenotype of dyscalculia (Carvalho & Haase, 2019).

## Summary and future outlook

Research on maths learning difficulties has made tremendous progress. However, compared to other neurodevelopmental disorders, a lot of evidence is still lacking. Deficits in basic understanding and processing of number have been identified as a hallmark of dyscalculia. But obviously, monocausal theories of neurodevelopmental disorders have been largely replaced by multideficit models (Pennington, 2006). This new perspective results from increasing evidence that specific developmental disorders are less specific than assumed by older definitions (Landerl, 2015; Landerl et al., 2013; Peters & Ansari, 2019). Thus we need a better understanding of the dynamic interactions between the potentially contributing factors that have been reviewed in this chapter. Longitudinal studies focusing on this complex interplay will be essential in order to better understand developmental trajectories. Of particular interest are studies that start at a very young age, ideally before the onset of any formal maths instruction, as they can be informative on early precursors of maths abilities and deficits (see Kucian, 2021, for an overview of current evidence).

Such longitudinal studies should be large and population based to inform us about inter- as well as intraindividual differences across the full range of ability levels (Peters & Ansari, 2019). However, given the prevalence rates of dyscalculia, such studies would have to be very large in order to cover the immense heterogeneity that we observe particularly in the lower end of the distribution of maths skills. Hence, a more focused approach may be helpful in order to gain specific information on early trajectories of dyscalculia: In the field of dyslexia research, a number of cutting-edge research projects exploited the knowledge that dyslexia is highly heritable by following samples of preschool children who had an affected first-degree relative (e.g., Lyytinen, 1997; Nash, Hulme, Gooch, & Snowling, 2014). Given our current knowledge about the heritability of dyscalculia, it would be worthwhile to expand this approach to the investigation of early development of domain-specific as well as domain-general precursors of arithmetic skills.

The at-risk design allows identifying cognitive "endophenotypes" (Carvalho & Haase, 2019) which are defined as markers associated with a genetic liability for the disorder. These markers can be observed not only in individuals developing the disorder but also in at-risk children who go on to show age-appropriate arithmetic skills. Investigation of the early developmental trajectories of children at risk for arithmetic deficits will improve our understanding of the complex interplay of cognitive, neurobiological, and environmental factors that cause children to struggle in their arithmetic development. Importantly, such studies will also allow us to identify protective factors in those children who will not develop arithmetic deficits in spite of familial risk. Such studies will also be helpful to devise task batteries for early identification so that programs of prevention can be efficiently implemented.

## References


Ansari, D. (2007). Does the parietal cortex distinguish between "10," "ten," and ten dots? *Neuron, 53*, 165–167.

Ansari, D. (2008). Effects of development and enculturation on number representation in the brain. *Nature Reviews. Neuroscience, 9*, 278–291.

Ansari, D., & Vogel, S. E. (2013). Cognitive neuroscience of numerical cognition. In K. Ochsner, & S. Kosslyn (Eds.), *Oxford handbook of cognitive neuroscience, volume 2: The cutting edges*. Oxford: Oxford University Press.

Arsalidou, M., & Taylor, M. J. (2011). Is 2+2=4? Meta-analyses of brain areas needed for numbers and calculations. *NeuroImage, 54*, 2382–2393.

Ashcraft, M. H. (2002). Math anxiety: Personal, educational, and cognitive consequences. *Current Directions in Psychological Science, 11*, 181–185.

Ashcraft, M. H., & Kirk, E. P. (2001). The relationships among working memory, math anxiety, and performance. *Journal of Experimental Psychology: General, 130*, 224–237.

Ashkenazi, S., Rosenberg-Lee, M., Tenison, C., & Menon, V. (2012). Weak task-related modulation and stimulus representations during arithmetic problem solving in children with developmental dyscalculia. *Developmental Cognitive Neuroscience, 2*, S152–S166.

Bellon, E., Fias, W., & De Smedt, B. (2019). More than number sense: The additional role of executive functions and metacognition in arithmetic. *Journal of Experimental Child Psychology, 182*, 38–60.

Berteletti, I., Prado, J., & Booth, J. R. (2014). Children with mathematical learning disability fail in recruiting verbal and numerical brain regions when solving simple multiplication problems. *Cortex, 57*, 143–155.

Brissiaud, R., & Sander, E. (2010). Arithmetic word problem solving: A situation strategy framework. *Developmental Science, 13*, 92–107.

Burgoyne, K., Witteveen, K., Tolan, A., Malone, S., & Hulme, C. (2017). Pattern understanding: Relationships with arithmetic and reading. *Child Development Perspectives, 11*, 239–244.

Butterworth, B. (1999). *The mathematical brain*. London: MacMillan.

Butterworth, B. (2010). Foundational numerical capacities and the origins of dyscalculia. *Trends in Cognitive Science, 14*, 534–541.

Butterworth, B. (2011). Foundational numerical capacities and the origins of dyscalculia. In S. Dehaene, & E. M. Brannon (Eds.), *Space, time, and number in the brain: Searching for foundations of mathematical thought* (pp. 249–265). London, UK: Academic Press.

Carrion-Castillo, A., Franke, B., & Fisher, S. E. (2013). Molecular genetics of dyslexia: An overview. *Dyslexia, 19*, 214–240.

Carvalho, M. R. S., & Haase, V. G. (2019). Genetics of dyscalculia 2: In search of endophenotypes. In A. Fritz, V. G. Haase, & P. Räsänen (Eds.), *International handbook of mathematical learning difficulties: From the laboratory to the classroom* (pp. 345–365). Cham, CH: Springer.

Chrisomalis, S. (2010). *Numerical notation: A comparative history*. Cambridge: Cambridge University Press.

Clayton, F. J., Copper, C., Steiner, A. F., Banfi, C., Finke, S., Landerl, K., & Göbel, S. M. (2020). Two-digit number writing and arithmetic in year 1 children: Does number word inversion matter. *Cognitive Development, 56*. https://doi.org/10.1016/j.cogdev.2020.100967, 100967.

Cohen, L. D., & Rubinsten, O. (2021). The complex pathways toward the development of math anxiety and links with achievements. In W. Fias, & A. Henik (Eds.), *Heterogeneous Contributions to Numerical Cognition: Learning and Education in Mathematical Cognition* (pp. 311–326). Amsterdam, NL: Elsevier.

Cragg, L., & Gilmore, C. (2014). Skills underlying mathematics: The role of executive function in the development of mathematics proficiency. *Trends in Neuroscience and Education, 3*, 63–68.

Davis, N., Cannistraci, C. J., Rogers, B. P., Gatenby, J. C., Fuchs, L. S., Anderson, A. W., & Gore, J. C. (2009). Aberrant functional activation in school age children at-risk for mathematical disability: A functional imaging study of simple arithmetic skill. *Neuropsychologia, 47*, 2470–2479.

de Hevia, M. D., & Spelke, E. S. (2010). Number-space mapping in human infants. *Psychological Science, 21*, 653–660.

De Smedt, B., Noël, M. P., Gilmore, C., & Ansari, D. (2013). How do symbolic and non-symbolic numerical magnitude processing skills relate to individual differences in children's mathematical skills? A review of evidence from brain and behavior. *Trends in Neuroscience and Education, 2*(2), 48–55.

Dehaene, S. (1992). Varieties of numerical abilities. *Cognition, 44*, 1–42.

Dehaene, S., Piazza, M., Pinel, P., & Cohen, L. (2003). Three parietal circuits for number processing. *Cognitive Neuropsychology, 20*, 487–506.

Dowker, A., Sarkar, A., & Looi, C. Y. (2016). Mathematics anxiety: What have we learned in 60 years? *Frontiers in Psychology, 7*, 508.

Fazio, L. K., Bailey, D. H., Thompson, C. A., & Siegler, R. S. (2014). Relations of different types of numerical magnitude representations to each other and to mathematics achievement. *Journal of Experimental Child Psychology, 123*, 53–72.

Finke, S., Freudenthaler, H. H., & Landerl, K. (2020). Symbolic processing mediates the relation between non-symbolic processing and later arithmetic performance. *Frontiers in Psychology, 11*, 549.

Froyen, D., Van Atteveldt, N., Bonte, M., & Blomert, L. (2008). Cross-modal enhancement of the MMN to speech-sounds indicates early and automatic integration of letters and speech-sounds. *Neuroscience Letters, 430*, 23–28.

Froyen, D., Willems, G., & Blomert, L. (2011). Evidence for a specific cross-modal association deficit in dyslexia: An electrophysiological study of letter-speech-sound processing. *Developmental Science, 14*, 635–664.

Gathercole, S. E., Brown, L., & Pickering, S. J. (2003). Working memory assessment at school entry as longitudinal predictors of National Curriculum attainment levels. *Educational and Child Psychology, 20*, 109–122.

Göbel, S. M., Watson, S. E., Lervåg, A., & Hulme, C. (2014). Children's arithmetic development: It is number knowledge, not the approximate number sense, that counts. *Psychological Science, 25*, 789–798.

Goffin, C., Sokolowski, H. M., Slipenkyj, M., & Ansari, D. (2019). Does writing handedness affect neural representation of symbolic number? An fMRI adaptation study. *Cortex, 121*, 27–43.

Grabner, R. H., Ansari, D., Koschutnig, K., Reishofer, G., Ebner, F., & Neuper, C. (2009). To retrieve or to calculate? Left angular gyrus mediates the retrieval of arithmetic facts during problem solving. *Neuropsychologia, 47*, 604–608.

Gunderson, E. A., Ramirez, G., Beilock, S. L., & Levine, S. C. (2012). The relation between spatial skill and early number knowledge: The role of the linear number line. *Developmental Psychology, 48*, 1229–1241.

Haberstroh, S., & Schulte-Körne, G. (2019). The diagnosis and intervention of dyscalculia. *Deutsches Ärzteblatt International, 116*, 107–114.

Hannula, M. M., & Lehtinen, E. (2005). Spontaneous focusing on numerosity and mathematical skills of young children. *Learning and Instruction, 15*, 237–256.

Haworth, C. M., Kovas, Y., Petrill, S. A., & Plomin, R. (2007). Developmental origins of low mathematics performance and normal variation in twins from 7 to 9 years. *Twin Research and Human Genetics, 10*, 106–117.

Hegarty, M., & Waller, D. (2004). A dissociation between mental rotation and perspective-taking spatial abilities. *Intelligence, 32*, 175–191.

Holloway, I. D., Battista, C., Vogel, S. E., & Ansari, D. (2012). Sematic and perceptual processing of number symbols: Evidence from a cross-linguistic fMRI adaptation study. *Journal of Cognitive Neuroscience, 25*, 388–400.

Imbo, I., Vanden Bulcke, C., De Brauwer, J., & Fias, W. (2014). Sixty-four or four-and-sixty? The influence of language and working memory on children's number transcoding. *Frontiers in Psychology*, *5*, 313.
Iuculano, T., Rosenberg-Lee, M., Richardson, J., Tenison, C., Fuchs, L., Supekar, K., & Menon, V. (2015). Cognitive tutoring induces widespread neuroplasticity and remediates brain function in children with mathematical learning disabilities. *Nature Communications*, *6*, 1–10.
Johnson, M. H. (2001). Functional brain development in humans. *Nature Reviews. Neuroscience*, *2*, 475–483.
Jolles, D., Ashkenazi, S., Kochalka, J., Evans, T., Richardson, J., Rosenberg-Lee, M., … Menon, V. (2016). Parietal hyper-connectivity, aberrant brain organization, and circuit-based biomarkers in children with mathematical disabilities. *Developmental Science*, *19*, 613–631.
Kaufmann, L., Vogel, S. E., Starke, M., Kremser, C., Schocke, M., & Wood, G. (2009). Developmental dyscalculia: Compensatory mechanisms in left intraparietal regions in response to nonsymbolic magnitudes. *Behavioral and Brain Functions*, *5*, 35.
Kaufmann, L., Wood, G., Rubinsten, O., & Henik, A. (2011). Meta-analyses of developmental fMRI studies investigating typical and atypical trajectories of number processing and calculation. *Developmental Neuropsychology*, *36*, 763–787.
Kucian, K. (2021). Developmental course of numerical learning problems in children and how to prevent dyscalculia: A summary of the longitudinal examination of children from kindergarten to secondary school. In W. Fias, & A. Henik (Eds.), *Heterogeneous Contributions to Numerical Cognition: Learning and Education in Mathematical Cognition* (pp. 229–251). Amsterdam, NL: Elsevier.
Kucian, K., Ashkenazi, S. S., Hänggi, J., Rotzer, S., Jäncke, L., Martin, E., & von Aster, M. (2014). Developmental dyscalculia: A dysconnection syndrome? *Brain Structure and Function*, *219*(5), 1721–1733. https://doi.org/10.1007/s00429-013-0597-4.
Kucian, K., Kaufmann, L., & van Aster, M. (2015). Chapter 39—Brain correlates of numerical disabilities. In R. C. Cohen-Kadosh, & A. Dowker (Eds.), *The Oxford handbook of numerical cognition* (pp. 716–729). Oxford University Press.
Kucian, K., Kohn, J., Hannula-Sormunen, M. M., Richtmann, V., Grond, U., Käser, T., … von Aster, M. G. (2012). Kinder mit Dyskalkulie fokussieren spontan weniger auf Anzahligkeiten [Children with dyscalculia show less spontaneous focusing on numerosities]. *Lernen und Lernstörungen*, *1*, 241–253.
Kucian, K., Loenneker, T., Dietrich, T., Dosch, M., Martin, E., & von Aster, M. (2006). Impaired neural networks for approximate calculation in dyscalculic children: A functional MRI study. *Behavioral and Brain Functions*, *2*, 31.
Landerl, K. (2013). Development of numerical processing in children with typical and dyscalculic arithmetic skills—A longitudinal study. *Frontiers in Psychology*, *4*, 459.
Landerl, K. (2015). How specific is the specific disorder of arithmetic skills? In S. Chin (Ed.), *The Routlege international handbook for mathematical difficulties and dyscalculia* (pp. 115–124). New York: Routledge.
Landerl, K. (2019). Neurocognitive perspectives on numerical development. In A. Fritz-Stratmann, V. Haase, & P. Räsännän (Eds.), *The international handbook of mathematical learning difficulties* (pp. 9–24). Cham: Springer.
Landerl, K., Bevan, A., & Butterworth, B. (2004). Developmental dyscalculia and basic numerical capacities: A study of 8-9 year old students. *Cognition*, *93*, 99–125.
Landerl, K., Fussenegger, B., Moll, K., & Willburger, E. (2009). Dyslexia and dyscalculia: Two learning disorders with different cognitive profiles. *Journal of Experimental Child Psychology*, *103*, 309–324.
Landerl, K., Göbel, S., & Moll, K. (2013). Core deficit and individual manifestations of developmental dyscalculia (DD): The role of comorbidity. *Trends in Neuroscience and Education*, *2*, 38–42.
Lee, K., Ng, S., Bull, R., Pe, M., & Ho, R. (2011). Are patterns important? An investigation of the relationships between proficiencies in patterns, computation, executive functioning, and algebraic word problems. *Journal of Educational Psychology*, *103*, 269–281.

LeFevre, J., Fast, L., Skwarchuk, S., Smith-Chant, B. L., Bisanz, J., Kamawar, D., et al. (2010). Pathways to mathematics: Longitudinal predictors of performance. *Child Development, 81*, 1753–1767.
Lyons, I. M., & Ansari, D. (2015). Foundations of children's numerical and mathematical skills: The roles of symbolic and nonsymbolic representations of numerical magnitude. *Advances in Child Development and Behavior, 48*, 93–116.
Lyons, I. M., Bugden, S., Zheng, S., De Jesus, S., & Ansari, D. (2018). Symbolic number skills predict growth in nonsymbolic number skills in kindergarteners. *Developmental Psychology, 54*, 440–457.
Lyons, I. M., Price, G. R., Vaessen, A., Blomert, L., & Ansari, D. (2014). Numerical predictors of arithmetic success in grades 1-6. *Developmental Science, 17*, 714–726.
Lyons, I. M., Vogel, S. E., & Ansari, D. (2016). On the ordinality of numbers: A review of neural and behavioural studies. *Progress in Brain Research, 227*, 187–221.
Lyytinen, H. (1997). In search of the precursors of dyslexia: A prospective study of children at risk for reading problems. In C. Hulme, & M. Snowling (Eds.), *Dyslexia: Biology, cognition and intervention*. London: Whurr.
MacKay, K. J., & De Smedt, B. (2019). Patterning counts: Individual differences in children's calculation are uniquely predicted by sequence patterning. *Journal of Experimental Child Psychology, 177*, 152–165.
Matejko, A. A., Hutchison, J., & Ansari, D. (2019). Developmental specialization of the left intraparietal sulcus for symbolic ordinal processing. *Cortex, 114*, 41–53.
Mazzocco, M. M. M., Feigenson, L., & Halberda, J. (2011). Preschoolers' precision of the approximate number system predicts later school mathematics performance. *PLoS One, 6*(9), e23749.
Mazzocco, M. M. M., & Thompson, R. E. (2005). Kindergarten predictors of math learning disability. *Learning Disabilities Research and Practice, 20*, 142–155.
McCaskey, U., von Aster, M., Maurer, U., Martin, E., O'Gorman Tuura, R., & Kucian, K. (2018). Longitudinal brain development of numerical skills in typically developing children and children with developmental dyscalculia. *Frontiers in Human Neuroscience, 11*, 629.
McCaskey, U., von Aster, M., O'Gorman, R., & Kucian, K. (2020). Persistent differences in brain structure in developmental dyscalculia: A longitudinal morphometry study. *Frontiers in Human Neuroscience, 14*, 272.
McCaskey, U., von Aster, M. G., O'Gorman Tuura, R., & Kucian, K. (2017). Adolescents with developmental dyscalculia do not have a generalized magnitude deficit –Processing of discrete and continuous magnitudes. *Frontiers in Human Neuroscience, 11*, 102.
McMullen, J., Chan, J. Y., Mazzocco, M. M. M., & Hannula-Sormunen, M. M. (2019). Spontaneous mathematical focusing tendencies in mathematical development and education. In A. Norton, & M. W. Alibali (Eds.), *Constructing number: Merging perspectives from psychology and mathematics education* (pp. 69–86). Cham, CH: Springer Nature.
McMullen, J., Hannula-Sormunen, M. M., & Lehtinen, E. (2018). Spontaneous focusing on quantitative relations as a predictor of rational number and algebra knowledge. *Contemporary Educational Psychology, 51*, 356–365.
Menon, V. (2015). Arithmetic in the child and adult brain. In R. Cohen Kadosh, & A. Dowker (Eds.), *The Oxford handbook of numerical cognition*. Oxford: Oxford University Press.
Michels, L., O'Gorman, R., & Kucian, K. (2018). Functional hyperconnectivity vanishes in children with developmental dyscalculia after numerical intervention. *Developmental Cognitive Neuroscience, 30*, 291–303.
Mix, K. S. S., & Battista, M. T. (2018). *Visualizing mathematics. The role of spatial reasoning in mathematical thought*. Cham, CH: Springer.
Miyake, A., Friedman, N. P., Emerson, M. J., Witzki, A. H., Howerter, A., & Wager, T. D. (2000). The unity and diversity of executive functions and their contributions to complex frontal lobe tasks: A latent variable analysis. *Cognitive Psychology, 41*, 49–100.

Moeller, K., Pixner, S., Zuber, J., Kaufmann, L., & Nuerk, H.-C. (2011). Early place-value understanding as a precursor for later arithmetic performance – A longitudinal study on numerical development. *Research in Developmental Disabilities*, *32*, 1837–1851.

Moeller, K., Zuber, J., Olsen, N., Nuerk, H.-C., & Willmes, K. (2015). Intransparent German number words complicate transcoding – A translingual comparison with Japanese. *Frontiers in Psychology*, *6*, 740.

Mussolin, C., Mejias, S., & Noël, M. (2010). Symbolic and nonsymbolic number comparison in children with and without dyscalculia. *Cognition*, *115*, 10–25.

Nash, H. M., Hulme, C., Gooch, D., & Snowling, M. J. (2014). Preschool language profiles of children at family risk of dyslexia: Continuities with specific language impairment. *Journal of Child Psychology and Psychiatry*, *55*, 1234–1243.

OECD (The Organisation for Economic Co-operation and Development). (2013). *PISA 2012 results: Ready to learn (volume III): Students' engagement, drive and self-beliefs*. Paris: OECD Publishing.

Peng, P., Namkung, J., Barnes, M., & Sun, C. (2016). A meta-analysis of mathematics and working memory: Moderating effects of working memory domain, type of mathematics skill, and sample characteristics. *Journal of Educational Psychology*, *108*, 455–473.

Pennington, B. F. (2006). From single to multiple deficit models of developmental disorders. *Cognition*, *101*, 385–413.

Peters, L., & Ansari, D. (2019). Are specific learning disorders truly specific, and are they disorders? *Trends in Neuroscience and Education*, *17*, 100115.

Peters, L., & De Smedt, B. (2018). Arithmetic in the developing brain: A review of brain imaging studies. *Developmental Cognitive Neuroscience*, *30*, 265–279.

Piazza, M., Facoetti, A., Trussardi, A. N., Berteletti, I., Conte, S., Lucangeli, D., … Zorzi, M. (2010). Developmental trajectory of number acuity reveals a severe impairment in developmental dyscalculia. *Cognition*, *116*, 33–41.

Piazza, M., Izard, V., Pinel, P., Le Bihan, D., & Dehaene, S. (2004). Tuning curves for approximate numerosity in the human intraparietal sulcus. *Neuron*, *44*, 547–555.

Piazza, M., Pinel, P., Le Bihan, D., & Dehaene, S. (2007). A magnitude code common to numerosities and number symbols in human intraparietal cortex. *Neuron*, *53*, 293–305.

Price, G. R., Holloway, I., Räsänen, P., Vesterinen, M., & Ansari, D. (2007). Impaired parietal magnitude processing in developmental dyscalculia. *Current Biology*, *17*, R1042–R1043.

Ranpura, A., Isaacs, E., Edmonds, C., Rogers, M., Lanigan, J., Singhal, A., … Butterworth, B. (2013). Developmental trajectories of grey and white matter in dyscalculia. *Trends in Neuroscience and Education*, *2*, 56–64.

Rathé, S., Torbeyns, J., De Smedt, B., & Verschaffel, L. (2019). Spontaneous focusing on Arabic number symbols and its association with early mathematical competencies. *Early Childhood Research Quarterly*, *48*, 111–121.

Rittle-Johnson, B., Zippert, E. L., & Boyce, K. L. (2019). The roles of patterning and spatial skills in early mathematics development. *Early Childhood Research Quarterly*, *46*, 166–178.

Rosenberg-Lee, M., Ashkenazi, S., Chen, T., Young, C. B., Geary, D. C., & Menon, V. (2015). Brain hyper-connectivity and operation-specific deficits during arithmetic problem solving in children with developmental dyscalculia. *Developmental Science*, *18*, 351–372.

Rotzer, S., Kucian, K., Martin, E., Aster, M., von Klaver, P., & Loenneker, T. (2008). Optimized voxel-based morphometry in children with developmental dyscalculia. *NeuroImage*, *39*, 417–422.

Rousselle, L., & Noël, M.-P. (2007). Basic numerical skills in children with mathematics learning disabilities: A comparison of symbolic vs non-symbolic number magnitude processing. *Cognition*, *102*, 361–395.

Rykhlevskaia, E., Uddin, L. Q., Kondos, L., & Menon, V. (2009). Neuroanatomical correlates of developmental dyscalculia: Combined evidence from morphometry and tractography. *Frontiers in Human Neuroscience*, *3*(November), 51.

Sasanguie, D., & Vos, H. (2018). About why there is a shift from cardinal to ordinal processing in the association with arithmetic between first and second grade. *Developmental Science, 21*, e12653.

Schneider, M., Beeres, K., Coban, L., Merz, S., Schmidt, S. S., Stricker, J., et al. (2017). Associations of non-symbolic and symbolic numerical magnitude processing with mathematical competence: A meta-analysis. *Developmental Science, 20*, e12372.

Shalev, R. S., Manor, O., Kerem, B., Ayali, M., Badichi, N., Friedlander, Y., & Gross-Tsur, V. (2001). Developmental dyscalculia is a familial learning disablility. *Journal of Learning Disabilities, 34*, 59–65.

Simmons, F. R., & Singleton, C. (2008). Do weak phonological representations impact on arithmetic development? A review of research into arithmetic and dyslexia. *Dyslexia, 14*, 77–94.

Simmons, F. R., Willis, C., & Adams, A. M. (2012). Different components of working memory have different relationships with different mathematical skills. *Journal of Experimental Child Psychology, 111*, 139–155.

Simos, P. G., Kanatsouli, K., Fletcher, J. M., Sarkari, S., Juranek, J., Cirino, P., … Papanicolaou, A. C. (2008). Aberrant spatiotemporal activation profiles associated with math difficulties in children: A magnetic source imaging study. *Neuropsychology, 22*, 571–584.

Sommerauer, G., Grass, K.-H., Grabner, R. H., & Vogel, S. E. (2019). The semantic control network mediates the relationship between symbolic numerical order processing and arithmetic performance in children. *BioRxiv*. https://doi.org/10.1101/791012, 791012.

Steiner, A. F., Banfi, C., Finke, S., Kemény, F., Clayton, F. J., Göbel, S. M., & Landerl, K. (2021). Twenty-four or four-and-twenty: Language modulates cross-modal matching for multi-digit numbers in children and adults. *Journal of Experimental Child Psychology, 202*. https://doi.org/10.1016/j.jecp.2020.104970, 104970.

Szucs, D., Devine, A., Soltesz, F., Nobes, A., & Gabriel, F. (2014). Cognitive components of a mathematical processing networtk in 9-year old children. *Developmental Science, 17*, 506–524.

Van der Ven, S. H. G., Klaiber, J. D., & Van der Maas, H. L. J. (2017). Four and twenty blackbirds: How transcoding ability mediates the relationship between visuospatial working memory and math in a language with inversion. *Educational Psychology, 37*, 487–505.

Van der Ven, S. H. G., Kroesbergen, E. H., Boom, J., & Leseman, P. M. M. (2012). The development of executive functions and early mathematics: A dynamic relationship. *British Journal of Educational Psychology, 82*, 100–119.

van Marle, K., Chu, F. W., Li, Y., & Geary, D. C. (2014). Acuity of the approximate number system and preschoolers' quantitative development. *Developmental Science, 17*, 492–505.

Verdine, B. N., Irwin, C. M., Golinkoff, R. M., & Hirsh-Pasek, K. (2014). Contributions of executive function and spatial skills to preschool mathematics achievement. *Journal of Experimental Child Psychology, 126*, 37–51.

Vogel, S. E., Goffin, C., & Ansari, D. (2015). Developmental specialization of the left parietal cortex for the semantic representation of Arabic numerals: An fMR-Adaptaton study. *Developmental Cognitive Neuroscience, 12*, 61–73.

Vogel, S. E., Goffin, C., Bohnenberger, J., Koschutnig, K., Reishofer, G., Grabner, R. H., & Ansari, D. (2017). The left intraparietal sulcus adapts to symbolic number in both the visual and auditory modalities: Evidence from fMRI. *NeuroImage, 153*, 16–27.

Vogel, S. E., Haigh, T., Sommerauer, G., Spindler, M., Brunner, C., Lyons, I. M., & Grabner, R. H. (2017). Processing the order of symbolic numbers: A reliable and unique predictor of arithmetic fluency. *Journal of Numerical Cognition, 3*, 288–308.

Vogel, S. E., Koren, N., Falb, S., Haselwander, M., Spradley, A., Schadenbauer, P., … Grabner, R. H. (2019). Automatic and intentional processing of numerical order and its relationship to arithmetic performance. *Acta Psychologica, 193*, 30–41.

Vogel, S. E., Matejko, A. A., & Ansari, D. (2016). Imaging the developing human brain using functional and structural magnetic resonance imaging: Methodological and practical guidelines. In J. Prior, & J. Van Herwegen (Eds.), *Practical research with children* (pp. 46–69). Routledge.

Vukovic, R. K., & Lesaux, N. K. (2013). The relationship between linguistic skills and arithmetic knowledge. *Learning and Individual Differences*, *23*, 87–91.

Wijns, N., Torbeyns, B., De Smedt, B., & Verschaffel, L. (2019). Young children's patterning competencies and mathematical development: A review. In K. M. Robinson, H. P. Osana, & D. Kotsopoulos (Eds.), *Mathematical leaning and cognition in early childhood: Integrating interdisciplinary research into practice* (pp. 139–161). Cham, CH: Springer.

Wilkey, E. D., Barone, J. C., Mazzocco, M. M. M., Vogel, S. E., & Price, G. R. (2017). The effect of visual parameters on neural activation during nonsymbolic number comparison and its relation to math competency. *NeuroImage*, *159*, 430–442.

Wilson, A. J., & Dehaene, S. (2007). Number sense and developmental dyscalculia. In D. Coch, G. Dawson, & K. Fischer (Eds.), *Human behavior, learning, and the developing brain: Typical development* (pp. 212–238). New York: Guilford Press.

World Health Organization. (2020, August 18). *International classification of diseases for mortality and morbidity statistics* (11th revision). Retrieved from https://icd.who.int/browse11/l-m/en.

Xu, F., & Spelke, E. S. (2000). Large number discrimination in 6-month-old infants. *Cognition*, *74*, B1–B11.

CHAPTER 16

# Early difficulties in numerical cognition

**Avishai Henik**
Department of Psychology, Ben-Gurion University of the Negev, Beer-Sheva, Israel

It is sometime interesting and revealing to examine old reports and results in light of current issues and debates. Such examination might focus on aspects that were deemed to be not prominent or important early on but became central to debates years later. A *Science* article by Mehler and Bever (1967) on number conservation stirred a lot of discussion 50 years ago, was taken as an indication for the existence of a number sense over 20 years ago (Dehaene, 1997), but might tell a different story in light of the recent debate regarding the contribution of noncountable/perceptual characteristics to the development of numerical abilities (Gebuis, Cohen Kadosh, & Gevers, 2016; Leibovich, Katzin, Harel, & Henik, 2017). We discuss these issues in what follows.

## Number conservation

Imagine the following experiment. You arrange two rows of objects, say a row of glasses and a row of bottles. Fig. 1 presents two different arrangements. In both, the bottles are arranged in the same way. In contrast, on the left, the glasses and the bottles are equally spaced and on the right the glasses are more spaced than the bottles.

If you are asked to judge whether the two arrays of objects on the left are the same or one has more objects (i.e., more bottles or more glasses), the answer is clear. Children and adults say that the glasses and bottles are of the same number. Now they are shown the arrangement on the right and asked the same question. Here is where young children and adults differ. Whereas the adults say that in terms of numbers nothing changed (i.e., spacing is irrelevant to the question), children tend to say that there are more glasses. It seems that they attend to the irrelevant spacing and spacing affects their response. Results of similar experiments (but see below Piaget's response

*Heterogeneous Contributions to Numerical Cognition*
https://doi.org/10.1016/B978-0-12-817414-2.00016-6

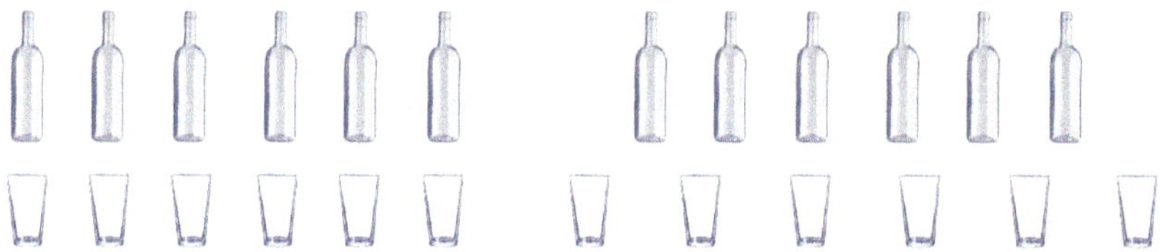


**Fig. 1** Example of arrangements of bottles and glasses.

article in 1968) were taken by Jean Piaget and his colleagues as support for their thoughts about the child's concept of number. In general, they suggested that children do not have an innate sense of number. Rather, children have to learn and wrestle with several difficulties before they can master numbers and create the sense of number. This particular example suggested to Piaget and his colleagues that children lack "number conservation" (that is, the number of items stays the same in spite of the change in spacing). Namely, they do not understand that the number of objects stays the same even if various changes, irrelevant to the number of objects, are introduced. Many experiments have suggested that at the age of 4 years old, children lack number conservation. In this example, the irrelevant feature is spacing. Another irrelevant feature, obvious to us all, is size; the bottles are larger than the glasses and there are many other features that should be ignored when judging numbers.

## Piaget's error?

Dehaene (1997) suggested that babies are able to sense numerosities. When presented with two arrays of objects, they can indicate which array has more objects. Needless to say, babies cannot compare any sizes of arrays but they can indicate that an array with four objects has more objects than an array with two objects. Dehaene suggested that we are born with a sense of number. In contrast, Piaget (1952) suggested that it takes time and effort for the child to acquire the number concept. In his book titled, "The Number Sense" (1997), Dehaene suggested that Piaget was wrong. Under the subtitle, "Piaget's Errors," he described a seminal research by Mehler and Bever (1967) suggesting that children's responses to the above question may change if the context or the question is changed. Mehler and Bever studied children between the ages 2 years 4 months (2–4) and 4 years 7 months (4–7). Children were allocated to seven groups according to their age. Each child was presented with two rows of objects. In one trial, the two rows had the same number of objects equally spaced. In the next trial, the experimenter modified one of the rows so that it was composed of six items that were less spaced than the other row with the smaller number of items (see Fig. 2). In separate trials, the objects could be clay pellets or candies (M&M's). In the first trial (equal number and spacing), the child was asked whether the two rows were the same. In the second trial, the question was changed. In the trials with the clay pellets, the child was asked which row had "more" and in the trial with the M&M's, responses were nonverbal, that is, the child was asked to "... take the row you want to eat, and eat all the M&M's in that row" (p. 141).

The researchers examined success in choosing or naming the row that had more. Such a choice was taken to be indicative of number conservation. There seems to be a decrease in conservation responses with age. At the age of 2–6, conservation seems to be high (between 80% and 100%) and it drops down up to the age of 3–8 or 3–11 and then starts to rise again (see Fig. 3). According to these results, children get worse as they get older and then they improve again. At the age of 4–6, children show conservation for both verbal and nonverbal responding, and the increase in proportion of conservation responses (from age 4 years old) is significant for both types of responses. Interestingly, the changes in proportion of conservation responses (i.e., the decrease and increase in proportions) are dramatic for the clay pellets (i.e., "more" responses, see the smooth line in Fig. 3) and are much smaller in the candy choice responses (i.e., the dashed line in Fig. 3 is rather shallow). Mehler and Bever (1967) examined consistency in responding, namely, children who selected the more numerous row in one trial and the less numerous (more spaced) row in the other trial. They reported that "...if a child gave inconsistent responses, it is more likely that the single conserving response was to the M&M's than to the clay pellets ..." (p. 142).

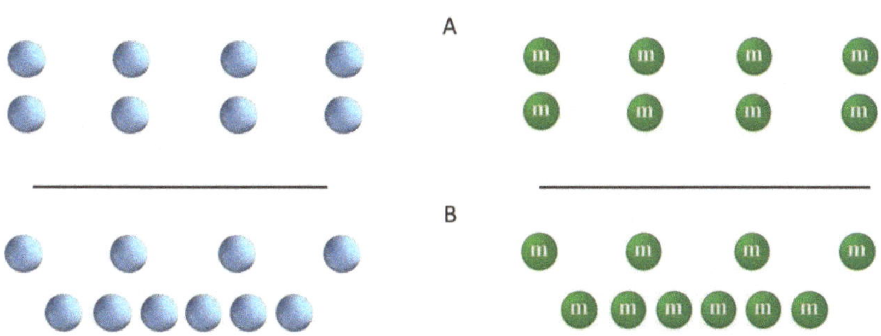


**Fig. 2** Example of clay pellet/M&M arrangements: (A) equally spaced in two rows and (B) with more items condensed into a smaller space in the second row.

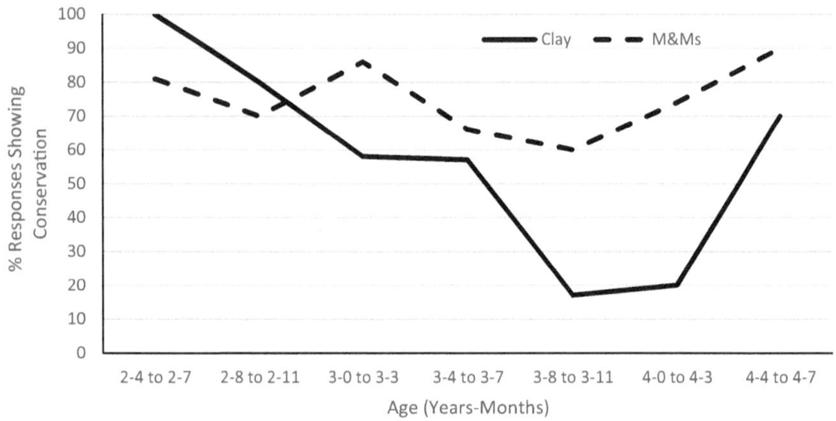


**Fig. 3** Approximate proportion by age of responses choosing the row with more members in the arrangement of Fig. 2B. *Redrawn from Mehler, J., & Bever, T. G. (1967). Cognitive capacity of very young children. Science, 158, 141–142.*

The authors' (Mehler & Bever, 1967) summary and conclusions are very interesting: "Our results indicate that the inability to conserve quantity is a temporary phase in the developing child. The child does not gradually acquire quantity conservation during his 4th year; rather, he reacquires it … We think that the temporary inability to solve the conservation problem reflects a period of overdependence on perceptual strategies. These strategies develop on the basis of experience with correlations of apparent shapes and actual quantity. Surely, it is a general rule that longer arrays usually have 'more' components, and a reasonable perceptual expectancy would reflect this … Eventually, the child develops a more sophisticated integration of the logical operation with his perceptual strategies which allow him to count the individual members of an array. He then has the capacity to ignore his perceptual expectancies in those critical instances in which they are not confirmed" (p. 142).

Dehaene (1997) suggested that these results, and in particular the performance of the youngest children, indicate that they have a number sense and are able to present number conservation. He wrote: "Only the older children failed to conserve the number of marbles … But the cognitive abilities of three- and four-year-olds are certainly not less well-developed than those of two-years-olds. Hence Piagetian tests cannot measure children's true numerical competence. For some reason, these tests seem to confuse older children to such an extent that they become unable to perform …" (p. 45). Confusion might be a problem but it is possible that Mehler and

Bever's (1967) results tell another story. We will come to this story after discussing a few more reactions to the Mehler and Bever article.

Mehler and Bever's (1967) work stimulated a debate and triggered quite a few publications. Beilin (1968) conducted a study using candies. He concluded that perceptual aspects of the stimuli modulated performance. Specifically, changes in the arrangement of candies affected children's decisions: "Children between the ages of 3 years and 4 years 7 months … are not able to conceptualize equality or inequality when objects are misaligned or spatially transformed" (p. 920). He added that "…. transformation leads to the incorrect 'perceptual' inference that the longer or shorter row is different in number from the standard when it is in fact equal" (p. 921). In addition, he suggested that the decline and rise of conservation responses reported by Mehler and Bever were not confirmed. Piaget (1968) himself was not impressed by Mehler and Bever's results and asserted that the results "… have nothing to do with conservation" (p. 976). This is because conservation means that the number of objects stays the same in spite of a change in some (irrelevant) characteristic of the enumerated objects. Piaget suggested that children reached a decision in the situation reflected in Fig. 2B on the basis of crowding or density, that is, the more crowded row has more pellets. On the basis of new experiments, Piaget (1968) suggested that very young children do not respond to numerosity but rather crowding affects their performance and from 3 years and 6 months, length starts to dominate children's performance. In their reply, Mehler and Bever (1968) discussed additional aspects pointed out during this debate. For example, several researchers questioned young children's ability to understand the term "more" because several tasks asked children to select the row with more pellets. They also reported that when asked the same question several times by the same experimenter, children tended to think that they might have provided the wrong answer and, in turn, changed their response. Bever, Mehler, and Epstein (1968) presented more studies, in which they attempted to control for various perceptual aspects of the stimuli, in particular length and spacing. They concluded that cognitive abilities that may exist in very young children (e.g., 2 years old) are constrained and cannot be expressed efficiently until the child learns to focus on relevant aspects of the information around them and ignore irrelevant aspects.

The debate presented above is interesting in many respects. Some are directly related to numerical cognition and its development, and some are related to general issues in psychological research.

## The development of numerical cognition

All the researchers who participated in the above debate agree that numerosity (i.e., the number of items in an array) is associated with a number of perceptual noncountable dimensions. Two such dimensions were discussed above: spacing or density and length or contour of the array. For example, when two arrays of objects are compared, the more numerous arrays are also denser in many cases. Hence, in comparative judgment tasks, when participants are asked to decide which array is more numerous, these associated aspects (e.g., density, length) create a confound that needs to be controlled. However, such control is not simple at all and may prove to be impossible Gebuis et al., 2016; Gebuis & Reynvoet, 2012; Leibovich & Henik, 2013; Leibovich et al., 2017). In addition, in all these comparative judgment tasks, participants may be asked to respond to two different questions. One relates to numerosity—which array has more elements—and the other relates to the noncountable dimension—which array is grayer? The first asks participants to pay attention to numerosity and ignore all other aspects, and the second task asks the participants to pay attention to the noncountable dimension and ignore numerosity. A display that could serve in such an experiment is depicted in Fig. 4. This is a contrast between how many or how much. If we go back to Mehler and Bever's (1967) work, it is possible to suggest that children were asked two different questions or that for the children, the M&M task was understood differently than the clay pellet task. The clay pellet task might be understood as requesting an answer to *how many*, whereas the M&M task might be understood as *how much* (chocolate). The example of the two tasks that might

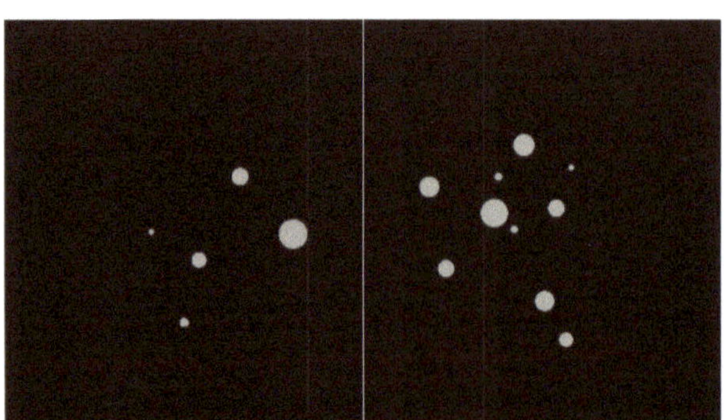

**Fig. 4** Two arrays of gray dots. In various studies using similar stimuli, participants are asked either which array has more dots (regardless of dot size, density, etc.) or which array is grayer (regardless of numerosity).

be applied to the two dot arrays presented in Fig. 4 is a good example that has been used in quite a few experiments. Mehler and Bever might be right in suggesting that the M&M task was more motivating. However, it is also possible that when one is asked to consume chocolate, he/she relates to how much chocolate (rather than how many M&M's) is available in one row compared to the other. In this context, the shallow M&M line (Fig. 3) suggests that this noncountable dimension dominates the perception and decision of the children, and it seems to prevail in the numerosity dimension. Moreover, because it might be easier for young children to judge noncountable dimensions of objects (e.g., density) than to judge numerosity, these noncountable dimensions may affect performance even when irrelevant. Because of the association between these dimensions and numerosity, these effects (or "use" of these noncountable dimensions) may go unnoticed by adults in many cases. However, the child has to learn numerosity and one of the struggles of the first years of life is to extract numerosity of discrete quantities (Mix, Levine, & Newcombe, 2016). To this end, the child has to learn to ignore the irrelevant noncountable dimensions, or in Mehler and Bever's terms, the child learns "to ignore his perceptual expectancies" (p. 142) and acquires the ability to count or compare numerosities. Whether Piaget (1968) was right in suggesting that very young children do not respond to numerosity but rather to crowding first and to length later, or that children may respond to numerosity but are still affected by the noncountable dimensions, it is clear that perception of noncountable, continuous aspects of the display affects performance of young children. This issue (effects of noncountable dimensions) has been the subject of much research in recent years but much more work is still necessary, especially in infants and children of preschool age.

## Density and experimental control

In many psychological experiments, the researcher manipulates a characteristic of the stimuli or the situation and examines whether these changes affect participants' performance. This is what Mehler and Bever (1967) did in their study. As can be seen in Fig. 2, they manipulated the arrangement of the clay pellets or the M&M's and examined if the children's evaluation of the numerosity (i.e., the number of clay pellets or the number of M&M's) changed. Much of the fierce discussion following their report can be summarized under the question: did they control all irrelevant variables (e.g., the perceptual features mentioned by many of the researchers) or not?

My suggestion that the M&M conditions are different from the clay pellet conditions is a critique of a lack of control. Mehler and Bever (1967) suggested that the use of the M&M's and the request to eat them only changed the motivation. Namely, the use of M&M's did not change anything major related to numerosity. In contrast, I suggested above that the M&M condition changed the situation and probably the question (i.e., pay attention to the amount of chocolate rather than pay attention to numerosity). This is intimately related to the issue of experimental control and thus might constitute an example for lack of experimental control; participants judged the amount of chocolate rather than compared numerosity as expected by the experimenters.

In many cases, experimental control works very well and we can rely on the results. That is, we can be confident that the relevant variable produced what we see and the irrelevant variable did not have any effect. However, it is possible that our efforts to introduce experimental control are useless or at the very least, inducing experimental control is very hard and maybe even impossible. In my opinion, some basic work in numerical cognition provides an excellent example of this state of affairs. A simple example appears in experiments that use comparative judgments.

If one examines the two dot arrays in Fig. 4, it is clear that the two aggregates of dots could vary in a number of ways; for example, density of the dots may change, the size of the individual dots may change, and so on. Why is density important? Because usually the denser an array appears, the more numerous it is. Needless to say, this is not always true but this seems to be a general rule that applies in many situations. If you see a very dense crowd, it is usually the case that this crowd has more people than the less dense crowd. Fig. 5 provides a good example. The upper group of silhouettes appears denser, and accordingly, it seems to have more people. Hence, when people are asked to judge which array of silhouettes has more people (i.e., numerosity), they may base their judgment on density rather than on numerosity. This is the issue of experimental control. If the experimenter wants to make sure that people compare the two arrays on numerosity and not on density, he/she needs to control for density. How can this be done? One possibility appears in Fig. 6. The figure presents three pairs of arrays, with one array of each pair having six dots and the other having four dots. In one pair, the six dots are denser than the four dots and we call this pair congruent because the more numerous array is also denser. In another pair, the array with the four dots is denser, that is, the two properties of the dots do not match. According to numerosity, the six-dot array is more numerous

**Fig. 5** The upper group of people looks much denser than the lower group. Accordingly, it seems that the upper group has more people. *Adapted from pictures designed by Kjpargeter/Freepik.*

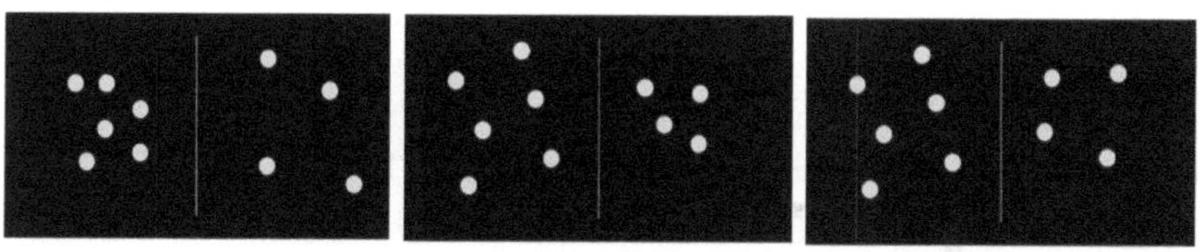

**Fig. 6** Examples for pairs of dot arrays in which numerosity and density vary. In the leftmost pair, the six dots are denser than the four dots—the congruent condition; in the middle pair, the array with the four dots is denser—the incongruent condition; and in the rightmost pair, density is the same in the two arrays.

but according to density, the four-dot array (less numerous) is denser, so this pair is called incongruent. If we run an experiment in which half of the trials are composed of congruent pairs and half of the trials are composed of incongruent pairs, density is not predictive of numerosity and participants should not use density rather than numerosity because they will consistently make mistakes on incongruent trials. Several researchers employed similar manipulations of congruency in order to control for continuous (noncountable) dimensions (Barth, La Mont, Lipton, & Spelke, 2005; Bonny & Lourenco, 2013; Nys & Content, 2012). However, quite a few researchers suggested that with the existing correlations between numerosity and other continuous variables like density, it is almost impossible to make sure that numerosity is experimentally controlled.

## Attention to numerosity

Going back to the Mehler and Bever (1967)–Piaget (1968) debate, we could agree that continuous properties of the stimuli or perceptual properties constitute a major difficulty in processing numerosity at a young age. Young children witness a difficult situation in which they have to learn to focus on numerosity and ignore other properties that many times correlate with numerosity. Moreover, it seems that attention is quite important in children of older ages also. A recent neuroimaging study (Wilkey & Price, 2019) showed an interesting association between numerosity judgment, congruency, and math achievement in 8–11-year-old participants. Participants were presented with congruent or incongruent arrays of dots and were asked to compare the numerosities of the two arrays. In addition, they were asked to perform an attention task (i.e., the flanker arrows task). Numerosity activated the expected intraparietal sulcus (IPS) and frontal brain areas, and also the right inferior frontal gyrus (rIFG). However, these activations did not correlate with mathematical achievement. Congruency also modulated activation in the rIFG. Most importantly, congruency activation in the rIFG predicted mathematical achievement. Please note that rIFG was activated in other studies that examined the ability to inhibit irrelevant information such as in the stop-signal task (Aron, 2008; Aron, Behrens, Smith, Frank, & Poldrack, 2007; Verbruggen & Logan, 2008) and its relation to attention (Chamberlain et al., 2009).

Interestingly, this discussion is related to Hannula-Sormunen and colleagues' work on spontaneous focusing on numerosity (SFON) (Hannula & Lehtinen, 2005; Hannula-Sormunen, McMullen, & Lehtinen, Chapter 9 in this book). In an earlier study, Hannula and Lehtinen (2005) examined children between the ages of 3 years and 6 months and 6 years old. SFON was tested using imitation tasks. For example, the experimenter put a number of envelopes, one by one, into a box and then turned to the child and said: "Now you do exactly like I did." Children's SFON level was scored (i.e., 0–3) according to their imitation behavior (correct or incorrect numerosity) and according to their various quantifying utterances (e.g., I will put in three envelopes) or actions (e.g., use of fingers to express numbers). The SFON tendency varied between children. Importantly, SFON scores correlated positively with tests of numerical ability (e.g., cardinality, counting of disarranged objects) (Hannula & Lehtinen, 2005). Recently, Ben-Shachar and Berger (2018) reported that SFON tendencies among preschool children were related to number discrimination (i.e., responding to the question:

which array has more dots?): children with better discrimination were more likely to attend to numerosity in the imitation tasks. Hannula-Sormunen and colleagues (Chapter 9 in this book) reviewed quite a few studies supporting the conclusion that those who spontaneously attend to numerosity show better achievements in various mathematical tests. Hence, it is possible that Piaget's number conservation and SFON tap a similar ability—paying attention to numerosity and at the same time ignoring distracting features (e.g., density, object identity) that are associated with numerosity and may hinder focusing on numerosity. Performance in tests that examine number conservation and SFON alike varies among children. Moreover, both improve with age and learning. Interestingly, Hannula, Räsänen, and Lehtinen (2007) reported positive correlations between SFON, subitizing, and counting in 4–5-year olds. Specifically, children showing better SFON scores also showed better subitizing and counting performance. Desoete (Chapter 14 in this book) presents results of a longitudinal study showing that subitizing of toddlers (2-year olds) was positively related to counting and calculation proficiency of 4-year olds. In another study, reported in the same chapter, counting and dot comparison at kindergarten (5–6-year olds) predicted arithmetic performance at school (7–8-year olds).

Several authors (e.g., Butterworth, 2005; Geary, 1993; Gelman, 1978; Siegler & Braithwaite, 2017) have suggested that mastering the counting principles (e.g., the one-to-one principle, cardinality, abstraction) is a foundation for developing appropriate mathematical knowledge. Aunola, Leskinen, Lerkkanen, and Nurmi (2004) conducted a longitudinal study starting at kindergarten ages with the aim of predicting the development of math performance during the first two years of school. They found that math performance progressed cumulatively, and proficiency in counting at preschool was a very good predictor of both the level of math performance and its subsequent development; "The higher the level of counting ability children showed at the beginning of the preschool year, the higher was their level of math performance and the faster the rate of growth they showed in their subsequent skill development" (p. 708). For similar results, see Desoete (Chapter 14 in this book).

Similar to the attention to numerosity, counting necessitates focusing on specific features of a set of objects and ignoring other features. For example, the abstractness principle means that any collection of items is countable. A countable set could be made of widely different entities. In terms of the SFON imitation task, the experimenter could put in the box a heterogenous set of objects rather than the homogenous set of envelopes or candies. Here, the child needs to ignore the differences among the selected objects—this is similar to the need to ignore density or spacing in Piagetian experiments. The child needs to learn that enumeration can be applied to nonidentical things. Inhibition (or selective attention) is an example for an executive function (EF). EFs include several different functions (e.g., working memory, set switching, inhibition) that influence the ability to achieve a goal or perform a task. Performing a task requires focusing on relevant aspects of the stimuli (e.g., numerosity) and ignoring irrelevant but potentially interfering aspects (e.g., density). Several authors have suggested that EFs are important for children's mathematical performance (e.g., Archambeau & Gevers, 2018; Aunola et al., 2004; Blair & Razza, 2007; Gilmore & Cragg, 2018) and this is specifically true with respect to inhibition (Archambeau & Gevers, 2018; Blair & Razza, 2007; Goldfarb, 2018).

## Inhibition, education, and numerical cognition

Inhibitory functions are part and parcel of the cognitive processes that generate and control behaviors designed to provide specific solutions to dealing with environmental demands. The ability to inhibit prepotent or reflexive attentional and behavioral reactions and to stop unnecessary or inappropriate behavior develops throughout childhood. These developments free us from interference by otherwise dominant tendencies. This, in turn, enables us to exercise choice and intention over our actions. An excellent example is Adele Diamond's studies of reaching behavior of infants (Diamond, 1990). In one experiment, infants were presented with a Plexiglass box with a building block inside or outside the box. Infants were able to retrieve the object when a direct line of reach was possible (the object was in front of the box touching its front wall or in the middle of the box, away from its front wall). However, when the object was placed inside the box touching its front wall, infants were unsuccessful in retrieving it. To retrieve the object in the latter situation, there was a need to execute a sequence of two movements in order to avoid touching the front wall of the box—one away from the object and a second in the direction of the object. Moreover, when the infants touched the front of the box, they reflexively grasped it or withdrew their hands. Seven-month-old infants rarely continued their reaching toward the object. In contrast, 10-month-old infants were much less likely to show these reflexive behaviors upon touching the edge of the box. These infants were able to retrieve the object in these circumstances. It seems that the older infants developed the ability to execute a reach that required a change of direction and to inhibit reflexive reactions of the hand. This is an example for response inhibition that is present very early in life.

Inhibitory control improves during preschool years (Blair & Razza, 2007; Garon, Bryson, & Smith, 2008; Geeraerts et al., 2021; Petersen, Hoyniak, McQuillan, Bates, & Staples, 2016). Petersen and colleagues authored a review article based on a meta-analysis of 198 studies. The reviewed tasks involved conflict, namely, tasks that required responding according to a rule and inhibiting habitual responding. For example, in the bear/dragon task (Kochanska, Murray, Jacques, Koening, & Vandegeest, 1996), the child follows directions from the bear puppet but ignores commands from the dragon puppet. In the day/night task (Gerstadt, Hong, & Diamond, 1994), the child says "day" when the experimenter shows a picture of a moon and "night" when the experimenter shows a picture of a sun. There were 14 different tasks that were used in various studies and with different age groups. Inhibitory control improved with age but specific tasks reached a plateau at different ages. For example, the bear/dragon task reached a plateau at 4 years and 6 months of age, whereas the day/night task reached a plateau at 6 years of age. Interestingly, the development of inhibitory control is affected by parental attitudes and strategies. Berger (2011, Ch. 4) summarizes quite a few studies suggesting that providing children with opportunities for autonomous regulation and encouraging them to practice self-regulation lead to improved self-regulation. In contrast, intrusive or "… preemptive behavior (e.g., carrying out activities for their children rather than allowing the children to do the activities for themselves)" (p. 67) was not favorable for the development of self-regulation. In a recent study, Geeraerts and colleagues reached a similar conclusion, specifically regarding the development of inhibitory control. Berger suggested an interesting insight: "… the most effective maternal strategy was the teaching-based limit-setting pattern … These were mothers who were relatively clear about prohibition, using occasional commands, and did not avoid their children's persistence or potential for a control struggle … They provided reasons, explanations, or both about the prohibition … Mothers with a teaching-based pattern seemed to facilitate their children's internalization of self-control …" (p. 69). This suggestion that successful teaching-based maternal style facilitates children's self-control leads to the question, how does teaching and learning in other educational contexts affect the development of numerical cognition and EF in preschool years?

Mathematical performance, focusing, and inhibition (among other EFs) develop during preschool years. Interestingly, Merkley, Matusz, and Scerif

(2018) suggested that the effects of attention are not necessarily unidirectional (attention modulates math performance) but rather work in the other direction (math proficiency may modulate attention) as well. Moreover, it seems that learning and practice play major roles in the development of all these skills. Hence, it is important that education in early years of development supports learning and training of mathematical abilities and related skills (e.g., EFs). Dedicated teaching efforts of this sort would promote children's future successful performance. Much effort in early education, in many countries, is devoted to develop literacy at the expense of other skills. It seems that results provided by various researchers and reviewed in various chapters of this book suggest that there is a need to attend to mathematics as well as to EFs in early education. Moreover, it is possible that attention to students' slow progress and difficulties could help advance the low-achieving students and might reduce the widening gaps among classmates (Aunola et al., 2004).

## Conclusions

Early in their development, children face difficulties in developing their numerical ability. Numerosity is one feature of their environment. To develop their numerical skills, children need to wrestle with the requirement to focus on certain aspects of their surroundings and ignore others. Attention to numerosity seems to be important for the development of mathematical abilities. In turn, attention to numerosity is contingent on grasping the concept of numerosity as well as on attentional and executive function abilities. The struggle to learn and deal with competing demands characterizes early development and features the development of various skills like math and EFs. This development could be advanced and shaped by appropriate pedagogy.

## References

Archambeau, K., & Gevers, W. (2018). (How) are executive functions actually related to arithmetic abilities. In A. Henik, & W. Fias (Eds.), *Heterogeneity of function in numerical cognition* (pp. 337–357). San Diego: Academic Press.

Aron, A. R. (2008). Progress in executive-function research. *Current Directions in Psychological Science*, *17*, 124–129.

Aron, A. R., Behrens, T. E., Smith, S., Frank, M. J., & Poldrack, R. A. (2007). Triangulating a cognitive control network using diffusion-weighted magnetic resonance imaging (MRI) and functional MRI. *The Journal of Neuroscience*, *27*, 3743–3752.

Aunola, K., Leskinen, E., Lerkkanen, M.-K., & Nurmi, J.-E. (2004). Developmental dynamics of math performance from preschool to grade 2. *Journal of Educational Psychology, 96*, 699–713. https://doi.org/10.1037/0022-0663.96.4.699.

Barth, H., La Mont, K., Lipton, J., & Spelke, E. S. (2005). Abstract number and arithmetic in preschool children. *Proceedings of the National Academy of Sciences of the United States of America, 102*, 14116–14121.

Beilin, H. (1968). Cognitive capacities of young children: A replication. *Science, 162*, 920–921.

Ben-Shachar, M. S., & Berger, A. (2018). An introduction to attention and its implication for numerical cognition. In A. Henik, & W. Fias (Eds.), *Heterogeneity of function in numerical cognition* (pp. 93–110). San Diego: Academic Press.

Berger, A. (2011). *Self-regulation: Brain, cognition, and development*. Washington, DC, USA: American Psychological Association.

Bever, T. G., Mehler, J., & Epstein, J. (1968). What children do in spite of what they know. *Science, 162*, 921–925.

Blair, C., & Razza, R. P. (2007). Relating effortful control, executive function, and false belief understanding to emerging math and literacy ability in kindergarten. *Child Development, 78*, 647–663.

Bonny, J. W., & Lourenco, S. F. (2013). The approximate number system and its relation to early math achievement: Evidence from the preschool years. *Journal of Experimental Child Psychology, 114*, 375–388. https://doi.org/10.1016/j.jecp.2012.09.015.

Butterworth, B. (2005). The development of arithmetical abilities. *Journal of Child Psychology and Psychiatry, 46*, 3–18.

Chamberlain, S. R., Hampshire, A., Müller, U., Rubia, K., del Campo, N., Craig, K., … Sahakian, B. J. (2009). Atomoxetine modulates right inferior frontal activation during inhibitory control: A pharmacological functional magnetic resonance imaging study. *Biological Psychiatry, 65*, 550–555.

Dehaene, S. (1997). *The number sense: How the mind creates mathematics*. Oxford, UK: Oxford University Press.

Diamond, A. (1990). Developmental time course in human infants and infant monkeys, and the neural bases of, inhibitory control in reaching. *Annals of the New York Academy of Sciences, 608*, 637–676. https://doi.org/10.1111/j.1749-6632.1990.tb48913.x.

Garon, N., Bryson, S. E., & Smith, I. M. (2008). Executive function in preschoolers: A review using an integrative framework. *Psychological Bulletin, 134*, 31–60. https://doi.org/10.1037/0033-2909.134.1.31.

Geary, D. C. (1993). Mathematical disabilities: Cognitive, neuropsychological, and genetic components. *Psychological Bulletin, 114*, 345–362.

Gebuis, T., Cohen Kadosh, R., & Gevers, W. (2016). Sensory-integration system rather than approximate number system underlies numerosity processing: A critical review. *Acta Psychologica, 171*, 17–35.

Gebuis, T., & Reynvoet, B. (2012). The interplay between nonsymbolic number and its continuous visual properties. *Journal of Experimental Psychology: General, 141*, 642–648. https://doi.org/10.1037/a0026218.

Geeraerts, S. B., Endendijk, J. J., Dekovic, M., Huijding, J., Deater-Deckard, K., & Mesman, J. (2021). Inhibitory control across the preschool years: Developmental changes and associations with parenting. *Child Development, 92*(1), 335–350. https://doi.org/10.1111/cdev.13426.

Gelman, R. (1978). Counting in the preschooler: What does and does not develop. In R. S. Siegler (Ed.), *Children's thinking: What develops?* (pp. 213–240). Hillsdale, NJ: Erlbaum.

Gerstadt, C. L., Hong, Y. J., & Diamond, A. (1994). The relationship between cognition and action: Performance of children 3 1/2–7 years old on a Stroop-like day-night test. *Cognition, 53*, 129–153. https://doi.org/10.1016/0010-0277(94)90068-x.

Gilmore, C., & Cragg, L. (2018). The role of executive function skills in the development of children's mathematical competencies. In A. Henik, & W. Fias (Eds.), *Heterogeneity of function in numerical cognition* (pp. 263–286). San Diego: Academic Press.
Goldfarb, L. (2018). Cognitive interferences and their development in the context of numerical tasks: Review and implications. In A. Henik, & W. Fias (Eds.), *Heterogeneity of function in numerical cognition* (pp. 245–262). San Diego: Academic Press.
Hannula, M. M., & Lehtinen, E. (2005). Spontaneous focusing on numerosity and mathematical skills in young children. *Learning and Instruction, 15*, 237–256.
Hannula, M. M., Räsänen, P., & Lehtinen, E. (2007). Development of counting skills: Role of spontaneous focusing on numerosity and subitizing-based enumeration. *Mathematical Thinking and Learning, 9*, 51–57.
Kochanska, G., Murray, K., Jacques, T.Y., Koening, A. L., & Vandegeest, K.A. (1996). Inhibitory control in young children and its role in emerging internalization. *Child Development, 67*, 490–507.
Leibovich, T., & Henik, A. (2013). Magnitude processing in non-symbolic stimuli. *Frontiers in Psychology, 4*, 375. https://doi.org/10.3389/fpsyg.2013.00375.
Leibovich, T., Katzin, N., Harel, M., & Henik, A. (2017). From "sense of number" to "sense of magnitude": The role of continuous magnitudes in numerical cognition. *Behavioral and Brain Sciences, 40*. https://doi.org/10.1017/S0140525X16000960, e164.
Mehler, J., & Bever, T. G. (1967). Cognitive capacity of very young children. *Science, 158*, 141–142.
Mehler, J., & Bever, T. G. (1968). Reply by J. Mehler and T. G. Bever. *Science, 162*, 979–981.
Merkley, R., Matusz, P. J., & Scerif, G. (2018). The control of selective attention and emerging mathematical cognition: Beyond unidirectional influences. In A. Henik, & W. Fias (Eds.), *Heterogeneity of function in numerical cognition* (pp. 111–126). San Diego: Academic Press.
Mix, K. S., Levine, S. C., & Newcombe, N. S. (2016). Development of quantitative thinking across correlated dimensions. In A. Henik (Ed.), *Continuous issues in numerical cognition: How many or how much* (pp. 4–36). San Diego: Academic Press.Nys and Content, 2012Nys, J., & Content, A. (2012). Judgement of discrete and continuous quantity in adults: Number counts! *Quarterly Journal of Experimental Psychology, 65*, 675–690. https://doi.org/10.1080/17470218.2011.619661.
Petersen, I. T., Hoyniak, C. P., McQuillan, M. E., Bates, J. E., & Staples, A. D. (2016). Measuring the development of inhibitory control: The challenge of heterotypic continuity. *Developmental Review, 40*, 25–71. https://doi.org/10.1016/j.dr.2016.02.001.
Piaget, J. (1952). *The child's conception of number*. New York, NY: Norton.
Piaget, J. (1968). Quantification, conservation, and nativism. *Science, 162*, 976–979.
Siegler, R. S., & Braithwaite, D. W. (2017). Numerical development. *Annual Review of Psychology, 68*, 187–213. https://doi.org/10.1146/annurev-psych-010416-044101.
Verbruggen, F., & Logan, G. D. (2008). Response inhibition in the stop-signal paradigm. *Trends in Cognitive Sciences, 12*, 418–424.
Wilkey, E. D., & Price, G. R. (2019). Attention to number: The convergence of numerical magnitude processing, attention, and mathematics in the inferior frontal gyrus. *Human Brain Mapping, 40*, 928–943. https://doi.org/10.1002/hbm.24422.

# Index

Note: Page numbers followed by *t* indicate tables.

Kolibri 360 GmbH
Ferdinand-Jühlke-Straße 7
99095 Erfurt, Deutschland
produktsicherheit@kolibri360.de